T0310063

Vulnerability of Land Systems in Asia

Vulnerability of Land Systems in Asia

Edited by

Ademola K. Braimoh

The World Bank

He Qing Huang

Institute of Geographic Sciences and Natural Resources Research, Chinese Academy of Sciences

Registered office: John Wiley & Sons, Ltd, The Atrium, Southern Gate, Chichester, West Sussex, PO19 8SQ, UK

Editorial offices: 9600 Garsington Road, Oxford, OX4 2DQ, UK
The Atrium, Southern Gate, Chichester, West Sussex, PO19 8SQ, UK
111 River Street, Hoboken, NJ 07030-5774, USA

For details of our global editorial offices, for customer services and for information about how to apply for permission to reuse the copyright material in this book please see our website at www.wiley.com/wiley-blackwell.

Library of Congress Cataloging-in-Publication Data is applied for.

ISBN: 978-1-118-85495-2

A catalogue record for this book is available from the British Library.

Wiley also publishes its books in a variety of electronic formats. Some content that appears in print may not be available in electronic books.

Set in 9/11pt Times Ten by Aptara Inc., New Delhi, India
Printed and bound in Singapore by Markono Print Media Pte Ltd

1 2015

Contents

Editors' Introductions

Dr Ademola Braimoh

Dr Ademola Braimoh is a Sustainability Scientist with expertise in human-environment relationships, land-use change, and vulnerability of ecosystem services to global environmental change. Dr Braimoh formerly worked as Scientist at the Global Water System Project, Postdoctoral Fellow of the Ecosystems and People Program at United Nations University, Professor of Land Change Science at the Center for Sustainability Science at Hokkaido University, and as Executive Director of the Global Land Project in Japan. At the World Bank, Dr Braimoh currently works at the Science–Policy interface of Climate Smart Agriculture, helping clients to realize increased productivity, enhanced resilience, reduced greenhouse gas emissions, and improved carbon sequestration in agricultural landscapes.

Dr He Qing Huang

Dr Huang obtained a Bachelor of Engineering degree from Wuhan University, China, in 1983, a Master of Engineering degree from Tsinghua University, Beijing, China, in 1986, and a PhD on Geosciences from the University of Wollongong, Australia, in 1997. During 1997–99, Dr Huang worked as a postdoctoral research fellow at the University of Wollongong in Australia. At the end of 1999, Dr Huang moved to the UK to work as a research scientist at the University of Glasgow (2000–02) and the University of Oxford (2002–05). In 2005, Dr Huang was appointed as a professor by the Institute of Geographic Sciences and Natural Resources Research (IGSNRR) of the Chinese Academy of Sciences, Beijing, China. During 2006–07, Dr Huang helped to establish a nodal office in IGSNRR for the Global Land Project (GLP) and was appointed as the executive director of the office by GLP in 2007.

Dr Huang has been working in the areas of sustainable use of land and water resources and the adjusting tendency of fluvial systems in response to climate change and human activities. In recent years, he has been leading several international collaborative research projects, studying the application of the newly emerged science of complexity to the field of sustainable use of land and water resources. For this, he has brought socio-economic and environmental scientists together and been developing agent-based simulation models in an attempt to bridge the divide between physical and human geographers. Having been working internationally and in a trans-disciplinary way, Dr Huang has a wide range of research interests, including sustainable management of land and water resources, land use and ecosystem interactions, fluvial processes and risk management of natural hazards (floods and droughts), earth surface process simulation, application of least action and entropy principles in geomorphologic systems, spatial data analysis, chaos and complexity science, advances in scientific research methods, and agent-based modelling of resources use and societal development. Dr Huang has so far published nearly 100 research papers in scientific journals.

List of Contributors

L. Alhamd, Research Center for Biology, Indonesian Institute of Sciences, Bogor, Indonesia

Myagmarsuren Altanbagana, Dryland Sustainability Institute and National University of Mongolia, Ulaanbaatar, Mongolia

Maria Anaya-Romero, Evenor-Tech, Spin-off from IRNAS-CSIC, Seville, Spain

Ademola K. Braimoh, The World Bank, Washington, DC, USA

Wen Chen, Institute of Atmospheric Physics, Chinese Academy of Sciences, Beijing, China

Togtokh Chuluun, Dryland Sustainability Institute and National University of Mongolia; Mongolian Development Institute, Ulaanbaatar, Mongolia

Xuefeng Cui, School of GeoSciences, University of Edinburgh, Edinburgh, UK; College of Global Change and Earth System Research, Beijing Normal University; State Key Laboratory of Earth Surface Processes and Resource Ecology, Beijing Normal University, Beijing, China

A. Damodaran, Indian Institute of Management Bangalore, Bangalore, India

Rajesh Daniel, Unit for Social and Environmental Research, Chiang Mai University, Chiang Mai, Thailand

S. Davaanyam, Dryland Sustainability Institute and National University of Mongolia, Ulaanbaatar, Mongolia

Terence Dawson, School of Geography, University of Southampton, Southampton, UK

Xiangzheng Deng, Institute of Geographic Sciences and Natural Resources Research, Chinese Academy of Sciences, Beijing, China

Peilei Fan, School of Planning, Design and Construction, Center for Global Change and Earth Observations, Michigan State University, East Lansing, USA

Jefferson Fox, East-West Center, Honolulu, Hawaii, USA

Thomas W. Giambelluca, Department of Geography, University of Hawaii, Manoa, Hawaii, USA

Stephan Harrison, Department of Geography, University of Exeter, Cornwall Campus, Penryn, Cornwall, UK

Shigeko Haruyama, Mie University, Graduate School and Faculty of Bioresource, Mie, Japan

He Qing Huang, Institute of Geographic Sciences and Natural Resources Research, Chinese Academy of Sciences, Beijing, China

Muyi Kang, School of Resources, Beijing Normal University, Beijing, China

Takashi Kohyama, Faculty of Environmental Earth Science, Hokkaido University, Hokkaido, Japan

Akihiko Kondoh, Center for Environmental Remote Sensing, Chiba University, Chiba, Japan

Ajay Kumar, Department of Geography, Delhi School of Economics, Delhi, India

Louis Lebel, Unit for Social and Environmental Research, Chiang Mai University, Chiang Mai, Thailand

Fen Li, Shenzhen Institute of Building Research Co., Ltd, Shenzhen, China

H.S. Limin, Center for International Management Tropical Peatland (CIMTROP), University of Palangka Raya, Indonesia

Gabriela Litre, Center for Sustainable Development, University of Brasilia, Brasilia, Brazil. Brazilian Research Network on Global Climate Change, National Institute for Space Research (Rede CLIMA-INPE)

Jiyuan Liu, Institute of Geographic Sciences and Natural Resources Research, Chinese Academy of Sciences, Beijing, China

Xiaoqian Liu, College of Urban and Environmental Sciences, Peking University, Beijing, China

Xing Lu, Yunnan University, Kunming, China

Yoshitaka Masuda, Nippon Telegraph and Telephone East Corporation, Tokyo, Japan

Dennis Ojima, Natural Resource Ecology Laboratory; Ecosystem Science and Sustainability Department, Colorado State University, Fort Collins, Colorado, USA

Paul Palmer, School of GeoSciences, University of Edinburgh, Edinburgh, UK

David G. Passmore, School of Geography, Politics and Sociology, Newcastle University, Newcastle upon Tyne, UK

Nina V. Pimankina, Institute of Geography, Academy of Sciences, Almaty, Republic of Kazakhstan

Jiaguo Qi, Department of Geography, Center for Global Change and Earth Observations, Michigan State University, East Lansing, USA

Alaric Rae, Department of Geography, Coventry University, Coventry, UK

Joeni S. Rahajoe, Research Center for Biology, Indonesian Institute of Sciences, Bogor, Indonesia

Diego De la Rosa, Institute for Natural Resources and Agrobiology (IRNAS), Spanish Science Research Council (CSIC), Seville, Spain

Mark Rounsevell, School of GeoSciences, University of Edinburgh, Edinburgh, UK

Bambang Hero Saharjo, Forest Fire Laboratory, Forest Protection Division, Department of Silviculture, Faculty of Forestry, Bogor Agricultural University, West Java, Indonesia

Fernando Teigao dos Santos, Instituto de Geografia e Ordenamento do Território (IGOT), Lisbon University, Lisbon, Portugal

Omer L. Sen, Istanbul Technical University, Eurasia Institute of Earth Sciences, Istanbul, Turkey

Igor Severskiy, Institute of Geography, Academy of Sciences, Almaty, Republic of Kazakhstan

Farzin Shahbazi, Department of Soil Science, Faculty of Agriculture, University of Tabriz, Tabriz, Iran

R.B. Singh, Department of Geography, Delhi School of Economics, Delhi, India

Troy Sternberg, School of Geography, Oxford University, Oxford, UK

M.S. Suneetha, United Nations University Institute of Advanced Studies, Yokohama, Japan

Shubhechchha Thapa, Global Land Project-Sapporo Nodal Office, Hokkaido University, Japan

B. Tserenchunt, Dryland Sustainability Institute and National University of Mongolia, Ulaanbaatar, Mongolia

Sandra Uthes, Leibniz-Centre for Agricultural Landscape Research (ZALF), Müncheberg, Germany

John B. Vogler, Center for Applied Geographic Information Science, University of North Carolina, Charlotte, USA

E.B. Walujo, Research Center for Biology, Indonesian Institute of Sciences, Bogor, Indonesia

Yunjie Wei, Institute of Geographic Science and Natural Resources Research, Chinese Academy of Sciences, Beijing, China

Vanessa Winchester, School of Geography and the Environment, University of Oxford, Oxford, UK

Huimin Yan, Institute of Geographic Sciences and Natural Resources Research, Chinese Academy of Sciences, Beijing, China

Li Yang, Institute of Geographic Science and Natural Resources Research, Chinese Academy of Sciences, Beijing, China

He Yin, College of Urban and Environmental Sciences, Peking University, Beijing, China

Yuan Jiang, State Key Laboratory of Earth Surface Processes and Resource Ecology, Beijing Normal University, China; School of Resources, Beijing Normal University, China

Wenze Yue, Department of Land Management, Zhejiang University, Hangzhou, China

Lin Zhen, Institute of Geographic Science and Natural Resources Research, Chinese Academy of Sciences, Beijing, China

Alan L. Ziegler, Department of Geography, National University of Singapore, Singapore

Preface

Humans have interacted with their environments for centuries. These interactions are driven by a complex suite of biophysical, institutional and socioeconomic drivers that determine the configuration of the land system. Human impacts on the environment are growing in magnitude, and development practitioners increasingly recognize the need to address the adverse consequences and the implications for sustainability. As the world faces up to the realities of a changing climate, coping with the negative impacts calls for fundamental changes in the way we manage our environment. This book offers a state-of-the-art overview of vulnerability of land systems to multiple interacting stressors in Asia. It identifies the stressors in various locales, the impacts on the land system and societal responses to meeting the challenges. The book is based on the sustainable principle that encourages the fusion of global change research and sustainable development. The chapters are written by international scholars with different disciplinary backgrounds. They have addressed the complex issues arising from human–environment interactions that cannot be satisfactorily dealt with by core disciplinary methods alone. The issues addressed range from the vulnerability of smallholder agriculture and urban systems to the impact of socioeconomic processes at the sub-regional level. Knowledge of the causal processes that affect land systems' vulnerability and capacity to cope with different perturbations is documented as well as the imperative of integrating vulnerability assessment into policies and decision-making.

Ademola K. Braimoh
He Qing Huang

1
Land Systems Vulnerability

Ademola K. Braimoh[1] and He Qing Huang[2]
[1] *The World Bank, Washington DC, USA*
[2] *Institute of Geographic Science and Natural Resources Research, Chinese Academy of Sciences, Beijing, China*

1.1 Introduction

Land-change science is an interdisciplinary field that seeks to understand the dynamics of the land system as a coupled human-environment system (CHES). A CHES is one in which the societal and biophysical subsystems of the land system are so intertwined that the system's condition, function, and responses to a perturbation depend on the synergy of the two subsystems (Turner *et al.*, 2010). Although humans have interacted with their biophysical environment since the beginning of human history, the extent and intensity of these interactions have increased spectacularly since the Industrial Revolution (Liu *et al.*, 2007; Steffen *et al.*, 2004). Therefore, the global change research community and development practitioners increasingly recognize the need to address the adverse consequences of changes taking place in the structure and function of the biosphere and the implications for human wellbeing (IPCC, 2007). Assessing land system dynamics requires detailed analysis of processes operating at different spatial and temporal scales, and the interactions between different drivers including policy shifts that often lead to emergent properties and nonlinear outcomes (Dearing *et al.*, 2010). As it becomes more obvious that, in future, human societies will be exposed to multiple interacting stressors, emphases on the means to understand and manage land systems are becoming stronger (Global Land Project, 2005; Turner *et al.*, 2007).

Land system vulnerability assessment is fundamental to the understanding of the link between global change, environmental sustainability and human wellbeing (Braimoh & Osaki, 2010; Leichenko & O'Brien 2008; Turner *et al.*, 2007). While the world faces up to the realities of a changing climate, coping with its adverse consequences constitutes an

Vulnerability of Land Systems in Asia, First Edition. Edited by Ademola K. Braimoh and He Qing Huang.

additional challenge to socioeconomic development. It is predicted that by 2050, agricultural production will need to increase by 60% to feed the world's population of over 9 billion people (Food and Agriculture Organization, 2012), whereas the accelerating pace of climate change is a further challenge to meeting the food security needs of the increasing population. Global surface temperatures have increased by 0.8°C since the late 19th century, with an average rate of increase of 0.15°C per decade since 1975 (IPCC, 2007). The Earth's mean temperature is projected to increase by 1.5–5.8°C during the 21st century (IPCC, 2001). Future global warming will exacerbate hydrologic scarcity and variability such that crops will have to grow in hotter and drier conditions. Higher temperatures and shorter growing seasons will reduce the yields of most food crops, and promote the spread of weeds and pests. Changes in precipitation patterns will also increase the likelihood of short-run crop failures and long-run productivity decline. Although there will be productivity gains in some crops in certain regions of the world, the overall impact of climate change on agriculture is expected to be negative, threatening global food security (Nelson *et al.*, 2009).

There is an urgent need to initiate measures that reduce vulnerability and increase the resilience of the CHES to climate change and other stressors. A recent study shows that between 2010 and 2050 the cost of adapting to a 2°C warmer world ranges between $75 billion and $100 billion per year, an amount equivalent to the Overseas Development Assistance received by the developing countries (World Bank, 2010b). Countries urgently need to shift development patterns or manage environmental resources in a manner that accounts for the potential impacts of global change.

Land system vulnerability assessment is fundamental to these issues. It identifies the consequences of interacting natural and socioeconomic stressors on exposed human groups and provides strategies for reducing exposure and sensitivity, whilst increasing adaptability to change. It answers questions such as how is the land system vulnerable to a given stressor, what is the level of vulnerability of different social groups, how are the changes induced by stressors alleviated by different conditions, what is the quality of coping capacity linked to different stressors, how do vulnerabilities of different places compare, and to what extent do vulnerable ecosystems produce vulnerable human conditions and vice-versa (Brooks *et al.*, 2005; Eakin & Luers, 2006; Leary *et al.*, 2008; Polsky *et al.*, 2007). Assessment of land system vulnerability requires approaches that treat the coupling of the land system as a human-environment system explicitly (Dearing *et al.*, 2010; Turner, 2010). Such integrated studies of coupled human-environment systems help to reveal new and complex patterns and processes not evident when studied by social or natural scientists separately (Liu *et al.*, 2007), deal with tradeoffs within and between the human and environmental subsystems (Braimoh *et al.* 2009; Carpenter *et al.* 2008; Rodriguez *et al.*, 2006), and facilitate the design of effective policies for ecological and socioeconomic sustainability (Braimoh & Huang, 2009; Tallis *et al.*, 2009).

Close to two-thirds of Asia's 4 billion people live in rural areas and depend on natural resources for their livelihood. The Asian continent is physiographically diverse – its northern part is located in the boreal climatic zone, the west and central part is predominantly arid, part of the eastern region is characterized by temperate rainforest, and the south is notably rich in agrobiodiversity, whereas the southeast is typified by tropical rainforests and monsoon climates with high rainfall (Galloway & Melillo, 1998; IPCC, 2007). Since the 1990s, Asia has witnessed impressive economic growth accompanied by diverse environmental management problems. Population growth and socioeconomic activities are important drivers of demand for ecosystem services, whereas climate change has adversely affected food and water security of the growing populace. Future climate change is predicted to profoundly affect agriculture, exacerbate water resource scarcity, and increase the threats to biodiversity as it compounds the pressure on ecosystem resources associated with urbanization and economic growth (Braimoh *et al.*, 2010; IPCC, 2007; World Bank, 2010a).

1.2 Overview of the book

This book integrates knowledge of the vulnerability of land systems to multiple stressors in Asia. It seeks to improve knowledge of the causal processes that affect land systems' vulnerability and capacity to cope with different perturbations. It also identifies factors that can help in integrating vulnerability assessment into policies and decision-making with the hope that countries with similar environmental conditions can learn from the Asian experience. The book is divided into three sections, namely 'Hazards and Vulnerability', 'Modeling and Impact Assessment', and 'Institutions'. The studies on hazards and vulnerability were carried out in the deserts of Mongolia and China, Rajasthan in India, and Kazakhstan, where changes in climate variability, ecosystem dynamics and social factors are interacting in remarkable ways to influence land-use systems and livelihood.

In Chapter 2, Sternberg shows that contrary to the prevailing concept of drought influencing severe winters, there are no connections between the two hazards. Rather, the major long-term correlation of drought is with human populations and their adaptation strategies. Public support in the form of risk management strategies should therefore take cognizance of the decoupling of extreme winters from droughts. This will improve the coping capacity of nomads in these arid environments. In Chapters 3 and 4, Ojima *et al.* show that the vulnerability of Mongolian rangelands to climate and land-use changes has increased since the transition to a market economy. To ensure a viable land-use system, adaptation options should include, among others, the development of cultural landscape restoration that incorporates community-based conservation and sustainable use of natural resources. In Chapter 5, Liu *et al.* construct a vulnerability profile for Middle Inner Mongolia by integrating biophysical and social datasets. The uneven development of the region is clearly reflected in the overall vulnerability across the landscape, but also indicates that irrespective of exposure to perturbations, actions that increase adaptive capacity at the local level can significantly mitigate vulnerability. In Chapter 6, Singh and Kumar's study highlights the importance of traditional water harvesting and appropriate tillage techniques for climate change adaptation in an area profoundly affected by wind erosion in India. In Chapter 7, Winchester *et al.* combine dendrogeomorphological techniques and sedimentological analysis with archival records to date debris flows in the Bolshaa Almatinka basin and Ozernaya valley floor in Kazakhstan. More and more younger trees on successively lower terraces in the Almatinka valley indicate numerous but decreasing flood magnitudes since the 1920s, whereas in the Ozernaya valley floor, the large-scale debris flows were caused by moraine-dammed glacial lake outbursts and intense summer rainstorms.

In the 'Modeling and Impact Assessment' section, various land-use models predict extensive anthropogenic modification of the landscape with grave consequences for ecosystem functioning. In Chapter 8, Fox *et al.* show that rapid intensification of agriculture and expansion of regional trade will affect 16% of montane mainland Southeast Asia by 2050. About 4% of the change could be due to an increase in rubber plantation, which markedly affects regional hydrology. By integrating Landsat Thematic Mapper data and a Net Primary Productivity (NPP) model, Yan *et al.* in Chapter 9 show that in the 1990s, cropland expansion led to a total agricultural productivity increase of 5% in China. However, interactions between agriculture and urban development implied that cropland expansion occurred mainly on poor quality land. The impacts of land-use change on agricultural production vary markedly across China with the largest effect occurring mainly in the Northeast Region. The long-term assessment of land-cover change in Amur River Basin by Haruyama *et al.* in Chapter 10 also confirms rapid cropland expansion in Northeast China, whereas in Chapter 11, the Parsimonious Land Use Model (PLUM) of Cui *et al.* linking population growth and socioeconomic development to changes in agricultural yield and environmental policies reasonably predicts changes in food consumption and consequent crop production in China. In Chapter 12, Shahbazi *et al.* illustrate the incorporation of soil use and protection concerns in land-use planning by applying the MicroLEIS decision support system for land evaluation in east Azerbaijan, Iran. Modules ranging from soil suitability identification, and plant species identification for land rehabilitation to soil capability classification were applied to guide sustainable agroecological planning. In Chapter 13, Rahajoe *et al.* analyze the impact of agricultural land change on environmental resources in Bawan Village, Indonesia, in the last few decades. Forest concessions led to the degradation of the forest ecosystem and conversion of clear-cut area to rubber plantation, which in turn influenced microclimate, biodiversity, and water resources. The authors suggest some policy interventions to reduce the vulnerability of the villagers and their environments to undesirable effects of land-use change. In Chapter 14, the study by Yue *et al.* reveals internal heterogeneity in Shanghai. Some districts exhibit positive urbanization trends, characterized by rapid economic growth, better environment conditions, and more appropriate land-use patterns, whereas others show negative urbanization trends, illustrated by slow economic growth, worse environment conditions, and less appropriate land-use patterns. Yue *et al.* identify economic restructuring and globalization, migration policy, and institutional factors as the major driving forces, and suggest addressing the increasing disparity between the urban districts for sustainable urban development.

Different institutions and decision-making structures intersect with the land systems in various ways to produce different outcomes for the maintenance of the land system (Global Land Project, 2005). Societal responses to global environmental change can be broadly classified into three: interventions designed to reduce uneven outcomes across individuals, households, or social groups; interventions devised to reduce vulnerability; and interventions intended to influence the causal processes of global change (Leichenko & O'Brien, 2008). Owing to the interconnections between vulnerability outcomes, contexts and driving factors, none of these interventions alone is sufficient in properly addressing vulnerability. Thus an integrated response is often recommended (Leichenko & O'Brien, 2008; Steffen *et al.*, 2004). The final section, 'Institutions', seeks to understand how various governance approaches and policies at different spatial levels help in prioritizing and sharing land-based ecosystem services between different stakeholders. In Chapter 15, Lebel and Daniel review the various approaches taken by communities to address ecosystem services governance challenges in Southeast Asian watersheds. Governance approaches taken by communities include spatial planning, zoning, rules and regulations, and incentives such as payment for ecosystem services (PES). Clearly, none of these approaches is a silver bullet in addressing environmental problems in all locales, as ecosystems may be mismanaged for several reasons, including the public good nature of many ecosystem services, lack of property rights, and capital market imperfections (Engel *et al.*, 2008; Tietenberg, 2006). The assessment by Li *et al.* of the socioeconomic impacts of the wetland restoration program in Chapter 16 indicate that while the farmer's willingness to accept eco-compensation (a form of PES) was 13,912 yuan (about US $2100) per household in Poyang, only 39% was actually compensated by the program. Without compensation, the region would suffer more degradation, necessitating the need to adjust the current compensation levels whilst also developing non-farm employment opportunities for the population. In Chapter 17, Thapa *et al.* further provide an overview, including the payment mechanisms, under the

Sloping Land Conversion Program, the largest payment for ecosystem services scheme in the world.

In Chapter 18 Saharjo discusses a community-based fire-free land preparation for the management of peat forests in Indonesia. The study indicates that the land preparation could potentially reduce greenhouse gas emissions whilst the fuel load can also be converted to soil amendments. The deficiencies in current climate finance mechanisms for climate adaptation are addressed in Chapter 19 by Damodaran. The author states that current climate finance mechanisms are not tuned to two important realities that affect vulnerable communities. The first reality is the fact that adaptation involves a gradient of measures ranging from simple natural response measures to more sophisticated techniques that entail high incremental costs that are unfavorable to farming systems in developing countries. The other reality is that an agent who undertakes planned adaptation measures that are designed to address possible climatic variability in the distant future faces the prospect of diminution of current income. Damodaran offers some suggestions to address these inadequacies to assist vulnerable agroecosystems in coping with climate change.

Increased understanding of global environmental change should be complemented with strategies that help reduce the vulnerability of people and places. However, effective science-policy linkage still remains a mirage. The role of communication in helping us meet the challenges of global environmental change is highlighted by Litre in Chapter 20. The author states that communication is vital in fostering effective science-policy dialogue by presenting complex scientific findings in a manner policy makers can use. The author also presents methods for building trust through two-way communication between researchers and policy makers.

In Chapter 21, Teigao dos Santos argues for a shift from 'command-and-control' to 'learning-and-adaptation' in planning for resilience. After exploring the relations and the implications of the resilience theories for the planning process, the author presents the SPARK (Strategic Planning Approach for Resilience Keeping) framework, a proactive approach designed to enhance the resilience of a system to changing circumstances.

Acknowledgements

Most of the chapters are drawn from a workshop on Vulnerability and Resilience of Land Systems in Asia held at Beijing in June 2009 for the implementation of the Global Land Project (http://www.globallandproject.org). We acknowledge the financial support of the Asia Pacific Network for Global-Change Research, the Ministry of Science and Technology of China (Grant No.: 2013DFA91700), the Chinese Academy of Sciences, the National Natural Science Foundation of China, and the Japan Ministry of Education through the Special Coordination Funds for Promoting Science and Technology. We are grateful for the support of Anette Reenberg, Tobias Langanke, and Lars Jorgensen of the Global Land Project International Project Office. We also acknowledge the untiring assistance of Kikuko Shoyama, Julius Agboola, Shubhechchha Thapa, and Eniola Fabusoro in the production of this book.

References

Braimoh, A., Huang, H.Q. (2009) Addressing issues for Land Change Science. Eos Transactions of the American Geophysical Union **90**: 334.

Braimoh, A.K., Osaki, M. (2010) Land-use change and environmental sustainability. Sustainability Science **5**: 5–7.

Braimoh, A.K., Agboola, J.I., Subramanian, S.M. (2009) The role of governance in managing ecosystem. service trade-offs. International Human Dimensions Program on global Environmental Change (IHDP) Update, Issue 3. pp. 22–5.

Braimoh, A.K., Suneetha, M.S., Elliott, W.S., Gasparatos, A. (2010). Climate and human related drivers of biodiversity decline in Southeast Asia. United Nations University Institute of Advanced Studies, Policy Report No 2, 50 pp.

Brooks, N., Adger, W.N., Kelly, P.M. (2005) The determinants of vulnerability and adaptive capacity at the national level and the implications for adaptation. Global Environmental Change **15**: 151–63.

Carpenter, S.R., Mooney, H.A., Agard, J. *et al.* (2009) Science for managing ecosystem services: beyond the millennium ecosystem assessment. Proceedings of the National Academy of Sciences of the USA **106**: 1305–12.

Dearing, J.A., Braimoh, A.K., Reenberg, A., Turner, B,L. II, van der Leeuw, S.E. (2010) Complex land systems: the need for long time perspectives in order to assess their future. Ecology and Society **15**: 21.

Eakin, H., Luers, A. (2006) Assessing the vulnerability of social-environmental systems. Annual Review of Environment and Resources **31**: 365–94.

Engel, S., Pagiola, S., Wunder, S. (2008) Designing payments for environmental services in theory and practice: An overview of the issues. Ecological Economics **65**: 663–74.

Food and Agriculture Organization (2012) The State of Food Insecurity in the World 2012. Economic growth is necessary but not sufficient to accelerate reduction of hunger and malnutrition. Rome: FAO.

Galloway, J.N., Melillo, J.M. (1998) *Asian Change in the Context of Global Climate Change.* New York: Cambridge University Press.

Global Land Project (2005) Science Plan and Implementation Strategy. International Geosphere Biosphere (IGBP) Report No. 53/International Human Dimensions Program on Global Environmental Change (IHDP) Report No. 19. Stockholm: IGBP

IPCC (2001) *Climate Change 2001: The Scientific Basis*. Cambridge University Press.

IPCC (2007) Climate change 2007. Climate change impacts, adaptation and vulnerability. Report of Working Group II. Geneva: IPCC.

Leary, N., Conde, C., Kulkarni, J., Nyong, A., Pulhin, J. (eds) (2008) *Climate Change and Vulnerability*. London: Earthscan.

Leichenko, R.M., O'Brien, K.L. (2008) *Environmental Change and Globalization: Double Exposures*. Oxford: Oxford University Press, 192 pp.

Liu, J., Dietz, T., Carpenter, S.R., *et al.* (2007) Complexity of coupled human and natural systems. Science **317**: 1513–16.

Nelson, G., Rosengrant, M.W., Koo, J., *et al.* (2009) Climate change: Impact on agriculture and costs of adaptation. Washington DC: International Food Policy Research Institute.

Polsky, C., Neff, R., Yarnal, B. (2007) Building comparable global change vulnerability assessments: The Vulnerability Scoping Diagram. Global Environmental Change **17**: 472–85.

Rodríguez, J.P., Beard, T.D. Jr, Bennett, E.M., *et al.* (2006) Trade-offs across space, time, and ecosystem services. Ecology and Society **11**: 28; available at: http://www.ecologyandsociety.org/vol11/iss1/art28/

Steffen, W., Sanderson, A., Tyson, P.D., *et al.* (2004) *Global Change and the Earth System: A Planet Under Pressure*. New York: Springer-Verlag.

Tallis, H., Goldman, R., Uhl, M., Brosi, B. (2009) Integrating conservation and development in the field: implementing ecosystem service projects. Frontiers in Ecology and Environment **7**: 12–20.

Tietenberg, T. (2006) *Environmental and Natural Resource Economics*, 6th edn. Boston: Addison-Wesley.

Turner, B.L. II (2010) Vulnerability and resilience: coalescing or paralleling approaches for sustainability science? Global Environmental Change **20**: 570–6.

Turner, B.L. II, Lambin, E.F., Reenberg, A. (2007) The emergence of land change science for global environmental change and sustainability. Proceedings of the National Academy of Sciences of the USA **104**: 20666–71.

World Bank (2010a) Development and climate change. World Development Report 2010. Washington DC: World Bank.

World Bank (2010b) The economics of adaptation to climate change. A synthesis report. Final Consultation Draft. Washington DC: World Bank.

I
Hazards and Vulnerability

2 Drought and Extreme Climate Stress on Human-Environment Systems in the Gobi Desert Mongolia

Troy Sternberg
School of Geography, Oxford University, Oxford, UK

2.1 Introduction

With the world's 'greatest concentration of areas of rapid land-cover changes, and in particular dryland degradation' (Lepers *et al.*, 2005, p. 122), Asia is now impacted by changing climates, increasing aridification, rapid socioeconomic and political transition, and evolving land-use practices, with these affecting both the physical and human geography (Geist, 2005; Lioubimtseva *et al.*, 2005; Rossabi, 2005; Yang *et al.*, 2005; Zhang *et al.*, 2007). Within the region the Gobi Desert environment, shared by southern Mongolia and northern China, is experiencing shifting biophysical and anthropogenic dimensions. A warming climate, increased precipitation volatility, and natural phenomena such as drought and *dzud* (extreme cold winter weather) impact the landscape (Batima *et al.*, 2005; Marin, 2010); economic, population, development, and land-use pressures affect human wellbeing. Such factors create ecological and livelihood dynamics that challenge system functionality. While natural processes are shared with other regions, ecosystem determinants are distinctly tied to local landscapes. Research embedded in the Asian environment is critical to assess today's dominant ecological concerns and address the implications of these concerns for the future (Fernandez-Gimenez, 2000; Sasaki *et al.*, 2008).

Vulnerability of Land Systems in Asia, First Edition. Edited by Ademola K. Braimoh and He Qing Huang.

Pastoralism is the dominant rural livelihood in Mongolia and is practiced in cold dryland conditions (Johnson *et al.*, 2006). Political and socioeconomic transition since 1990 has changed the herding model from a collective, state-supported approach to one featuring privatized livestock and individual responsibility. On the steppe grasslands environmental challenges impact herding livelihoods with the natural hazards of drought and *dzud* (extreme winter conditions) prevalent yet infrequently researched (Sternberg, 2008). These natural hazards are significant because they impact the physical environment that provides the resource base for pastoralists' livestock (Nandintsetseg *et al.*, 2007). As livelihoods are stressed by hazard episodes, questions of vulnerability and resilience are framed by the landscape and herders' ability to mitigate risks. The common perception at the governmental and international levels is that drought and *dzud* are linked, though clarification of cause and effect is lacking (Begzsuren *et al.*, 2004; World Bank, 2004). Clarification of hazard interaction and systemic impact is essential to understanding and addressing pastoral environments in the Gobi.

2.1.1 Social ecological systems

In the Gobi Desert the physical environment cannot be examined in isolation from herder action and impact (Brand, 2009; Sternberg, 2009; Turner *et al.*, 2007). This is essential when considering the low resource base of steppe grasslands where subtle fluctuations in biophysical conditions or human action can impact livelihood and grassland productivity. The intertwined roles of humanity and nature have shaped the steppe – failure to recognize this would render any conclusion incomplete (Shinoda *et al.*, 2007). For this reason theoretical linkage of the two forces is essential. Using social-ecological system (SES) theory this chapter examines if rangelands can experience disturbance, such as grazing or drought, yet maintain functioning and the role flexibility in biophysical and human systems that may enable continued range productivity (Walker & Abel, 2002). The strength of an SES foundation is the combination of biophysical and social components to underpin research questions through concepts including vulnerability, adaptation, and resilience (Gallopin, 2006; Walker *et al.*, 2004). Derived from life sciences and social sciences, as an analytical tool SES can bridge the natural-social divide through cross-discipline application to address more fully environmental dynamics on the steppe (Gallopin *et al.*, 2001). The strength of these analytical strategies is their relevance to the Gobi Desert (Batima *et al.* 2005; Chuluundorj, 2006; Walker & Abel, 2002) and ability to geographically integrate human and physical factors.

Vulnerability is composed of a system's exposure and susceptibility to external factors, sensitivity to disturbance, and coping capacity (Adger, 2006). Focus has been on human risk and the likelihood of an individual or group being exposed to and adversely impacted by a hazard (Cutter *et al.*, 2003). In this way it considers forces that have a limiting or detrimental effect on a system resulting from stresses or events beyond normal system variability that can originate internally or externally (Turner *et al.*, 2003). While including resource availability, it places society, particularly constraints and response capacity, at the centre of analysis rather than attempting to interpret physical conditions (Kelly & Adger, 2000; Smit & Wandel, 2006). Regionally, vulnerability can be applied to the role of pastoralists in reacting to external hazards and shifting socioeconomic forces that may limit their livelihoods. Vulnerability exists within a social context; here it relates to the impact of hazards on people and livelihoods, not on the physical environment.

Resilience theory probes the ability of a system to maintain basic structure and functioning when encountering disturbance and disruption. It allows examination of ecosystem adaptability to changing conditions regardless of cause, and asks if a system can return to a stable state (Falkenmark & Rockstrom, 2008; Holling, 1973; Walker & Salt, 2006). Resilience, in highlighting the systemic ability of an actor or environment to cope with alteration, conceives of a system returning to stability without lasting negative consequences (Chuluundorj, 2006; Pelling, 2003). This suggests a process that can encompass changing physical dynamics and social patterns and allow for recognition of multiple timescales within a system. Rather than drawing definitive conclusions, the theory presents conditions and states at a point in time, acknowledging that circumstances, such as those in the Gobi, are evolving rather than fixed. The ability to encompass variability and volatility within a system is the inherent strength of resilience and makes it relevant to the region (Marin, 2010).

2.1.2 Mongolian rangelands

Two natural hazards dominate in the Gobi region – drought and extreme winter *dzuds*. The former inhibits vegetation growth, reduces livestock summer weight gain, and leads to livestock concentration around water points and overgrazing; the latter happen when severe environmental conditions restrict the ability of animals to forage and threaten livestock survival (Table 2.1) (Suttie, 2005). Often detrimental environmentally, socially, and economically, drought affects more people globally than any other type of hazard yet remains a poorly understood weather phenomenon (Keyantash & Dracup, 2002; Wu *et al.*, 2001). Meteorological drought is generally regarded as significant negative variation from mean precipitation that can develop into a severe climatic event affecting water-related processes and vulnerable communities (Keyantash & Dracup, 2002; Sonmez *et al.*, 2005; World Meteorological Organization, 1986). A *dzud* occurs when extreme winter cold, snow, and/or ice limit forage potential, often resulting in high livestock mortality rates from starvation. Temperatures in the countryside can reach –40°C at night, causing a snow or ice 'crust' that prevents livestock from consuming vegetation. A result of local conditions, there is not an equivalent English word. Unique to Inner Asian pastoralism, limited reports of *dzuds* suggest they take place from once in three to once in seven years, often in combination with drought (Begzsuren *et al.*, 2004; World Bank, 2002).

Low precipitation and high climatic variability in this dryland environment impact the landscape and affect pastoralism, the dominant rural lifestyle. Serious climatic events occur with increasing frequency on Mongolian grasslands that are identified as highly degraded,

Table 2.1 Recent drought and *dzud* events in Mongolia

Year	Type of disaster
1944–45	*Dzud* + drought
1954–55	*Dzud*
1956–57	*Dzud*
1967–68	*Dzud* + drought
1976–77	*Dzud*
1986–87	*Dzud*
1993–94	*Dzud*
1996–97	*Dzud*
1999–00	*Dzud* + drought
2000–01	*Dzud* + drought

Data from UN records cited in Reading *et al.* (2006).

with this resulting in an intensification of natural hazards (Bold, 2001; Fernandez-Gimenez &Allen-Diaz, 1999; UNEP, 2002; Yang *et al.*, 2004). This stresses pastoral livelihoods, practiced in open-range grasslands, that must adjust to changing conditions in a harsh landscape.

Articles and reports, stressing drought frequency (Johnson *et al.*, 2006; World Bank, 2002) and the severity of *dzud,* state that summer droughts in 1999 and 2000 caused serious *dzuds* in 1999–2000 and 2000–2001 (Asian Development Bank, 2005; Batima *et al.*, 2005; Nandintsetseg *et al.*, 2007). These authors recognize how water and forage resources affect livestock conditions and herder wellbeing and then infer a causal role of summer droughts in initiating *dzud* events. However, a relationship between drought and *dzud* has not been clearly documented. Better understanding of these natural factors is essential for human wellbeing and improved pasture management. This chapter first establishes a drought record for the region. It then assesses drought interaction with extreme winter events, human population, and livestock numbers during the 1999–2001 disaster in Omnogovi Province. The study then evaluates systemic vulnerability and resilience in the region.

2.2 Methods

2.2.1 Study area

Five meteorological stations (Arvaheer, Bulgan, Dalanzadgad, Mandalgovi, Saixan Ovoo) that reflect the steppe to desert-steppe zone encompassing >50% of the country were selected in south-central Mongolia (Figure 2.1). This region, situated between the Hangai Mountains to the north and the Chinese border to the south, consists of Dundgovi, Ovorhangai, and Omnogovi Provinces, which combined cover approximately 300,000 km^2. The terrain comprises rolling gravel plains at an elevation of 1000 to 2000 m a.s.l. (Hilbig, 1995). In this ecosystem rainfall and coefficient of variation (CV) are telling statistics as they confirm substantial precipitation variability over the study period, particularly in the key summer months (Table 2.2). Such high precipitation variability implies a non-equilibrium environment as noted by prior study (Begzsuren *et al.*, 2004; Munkhtsetseg, 2007; Retzer & Reudenbach, 2005). The area has a harsh continental climate with distinct seasons and large daily and annual temperature fluctuation. The proportions of the population engaged in weather-dependent pastoral livelihoods are: in Omnogovi 49%, in Ovorhangai 57%, and in Dundgovi 63% (Mongolian Statistical Yearbook, 2006).

2.2.2 Methodology

The Standardized Precipitation Index (SPI) was used to assess anomalous and extreme precipitation and drought events in the Gobi Desert from 1970 to 2006 (Hayes *et al.*, 1999;

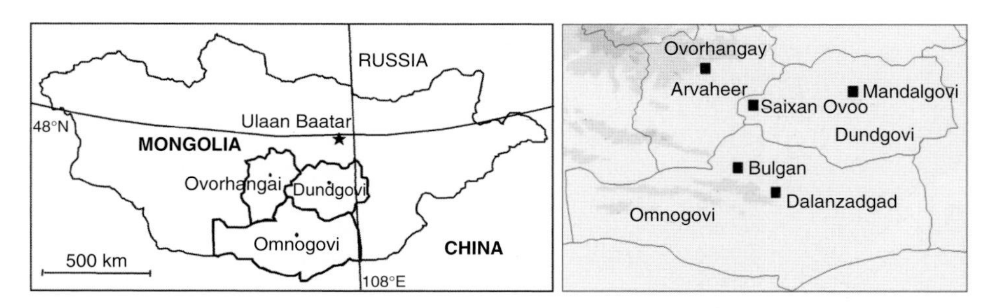

Figure 2.1 Map of Mongolia with study provinces and locations of meteorological stations

Table 2.2 Precipitation and coefficient of variation records, 1970–2006

	Precipitation		Coefficient of variation	
	Mean	Min./max.	Annual	Summer
Arvaheer	231	119/378	27.3	60.1
Bulgan	122	57/261	35	87.8
Dalanzadgad	122	51/235	33.5	71.8
Mandalgovi	147	72/243	31.6	60.8
Saixan Ovoo	115	59/268	39.5	76.2

McKee *et al.*, 1993; Sonmez *et al.*, 2005; Wu *et al.*, 2007). Based on meteorological data, the SPI gives a numeric value to precipitation that allows monitoring of drought at different timescales and enables comparison across regions and varied climatic zones. The index calculates drought initiation, magnitude, duration, and frequency while providing spatial and temporal flexibility; the SPI reflects the probability of precipitation at selected timescales measured by the number of standard deviations the observed value is from the long-term mean (Labedzki, 2007; Rouault & Richard, 2003). The SPI provides statistical consistency, identification of short- and long-term drought episodes, and is effective in areas where limited data availability can restrict drought quantification (Cancelliere *et al.*, 2007; Smakhtin & Hughes, 2007). Using meteorological data from the Institute of Hydrology and Meteorology, drought was examined at 3-, 6-, 12-, and 24-month timescales through August as this period coincides with high seasonal precipitation, with ≥60% falling during June, July, and August. Additional data were obtained from the government records.

2.3 Results

Wet and dry conditions across the region were established using the SPI. Results show cyclical fluctuation with broadly wetter conditions in the 1970s and 1990s, a notably drier period in the 1980s, and alternating wet-dry episodes in the 2000s, with 2006 being a particularly dry year (Figure 2.2). Drought event rates varied within and between sites and at different timescales. Site drought records fluctuated, with Arvaheer and Bulgan below estimated probability at each measurement and Dalanzadgad in drought 25% more of the time than Arvaheer. Moderate and extreme drought rates were slightly below average whereas severe levels were somewhat elevated (Table 2.3).

At moderate and severe levels drought was extant at multiple sites and different timescales. Saixan Ovoo SPI values were significantly related with all sites at all timescales ($P = 0.05$) whereas Mandalgovi, at 50%, was least related to other sites ($P = 0.05$). Patterns emerged between sites, with Mandalgovi appearing to follow latitude-influenced relationships with Arvaheer and Saixan Ovoo whereas Dalanzadgad had similarities with Bulgan, Saixan Ovoo, and Arvaheer. Bulgan's relationships were proximal with Dalanzadgad and Saixan Ovoo, the two nearest sites. Arvaheer's distribution was consistent with all but Bulgan. Numbers of inter-site correlations were the same at 3, 6, and 12 months and slightly increased at SPI 24-month.

2.3.1 Spatial continuity of droughts

The magnitude of drought events fluctuated between stations. Mapping shows spatial drought coverage at decadal intervals (Figure 2.3) and at different timescales in 2006

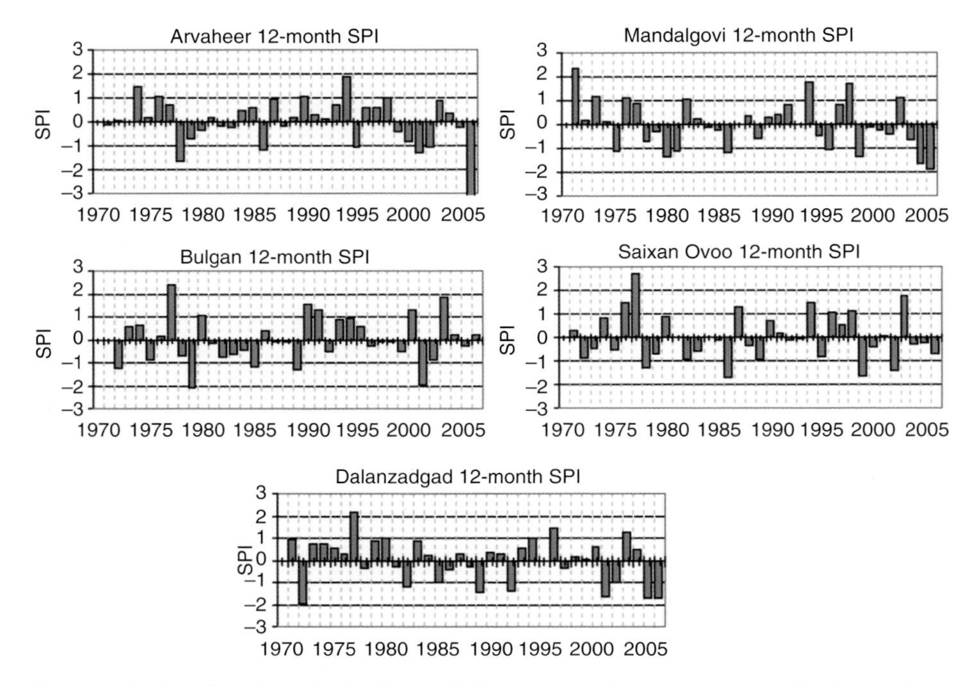

Figure 2.2 Standard Precipitation Index (SPI) at 12 months through August for five stations, 1970–2006

(Figure 2.4). Distribution is not spatially consistent nor are there uniform patterns over time. Within the same year drought identification and coverage are timescale-dependent. In 2006 Arvaheer was in extreme drought at all time periods whereas Mandalgovi and Dalanzadgad moved into extreme drought when examined over a longer duration. At the same time Bulgan and Saixan Ovoo fluctuated between mildly wet and dry episodes, only reaching moderate drought conditions.

2.3.2 *Dzud* of 1999–2001

Investigation of the 1999–2001 drought-*dzud* event, identified as Mongolia's worst natural disaster, focused on climate data for 1998 through 2003 in Omnogovi Province (Batima *et al.*, 2005). The 1999 and 2000 SPI values for Dalanzadgad and Bulgan show neither site experienced drought at any timescale. The 1999 12-month SPI values were mild, changing in 2000 to wet conditions (Figure 2.5). In July and August 2001 both sites became drought stricken at a 12-month scale. Dalanzadgad reached drought status at each timescale in 2002 whereas drought at Bulgan occurred at 3 and 24 months. From August 2002 through April 2003 the two sites experienced a 24-month drought, reaching extreme intensity in August. In May 2003 both sites shifted to wet conditions.

During 1999–2002 mean January temperature was generally warmer at both sites than the long-term averages; 2000–01 was ~3°C colder whereas 2002 was 4.8°C and 3.8°C warmer than historic norms at the two sites. Winter precipitation (December–February) in 1999–2001 was normal or slightly low at both sites (Sternberg *et al.*, 2009). High precipitation in December 2001 was followed by above-average temperatures and preceded

Table 2.3 Months in drought 1970–2006

	Months in Drought (% time in drought)			
Arvaheer	**3-month**	**6-month**	**12-month**	**24-month**
Moderate	32 (7.2)	36 (8.1)	37 (8.3)	16 (3.6)
Severe	18 (4.0)	9 (2.0)	11 (2.5)	26 (5.9)
Extreme	9 (2.0)	11 (2.5)	13 (2.9)	18 (4.1)
	Total 59 (13.3)	56 (12.6)	61 (13.7)	60 (13.5)
Bulgan				
Moderate	32 (7.2)	38 (8.6)	40 (9.0)	28 (6.3)
Severe	23 (5.1)	25 (5.6)	22 (4.9)	29 (6.5)
Extreme	10 (2.2)	6 (1.4)	8 (1.8)	5 (1.1)
	Total 65 (14.6)	69 (15.5)	70 (15.7)	62 (13.9)
Dalanzadgad				
Moderate	31 (7.0)	53 (11.9)	62 (13.9)	34 (7.7)
Severe	19 (4.3)	17 (3.8)	19 (4.3)	23 (5.1)
Extreme	10 (2.2)	7 (1.6)	8 (1.8)	12 (2.7)
	Total 60 (13.5)	77 (17.3)	89 (20.1)	69 (15.5)
Mandalgovi				
Moderate	37 (8.3)	36 (8.1)	34 (7.6)	44 (9.9)
Severe	25 (5.6)	24 (5.4)	19 (4.2)	18 (4.1)
Extreme	4 (0.9)	6 (1.4)	10 (2.2)	14 (3.1)
	Total 66 (14.8)	66 (14.9)	63 (14.1)	76 (17.1)
Saixan Ovoo				
Moderate	46 (10.3)	40 (9.0)	41 (9.2)	32 (7.2)
Severe	23 (5.1)	12 (2.7)	16 (3.6)	29 (6.5)
Extrerre	3 (0.6)	12 (2.7)	0 (0)	1 (0.2)
	Total 72 (16.2)	64 (14.4)	57 (12.8)	62 (13.9)

livestock losses in 2002. The human population fell by 50% in both districts in 2001 (Figure 2.6); recovery for human populations started in 2003 with livestock numbers rebounding in 2004 as vegetation cover increased.

2.3.3 *Dzud* and drought: non-drought years

Examination of other historic *dzuds* (1987–88, 1994–95, 1997–98) at both sites showed that drought was not present during these years. Five of the six *dzud* periods (3 Bulgan, 3 Dalanzadgad) had above-average rainfall with, in all cases, precipitation decline the year after the *dzud*. Livestock numbers increased in five of the six years, human populations increased during four of the years and decreased slightly (2%) in two years. In brief, an analysis of *dzuds* found there were no significant correlations between *dzud* years and livestock, precipitation, or January average temperatures.

Time series analysis was used to investigate the relation of non-drought years to livestock, precipitation, and January and July temperatures since 1981 at the 12-month timescale (through August). Results showed that livestock numbers and July temperatures were significantly related ($P = 0.01$) at both sites. Results did not find other significant

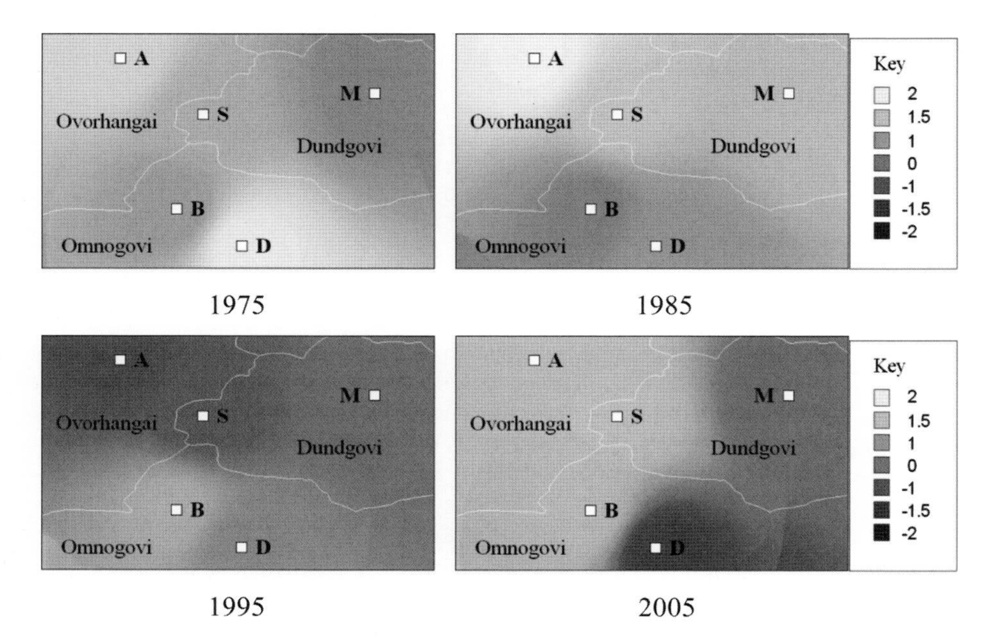

Figure 2.3 Spatial drought at 10-year intervals, calculated at 12-month scale (A = Arvaheer, B = Bulgan, D = Dalanzadgad, M = Mandalgovi, S = Saixan Ovoo)

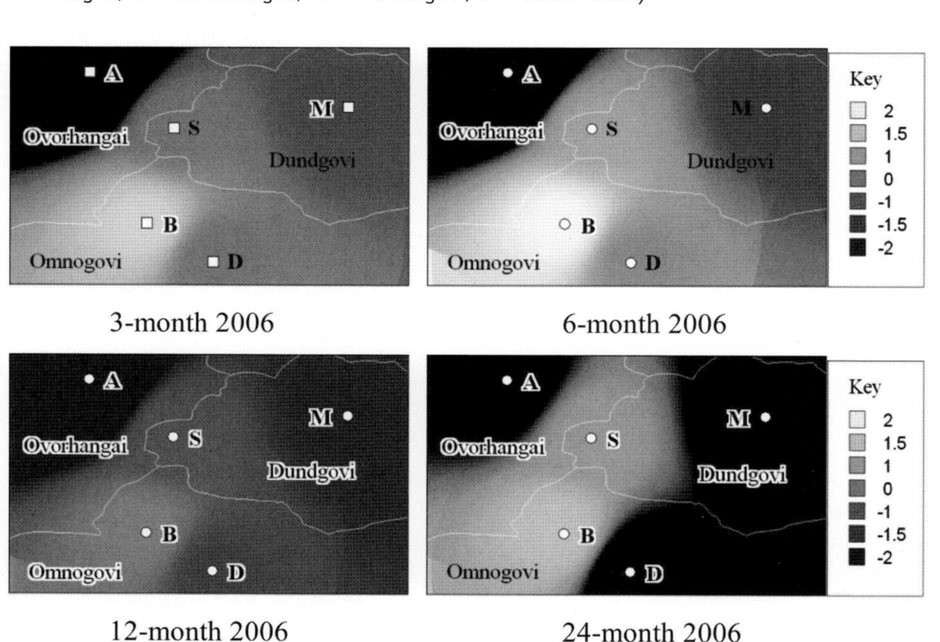

Figure 2.4 Drought at short- and long-term timescales, calculated at 12-month scale (A = Arvaheer, B = Bulgan, D = Dalanzadgad, M = Mandalgovi, S = Saixan Ovoo)

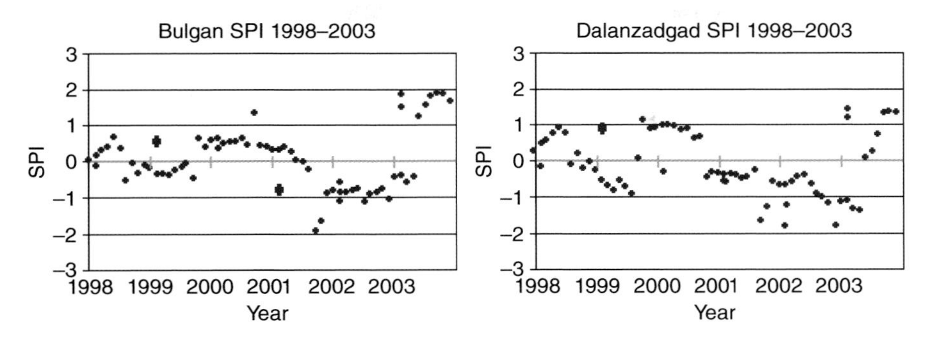

Figure 2.5 12-month SPI records, 1998–2003

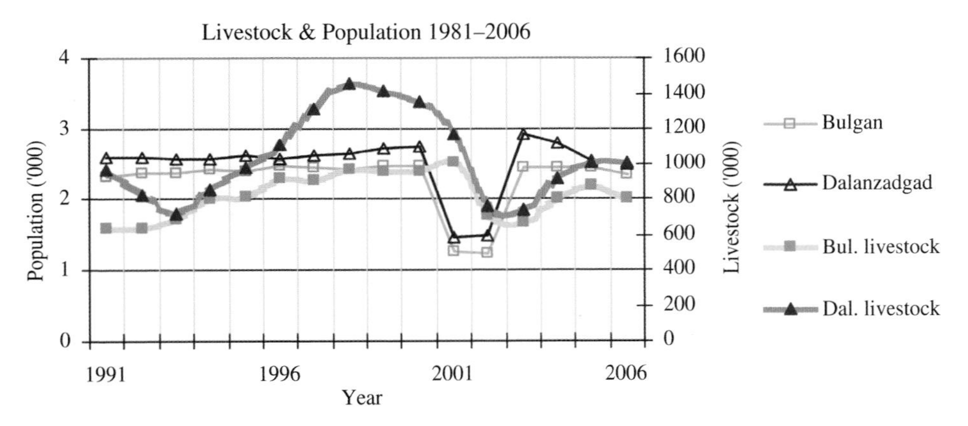

Figure 2.6 Human and livestock populations in Dalanzadgad and Bulgan, 1991–2006

associations. Non-drought years showed less correlation with environmental factors than drought years.

2.4 Discussion

Drought is a recurrent natural hazard in the Gobi region of Mongolia. The most salient factors are the variability of drought events over time, their site-specificity, and duration. The implications are several for human populations: unpredictability in pasture resources for pastoral livelihoods; the limitations of the climate regime on potential agricultural production in the steppe zone; the marginality of the desert-steppe areas experiencing drought; and the ongoing threat to livelihoods dependent on the natural environment for sustenance and survival. Additionally, the conventional understanding is that drought precedes or intensifies a *dzud* (World Bank, 2004). However, results found that *dzuds* were not preceded or concurrent with 12-month droughts, nor were years of extreme drought followed by identified *dzud* conditions. Similarly, in South Gobi Province the *dzuds* of 1987, 1994, 1997, 1999–2000, and 2000–2001 (Reading *et al.*, 2006) are also uncorrelated with drought. Rather than climatic or ecological variables, the strongest long-term correlation of drought is with pastoralist movements, with populations adapting to drought stress through mobility (Sternberg *et al.*, 2009). The frequency and intensity of drought here showed a moderate link with natural factors with little influence on livestock numbers. This study suggests that drought and *dzud* are de-coupled in South Gobi Province; thus the prevailing concept needs more thorough examination, particularly when *dzud* is defined by stock losses rather than meteorological data alone (Suttie, 2005).

Study findings question the oft-cited detrimental effect of drought as events neither magnify nor act as indicators of extreme winter conditions in the region (FAO, 2006; Adiyasuren, 1998; UNCCD, 2002; World Bank, 2002). Drought is a commonplace event and only one of several environmental challenges affecting this arid region. Accurate knowledge of regional drought conditions is essential information for herder decision-making and livelihood wellbeing. Beyond documenting historical patterns, the SPI can serve as a predictive tool for both the immediate future, through identifying precipitation shortfalls at selected timescales, and generating drought probability perspectives for the long-term (Cancelliere *et al.*, 2007).

2.4.1 Resilience

Results suggest variable climate patterns (thus fluctuating pasture conditions) and repeated drought events occur without significantly altering the physical landscape. Human resilience enables shifting herding forces and livestock numbers to be encompassed. In the region both physical and human systems adjust to often serious perturbations (drought, *dzud*, political and economic change) on the steppe, a process that allows the environment to maintain sufficient productivity in its dryland setting for pastoralism to continue as a viable livelihood. On the steppe continued land cover variability represents fluctuation rather than an identifiable move to an altered state.

The nature of pastoral livelihood strategies emphasizes efficient use of limited physical resources and adaptability to the country's transitioning socioeconomic and physical conditions. Today herders address economic pressures, evolving social perspectives, and a loss of government support with traditional mobility patterns, water access, and herd composition decisions that accommodate existing resources. Whereas a rigid system would have limited coping abilities, pastoralists' flexible livelihood approach maintains productivity in a fluid environment and a range of volatile conditions. Socioeconomic processes include a shift to goats for income generation, adjustment to a lack of infrastructure or emergency fodder through personal adaptation and expense, a move away from herding as a career choice, greater herder differentiation, and increased expenses. Concurrently natural and human-influenced disturbances abound in pastoralism (Suttie, 2005). Vegetation cover fluctuation, drought episodes, and drying surface water sources combine with reduced wells and increased livestock numbers to affect the grassland. Despite these challenges, the countryside has seen the expansion of herds and the continued dependence of a large percentage of the population on pastoralism as herders adjust livelihoods to maintain viability (Johnson *et al.*, 2006).

Falkenmark and Rockstrom (2008, p. 94) state that resilience reflects the capacity to 'adapt, recover, develop and remain flexible'. This is relevant in the Gobi as variable conditions are prevalent yet do not disrupt environmental functioning (Walker & Abel, 2002). In encompassing changing conditions systems are adaptive; this contrasts with vulnerability, which identifies a system as reactive and having limited ability to cope with perturbations. The Gobi environment and pastoralism include mechanisms (e.g. spatial fluctuation in vegetation cover, seasonal pastures, mobility) to reduce hazard impact on landscapes and livelihoods. The inherent systemic variability exemplified by drought patterns and the long record of regional pastoralism suggests a history of encountering and resolving disturbances. Thus vulnerability, the notion of exposure and susceptibility to external factors, is a less relevant model than resilience in the region, particularly when linking human-environment processes. Study findings suggest developed coping skills limit herder vulnerability in the Gobi and that natural and learned resilience is a pastoral strength in Mongolia.

Research on resilience in other dryland regions highlights the importance of maintaining ecological balance for landscape viability and pastoral coping skills, such as when dealing with drought (Reed & Dougill, 2002; Sorbo, 2003). Most grassland regions are identified as undergoing rapid change with each rangeland facing localized physical and institutional conditions (Walker & Abel, 2002). How Mongolia adapts to disturbance and reorganization will depend on forces (e.g. economics, precipitation) that shape the environment and policy that enables herding to be practiced at adequate spatial and temporal scales to maintain viability. Plummer and Armitage's (2007) assertion that resilience encompasses ecology, economics, and society is particularly relevant to findings in the Gobi Desert.

Observations on resilience segue into evaluating sustainability (Brand, 2009). Exhibited resilience in the steppe pasture systems can be regarded as an important part of sustainability; Maler (2008) states that resilience and sustainable development share a single conceptual framework. The Gobi region exemplifies this cohesion as findings suggest that pastoral livelihoods can continue to be productive without threatening environmental functionality (Turner, 2010). Challenges such as water access and economic pressures affect both sustainable development and future systemic resilience. In Walker and Abel's (2002) four stages of range development Mongolia is at the initial, extensive phase of livestock movement with low human and livestock density, self-reliance, and strong ecological-household linkage. At this level the steppe shows sustainability; evolution beyond current range dynamics will call for new evaluation.

Prior work on sustainability in Mongolia highlighted uncertainty about the impact of policy, institutional support, and socioeconomic pressures in the countryside. Scientists stressed continued mobility, management flexibility, and communal rangeland use, factors that contribute to resiliency, as key strategies as well as the need for herders to be involved with range management and regulation through a participatory process (Fratkin & Mearns, 2003; Sayre & Fernandez-Gimenez, 2003). Ojima and Chuluun (2008) identified ecologically-based land use, including redrawing administrative boundaries for better resource access, investment in infrastructure, building more wells to facilitate mobility and pasture access, and improved services as essential. Further, strengthening grazing regulation, related services, and economic viability are important for sustainable rangeland use (Johnson *et al.*, 2006).

A regional perspective on the social-ecological debate can improve knowledge of the Gobi region. This involves greater understanding of systems in northern China, a region with a similar pastoral history. Different development patterns and a shift to settled livestock raising in China exemplifies how political and economic systems can impact pastoralism in a shared landscape. China represents an alternate range model that involves a marked change from open grazing to fenced enclosures and intensified practices (Chen & Tang, 2005). It also identifies potential implications for ecosystem sustainability and livelihood wellbeing as dynamics focus on economic production in a well-organized, government-controlled pastoral system integrated into a larger society. Issues of pastoral viability, vulnerability, and long-term resilience encounter perturbations that contrast with Mongolia and can be evaluated with similar investigation. In China particular attention is given to poor management practices and high desertification rates in northern China that question the viability of this approach on the steppe (Liu *et al.*, 2005a; Normile, 2007; Sneath, 1998). Degradation has become a major steppe concern, and awareness of

Chinese conditions can inform debate and management in Mongolia. Though political and economic organization differs from Mongolia's governance system, the comparison highlights existing forms of livestock development on the plateau that may come to Mongolia in a modified form.

Research in northern China identifies several causes of recent land-cover change. These include inappropriate land use, over-cultivation in marginal areas, irrational water use, climate change, resource exploitation, population pressure, property rights/fencing, and loss of traditional management strategies (Jun Li *et al.*, 2007; Lee & Zhang, 2005; Ma *et al.*, 2007; Wang *et al.*, 2008). Several factors (water, mining, land tenure, customary practices) are relevant; examination of the Chinese paradigm can inform discussion and action in Mongolia. By studying its neighbour's land use patterns and basing decisions on steppe conditions and biophysical parameters the country can limit the need for future ecological restoration that is ongoing in China (Liu *et al.*, 2005b; Zha & Gao, 1997). Chinese researchers have concluded that 'Mongolian nomadic culture has advantages over agrarian culture in ecology and environmental care, sustainable utilization of grasslands, and in sustainable human social economic development in the region' (Zhang *et al.*, 2007, p. 19). This suggests that a combination of existing pastoral strengths with modern knowledge and skills can maintain ecological viability and socioeconomic wellbeing in Mongolia.

2.5 Conclusion

The study finds that drought and localized extreme weather events in Mongolia are unassociated. Development design and relief-aid efforts would be more productive if it were recognized that these hazards can be unconnected. Further, by acknowledging resilient strategies that are a part of pastoralists' traditional ability to accommodate drought, for example through migration, their need for external support when dealing with *dzuds* that threaten livestock survival would be decoupled from drought aid per se. Establishment of adequate risk management strategies could then improve herders' ability to cope with drought and *dzuds*.

Addressing environmental issues on the Mongolian plateau, an area shared with China, adds to the understanding of a region that receives much attention politically and economically but where documentation of physical processes has received less attention. This work presents material that can be compared with other pastoral areas and provides a factual basis for regional development and policy debate. Expanded research is required to aid the identification of drought risks and the degree of pastoralist exposure to precipitation deficiency over multiple time intervals in this arid region.

Acknowledgements

The author would like to thank the British Academy and the Royal Geographical Society for their research funding. The study is supported by an international collaborative project of the Ministry of Science and Technology of China (Grant No. 2013DF91700).

References

Adger, W. (2006) Vulnerability. *Global Environmental Change* **16**: 268–81.

Adiyasuren, T. (1998) *Environment and Development Issues in Mongolia*. Ulaanbaatar: Ministry of Nature and Environment.

Asian Development Bank (2005) *Mongolia: Country Environmental Analysis*. Manila: Asian Development Bank.

Batima, P., Natsagdorj, L., Gombluudev, P., Erdenetsetseg, B. (2005) Observed climate change in Mongolia. AIACC Working Paper No. 12. Available at: www.aiaccproject.org/working_papers/Working%20Papers/AIACC_WP_No013.pdf

Begzsuren, S., Ellis, J., Ojima, D., Coughenour, M., Chuluun, T. (2004) Livestock responses to droughts and severe winter weather in the Gobi Three Beauties National Park, Mongolia. *Journal of Arid Environments* **59**: 785–96.

Bold, B. (2001) *Mongolian Nomadic Society*. New York: St Martin's Press.

Brand, F. (2009) Critical natural capital revisited: ecological resilience and sustainable development. *Ecological Economics* **68**: 605–12.

Cancelliere, A., Di Mauro, G., Bonaccorso, B., Rossi, G. (2007) Drought forecasting using the Standardized Precipitation Index. *Water Resource Management* **21**: 801–19.

Chen, Y., Tang, H. (2005) Desertification in north China: background, anthropogenic impacts and failures in combating it. *Land Degradation and Development* **16**: 367–76.

Chuluundorj, O. (2006) A multi-level study of vulnerability of Mongolian pastoralists to natural hazards and its consequences on individual and household well-being. PhD thesis, University of Colorado, Denver.

Cutter, S., Boruff, B., Shirley, W. (2003) Social vulnerability to environmental hazards. *Social Science Quarterly* **84**: 242–59.

Falkenmark, M., Rockstrom, J. (2008) Building resilience to drought and desertification-prone savannas in sub-Saharan Africa: the water perspective. *Natural Resource Forum* **32**: 93–102.

FAO (Food and Agriculture Organization) (2006) Country Pasture/Forage Resource Profiles – Mongolia. Available at: http://www.fao.org/ag/AGP/AGPC/doc/Counprof/Mongolia/mongol1.htm

Fernandez-Gimenez, M. (2000) The role of Mongolian nomadic pastoralists' ecological knowledge in rangeland management. *Ecological Applications* **10**: 1318–26.

Fernandez-Gimenez, M., Allen-Diaz, B. (1999) Testing a non-equilibrium model of rangeland vegetation dynamics in Mongolia. *Journal of Applied Ecology* **36**: 871–85.

Fratkin, E., Mearns, R. (2003) Sustainability and pastoral livelihoods: lessons from East Africa and Mongolia. *Human Organization* **62**: 112–24.

Gallopin, G. (2006) Linkages between vulnerability, resilience, and adaptive capacity. *Global Environmental Change* **16**: 293–303.

Gallopín, G., Funtowicz, S., O'Connor, M., Ravetz, J. (2001) Science for the 21st century: from social contract to the scientific core. *International Social Science Journal* **168**: 219–29.

Geist, H. (2005) *The Causes and Progression of Desertification*. Burlington, VT: Ashgate.

Hayes, M., Svoboda, M., Wilhite, D., Vanyarkho, O. (1999) Monitoring the 1996 drought using the Standardized Precipitation Index. *Bulletin of the American Meteorology Society* **80**: 429–38.

Hilbig, W. (1995) *The Vegetation of Mongolia*. Amsterdam: SPB Academic Publishing.

Holling, C. (1973) Resilience and stability of ecological systems. *Annual Review of Ecology and Systematics* **4**: 1–23.

Johnson, D., Sheehy, D., Miller, D., Damiran, D. (2006) Mongolian rangelands in transition. *Secheresse* **17**: 133–41.

Jun Li, W., Ali, S., Zhang, Q. (2007) Property rights and grassland degradation: A study of the Xilingol Pasture, Inner Mongolia, China. *Journal of Environmental Management* **85**: 461–70.

Kelly, P., Adger, N. (2000) Theory and practice in assessing vulnerability to climate change and facilitating adaptation. *Climatic Change* **47**: 325–52.

Keyantash, J., Dracup, J. (2002) The quantification of drought: an evaluation of drought indices. *Bulletin of the American Meteorological Society* **83**: 1167–80.

Labedzki, L. (2007) Estimation of local drought frequency in central Poland using the standardized precipitation index SPI. *Irrigation and Drainage* **56**: 67–77.

Lee, H., Zhang, D. (2005) Perceiving land-degrading activities from the lay perspective in northern China. *Environmental Management* **36**: 711–25.

Lepers, E., Lambin, E., Janetos, A., *et al.* (2005) A synthesis of information on rapid land-cover change for the period 1981–2000. *BioScience* **55**: 115–24.

Lioubimtseva, E., Cole, R., Adams, J., Kapustin, G. (2005) Impacts of climate and land-cover changes in arid lands of Central Asia. *Journal of Arid Environments* **62**: 285–308.

Liu, Y., Wang, D., Gao, J., Deng, W. (2005a) Land use/cover changes, the environment and water resources in northeast China. *Environmental Management* **36**: 691–701.

Liu, H., Cai, X., Geng, L., Zhong, H. (2005b) Restoration of pastureland ecosystems: case study of western Inner Mongolia. *Journal of Water Resources Planning and Management* **131**: 420–30.

Ma, Y., Fan, S., Zhou, L., Dong, Z., Zhang, K., Feng, J. (2007) The temporal change of driving factors during the course of land desertification in arid region of North China: the case of Minqin County. *Environmental Geology* **51**: 999–1008.

Maler, K. (2008) Sustainable development and resilience in ecosystems. *Environmental and Resource Economics* **39**: 17–24.

Marin, A. (2010) Riders under storms: contributions of nomadic herders' observations to analysing climate change in Mongolia. *Global Environmental Change* **20**: 162–76.

McKee, T., Doeskan, N., Kleist, J. (1993) The relationship of drought frequency and duration to time scales. *Eighth Conference on Applied Climatology*. Anaheim, CA: American Meteorological Association, pp. 179–84.

Mongolian Statistical Yearbook (2006) Ulaanbataar: National Statistical Office of Mongolia.

Munkhtsetseg, E., Kimura, R., Wang, J., Shinoda, M. (2007) Pasture yield response to precipitation and high temperature in Mongolia. *Journal of Arid Environments* **70**: 94–110.

Nandintsetseg, B., Greene, J., Goulden, C. (2007) Trends in extreme daily precipitation and temperature near Lake Hovsgol, Mongolia. *International Journal of Climatology* **27**: 341–7.

Normile, D. (2007) Getting at the roots of killer dust storms. *Science* **317**: 314–16.

Ojima, D., Chuluun, T. (2008) Policy changes in Mongolia: implications for land use and landscapes. In: Galvin, K., Reid, R., Behnke, Jr R., Hobbs, N. (eds), *Fragmentation in Semi-Arid and Arid Landscapes: Consequences for Human and Natural Systems*. Dordrecht, The Netherlands: Springer, pp. 179–93.

Pelling, M. (2003) *The Vulnerability of Cities: Natural Disasters and Social Resilience*. London: Earthscan.

Plummer, R., Armitage, D. (2007) A resilience-based framework for evaluating adaptive co-management: Linking ecology, economics and society in a complex world. *Ecological Economics* **61**: 62–74.

Reading, R., Bedunah, D., Amgalanbaatar, S. (2006) Conserving biodiversity on Mongolian Rangelands: implications for protected area development and pastoral uses. *USDA Forest Service Proceedings* RMRS-P-39.

Reed, M., Dougill, A. (2002) Participatory selection process for indicators of rangeland condition in the Kalahari. *The Geographical Journal* **168**: 224–34.

Retzer, V., Reudenbach, C. (2005) Modelling the carrying capacity and coexistence of pika and livestock in the mountain steppe of the south Gobi, Mongolia. *Ecological Modelling* **189**: 89–104.

Rossabi, M. (2005) *Modern Mongolia: from Khans to Commissars to Capitalists*. Berkeley: University of California Press.

Rouault, M., Richard, Y. (2003) Intensity and spatial extension of drought in South Africa at different time scales. *Water South Africa* **29**: 489–500.

Sasaki, T., Okayasu, T., Jamsran, U., Takeuchi, K. (2008) Threshold changes in vegetation along a grazing gradient in Mongolian rangelands. *Journal of Ecology* **96**: 145–54.

Sayre, N., Fernandez-Gimenez, M. (2003) The genesis of range science, with implications for current development policies. In: Also, N., Palmer, A., Milton, S., *et al.* (eds), *Proceedings of the VIIth International Rangelands Congress, 26th July – 1st August 2003*. Durban, South Africa: The Congress, pp. 1976–85.

Shinoda, M., Ito, S., Nachinshonhor, G., Erdenetsetseg, D. (2007) Phenology of Mongolian grasslands and moisture conditions. *Journal of the Meteorological Society of Japan* **85**: 359–67.

Smakhtin, V., Hughes, D. (2007) Automated estimation and analyses of meteorological drought characteristics from monthly rainfall data. *Environmental Modelling & Software* **22**: 880–90.

Smit, B., Wandel, J. (2006) Adaptation, adaptive capacity, and vulnerability. *Global Environmental Change* **16**: 282–92.

Sneath, D. (1998) State policy and pasture degradation in inner Asia. *Science* **281**: 1147–8.

Sonmez, K., Komuscu, A., Erkhan, A., Turgu, E. (2005) An analysis of spatial and temporal dimension of drought vulnerability in Turkey using the Standard Precipitation Index. *Natural Hazards* **35**: 243–64.

Sorbo, G. (2003) Pastoral ecosystems and the issue of scale. *Ambio* **32**: 113–17.

Sternberg, T. (2008) Environmental challenges in Mongolia's dryland pastoral landscape. *Journal of Arid Environments* **72**: 1294–304.

Sternberg, T. (2009) Nomadic geography: pastoral environments in the Gobi Desert, Mongolia. PhD thesis, Oxford University, Oxford.

Sternberg, T., Middleton, N., Thomas, D. (2009) Pressurized pastoralism: what is the role of drought? *Transactions of the Institute of British Geographers* **34**: 364–77.

Suttie, J. (2005) Grazing management in Mongolia. In: Suttie, J., Reynolds, S., Batello, C. (eds), *Grasslands of the world. FAO Plant Production and Protection Series No. 34*. Rome: FAO.

Turner II, B. (2010) Sustainability and forest transitions in the southern Yucatan: the land architecture approach. *Land Use Policy* **27**: 170–9.

Turner II, B., Kasperson, R., Matson, P., *et al.* (2003) A framework for vulnerability analysis in sustainability science. *Proceedings of the National Academy of Sciences of the USA* **100**: 8074–9.

Turner II, B., Lambin, E., Reenberg, A. (2007) The emergence of land change science for global environmental change and sustainability. *Proceedings of the National Academy of Sciences of the USA* **104**: 20666–71.

UNCCD (United Nations Commission to Combat Desertification) (2002) Second report on the implementation of the UN Convention to Combat Desertification. Available at: www.unccd.int/RegionalReports/mongolia-eng2002.pdf

UNEP (United Nations Environmental Programme) (2002) *Mongolia: State of the Environment 2002*. Ulaanbataar: United Nations Environmental Program.

Walker, B., Abel, N. (2002) Resilient rangelands – adaptation in complex systems. In: Gunderson, L. Holling, C. (eds), *Panarchy: Understanding Transformations in Human and Natural Systems*. London: Island Press.

Walker, B., Holling, C.S., Carpenter, S.R., Kinzig, A. (2004) Resilience, adaptability and transformability in social-ecological systems. *Ecology and Society* **9**: 5 [online]: http://www.ecologyandsociety.org/vol9/iss2/art5/

Walker, B., Salt, D. (2006) *Resilience Thinking*. London: Island Press.

World Bank (2002) *Mongolia Environmental Monitor*. Ulaanbaatar: World Bank.

World Bank (2004) *Mongolia Environmental Monitor*. Ulaanbaatar: World Bank.

World Meteorological Organization (1986) Report on drought and countries affected by drought during 1974–1985. WCP-118, WMO/TD. No. 133.

Wu, H., Hayes, M., Weiss, A., Hu, Q. (2001) An evaluation of the Standardized Precipitation Index, the China-Z Index, and the Statistical Z-Score. *International Journal of Climatology* **21**: 745–58.

Wu, H., Svoboda, M., Hayes, M., Wilhite, D., Wen, F. (2007) Appropriate application of the Standardized Precipitation Index in arid locations and dry seasons. *International Journal of Climatology* **27**: 65–79.

Wang, X., Chen, F., Hasi, E., Li., J. (2008) Desertification in China: an assessment. *Earth-Science Reviews* **88**: 188–206.

Yang, X., Rost, K., Lehmkuhl, F., Zhenda, Z., Dodson, J. (2004) The evolution of dry lands in northern China and in the Republic of Mongolia since the last Glacial Maximum. *Quaternary International* **118–119** : 69–85.

Yang, X., Zhang, K, Jia, B., Ci, L. (2005) Desertification assessment in China: An overview. *Journal of Arid Environments* **63**: 517–31.

Zha, Y., Gao, J. (1997) Characteristics of desertification and its rehabilitation in China. *Journal of Arid Environments* **37**: 419–32.

Zhang, M., Borjigin, E., Zhang, H. (2007) Mongolian nomadic culture and ecological culture: on the ecological reconstruction in the agro-pastoral mosaic zone in Northern China. *Ecological Economics* **62**: 19–26.

3 Vulnerability and Resilience of the Mongolian Pastoral Social-Ecological Systems to Multiple Stressors

Dennis Ojima[1,2], Togtokh Chuluun[1,3,4] and Myagmarsuren Altanbagana[3,4]

[1]*Natural Resource Ecology Laboratory, Colorado State University, Fort Collins, Colorado, USA*
[2]*Ecosystem Science and Sustainability Department, Colorado State University, Fort Collins, Colorado, USA*
[3]*National University of Mongolia, Ulaanbaatar, Mongolia*
[4]*Dryland Sustainability Institute of Mongolia, Ulaanbaatar, Mongolia*

3.1 Introduction

Dramatic changes have occurred in pastoral systems of Mongolia during the past two decades. An assessment of the impact of these changes on the environment and quality of life is essential for sustainability of the region. An understanding of the complex interactions and impacts of various management strategies on the environment and human systems is critical for the development of sustainable adaptation strategies. Evaluation of the pastoral systems has been conducted in the region (Chuluun *et al.*, 2005a; Ojima *et al.*, 2004). Pastoral systems, where humans depend on livestock, exist largely in arid or semi-arid ecosystems where climate is highly variable. Thus, in many ways pastoral systems are adapted to climatic variability. It is plausible to assume a direct connection between

Vulnerability of Land Systems in Asia, First Edition. Edited by Ademola K. Braimoh and He Qing Huang.

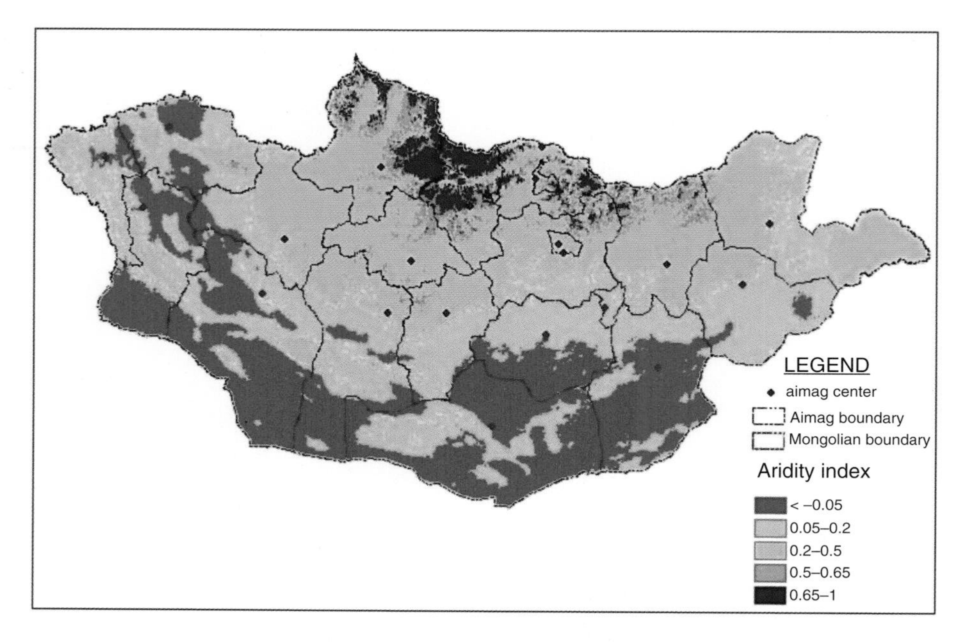

Figure 3.1 Aridity classification map of Mongolia

climate variability, ecosystem dynamics and a nomadic land-use system in Mongolia. Interactions between ecosystems and nomadic land-use systems have co-shaped them in mutually adaptive ways for hundreds of years, thus making both the Mongolian rangeland ecosystem and nomadic pastoral system resilient and sustainable. However, socio-economic and climate conditions of the past three decades are affecting these management systems in dramatic ways.

The majority of lands (about 90%) in Mongolia are drylands (Khaulenbek, 2009). Hyper-arid, arid and semi-arid zones occupy more than 80%, and a dry sub-humid zone occupies less than 10% of the territory of Mongolia (Figure 3.1). The deserts (and southern part of the desert-steppe) fall into the hyper-arid zones. The desert-steppe (northern part of the desert-steppe) and dry steppe fall into the arid zones. All other steppe ecosystems, such as steppe, mountain steppe, forest steppe and meadow steppes, are included in the semi-arid zones. This aridity greatly contributes to the sensitivity of this region to climate warming and variability in precipitation, and pastoral responses to this range of aridity are reflected in the mobility of nomadic patterns of pastoralists. Some of the hyper-arid zones are not used by herders, and herders in dry sub-humid zones are the least mobile, moving for short distances and only twice a year: between summer-fall and winter-spring pastures. Herders living in arid and semi-arid landscapes are more mobile with seasonal movements.

Mongolian land-use dynamics of the past two decades have also been impacted significantly by changing demographic, political and economic forces affecting pastoral exploitation. The response of livestock dynamics to political and economic changes has led to a tripling of goat numbers and a reduction in cattle numbers across Mongolia since its transition to a market economy based on cashmere with both socio-economic and ecological consequences. The general trend involves greater intensification of resource exploitation at the expense of traditional patterns of extensive range utilization. This set of drivers is orthogonal to the above described climate drivers. Thus we expect relationships between climate, land use and land cover to be crucially modified by the socio-economic forces mentioned above. Nevertheless, the complex relationship between climate variability and pastoral exploitation patterns will still form the environmental framework for overall patterns of land-use change.

The objective of this chapter is to identify current trends in key factors affecting land-use systems, ecosystem services and characteristics, and how these are affecting the vulnerability of the Mongolian pastoral systems to climate and other stresses being experienced in the region over the past three decades. In addition, we will discuss various coping strategies in response to these stresses.

3.2 The current situation

3.2.1 Climate conditions

Climate variability and change are being experienced as droughts, extreme winter conditions and warming. We have analyzed climate data and land cover changes to evaluate factors affecting land-use changes. Linkages between current trends in policy decisions and economic forces will be developed in the analysis of environmental and ecosystem dynamics. During the last 70 years the annual mean air temperature increased by 2.14°C in Mongolia (Figure 3.2, MNET, 2009: 'MARCC'). Winter temperature has increased by 3.61°C and spring-fall temperature by 1.4–1.5°C. However, the summer temperature has decreased by 0.3°C. Changes in warming have a spatial character: winter warming is more pronounced in the high mountains and mountain valleys, and less in the Gobi desert and the steppe. There is a slight increasing trend in annual precipitation over the last 60 years (Natsagdorj *et al.*, 2003). During 1940–98, annual precipitation increased by 6%, while summer precipitation

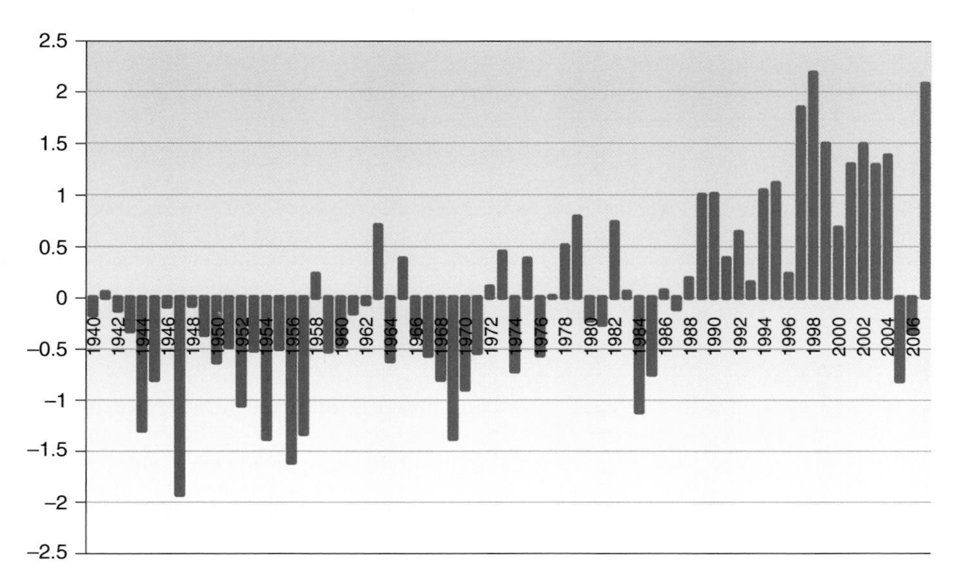

Figure 3.2 Mean annual temperature changes for Mongolia. Long-term average computed for the length of record starting in 1940 and ending in 2007

increased by 11% (mostly in August) and spring precipitation decreased by 17%, mostly in May.

Changes in seasonal precipitation (i.e. initiation of snowfall in the mountain regions) and availability of water have resulted in water shortages and additional stress to rangelands. Observations of seasonal changes in the initiation of plant growth across regions of Mongolia have been documented with satellite data analysis (Ojima *et al.*, 2004). The herders were very sensitive to water availability during both the warm and cold seasons and there was also ecosystem degradation as a result of overgrazing around the wells and the few remaining springs that had not already shrunk due to the impact of climate change. The herders were very sensitive to snow cover change as well. For instance, Nogoon Suuri, a herder of Hujirt sum, indicated that six springs had disappeared and that only a single watering point remained for 14 households with 3000 livestock. This only remaining spring was prone to freezing, leaving these households without water in early December 2006. As there is usually snow on the ground during this time of year, the herders dispersed, moving away from their winter camps so that they could use the snow as a winter water source.

3.2.2 Water resources

In the central Mongolia study sites, the disappearance of small streams, lakes and springs was also observed. Decreases in snowfall, increased tree cutting, the melting of permafrost, intensifying drying trends, destruction of riparian zone shrubs and swamps, and overgrazing all interacted in a non-linear way, resulting in the disappearance of water sources. Regional climate may be affected due to the albedo change that comes with land and snow cover changes. The summer of 2009 had large floods in Hujirt sum territory due to both heavy rainstorms and drought conditions.

According to the 2000 statistical information (NSO, 2000), the total number of water points in Mongolia was slightly fewer than 31,000 of which approximately 8000 were mechanical wells and the remainder were hand wells (almost 23,000). Of the approximately 21,000 water points located in pastureland approximately 25% are dysfunctional. Since 1990, the number of water points decreased by 28%, with about 75% of these being mechanical wells. Of the wells constructed between 1960 through 1990, only about 40% are still operational (UNDP, 2005).

Because the number of water points continues to decline from year to year, traditional nomadic pastoral patterns of seasonal grazing have been disrupted with the loss of these watering points. The increased grazing pressure around the remaining water points has resulted in overgrazing of pastures. In addition to these anthropogenic grazing effects, a warmer and drier climate has created conditions promoting expansion of deserts. From the beginning of the 20th century global warming intensified in northern latitudes, and the temperature in Mongolia has increased by 1.8°C since 1940 (Batima *et al.*, 2005). Forage availability, determined by remote sensing data from 1982 through 2002 in central parts of Mongolia, was affected by these climate effects (Ojima *et al.*, 2004). Ellis and colleagues (2002) showed that the steppe area adjacent to the Gobi region is especially vulnerable to climate changes, and with increased grazing pressures during the past decade these have led to desertification.

Riparian ecosystems appear to have keystone value in coupled pastoral social-ecological systems. The collapse of these critical ecosystems' ability to provide water would greatly impact the pastoral community, as water is the most valuable resource for both people and animals in drylands.

3.2.3 The nomadic system

The Mongolian nomadic pastoralists have traditionally adjusted their movements to environmental conditions in the region where they live. In regions with relatively greater climate variability and increased uncertainty, pastoral movements tend to be more chaotic and follow more opportunistic strategies to secure forage. These movements are associated with drier parts of the steppe and desert areas such as the Gobi desert and desert-steppe regions, where non-equilibrium ecosystem dynamics are observed (Ellis & Chuluun, 1993; Fernández-Giménez, 1999).

During the collectivization, or *negdel*, period between 1960 and 1990, herders were moving less frequently, and across longer distances with mechanized transport vehicles (e.g. trucks and tractors). However, herders still kept the traditional pastoral land-use concept. *Negdels*, or collectives, were dissolved with the privatization of livestock in the early 1990s. The pastoral collectives and the state assigned new administrative territories that cut across lines of traditional nomadic movements. Within these collectives were several smaller units or teams, which were the levels at which herding activities were allocated and comprised one or two households (Mearns, 1993). Although these traditional functions may have been subsumed by the collectives, the customary institutions did not disappear altogether, as is demonstrated by the fact that many are now re-emerging.

From 1960 through 1990 many aspects of the traditional nomadic culture were replaced by socialist practices. Herdsmen were commonly organized into collectives and were allowed only a small number of animals for private ownership (Sneath, 2003). The provincial and national government established a strategy for short-term, long-distance moves (*otor*) to safeguard against drought and rangeland overuse. This strategy also served as a mechanism to fatten the stock in the summer and fall seasons. The collectives made all decisions over allocations of animals, and specialization of tasks and species. Although these functions may have been subsumed by the collectives, the customary institutions did not disappear altogether, as is demonstrated by the fact that many *hot ails*[1] have re-emerged during the transition to a market economy (Janzen, 2005; Reading *et al.*, 2006; Schmidt *et al.*, 2002; Schmidt, 2006).

Since de-collectivization started in the early 1990s, pastoral movements have become less frequent and shorter due to the lack of subsidies to maintain transportation for long-distance travel and a preference to stay closer to settled areas (Fernández-Giménez, 2006; Janzen, 2005; Reading *et al.*, 2006). The higher concentration of livestock near settled areas and year-round use of riparian zones have led to deterioration of the rangelands (Chuluun *et al.*, 2005a; Janzen, 2005; Ojima *et al.*, 2004).

3.2.4 Livestock changes

Additional stresses on the system have arisen from changing livestock stocking rates due to market factors contributing to the expansion of cashmere production in the region and the increase in goat numbers since 1990. Since 1990, Mongolia has shifted to a free-market economy, which has led to changes in the livestock sector. In addition, during this transition period, unemployment increased because of stagnation of enterprises in the capital city and other civic centers, and poverty has increased in part due to accelerating inflation. Accompanying these trends is a decline in social services available in these civic centers. Numerous small administrative units or villages became less viable due to a lack of economic and

[1] A Mongolian term for groups of three or four herding families herding and moving together.

resource support from the central government. Although livestock was privatized in rural areas and the number of livestock increased, the livestock industry has suffered due to degradation of pasture and unfavorable climate conditions (i.e. summer droughts followed by winter storms, or *zuds*) in recent years (i.e. 1999 to 2002). The higher concentration of livestock near settled areas and year-round use of riparian zones have led to deterioration of the rangelands (Ojima *et al.*, 2004). As a result, generally livestock herd size has shrunk and poverty has increased.

Cultural patterns of livestock movement have changed during the past several decades, which has increased the vulnerability of these pastoral communities to climate change. Environmental degradation has increased markedly in Mongolia and is associated with increased livestock numbers. Since about 1995 the area of highly degraded land increased 1.8 times (MNE, 2001) and desertification in the arid and semi-arid regions of Mongolia increased by 3.4% during 1990–2004 (MNE, 2006). Acceleration in desertification has occurred in part because of human influences and in part due to a changing climate. As the number of livestock exceeded pasture carrying capacity, a number of impacts have been observed. These include pastureland degradation, a decline in plant production, ecosystem breakdown, and a shift of grasslands to more desert-like conditions. In some areas, soil deteriorated and sand migration increased as bushes and trees, which are the main factor for arresting sand migration, were cut for local use.

3.3 Analysis of vulnerability of critical ecosystem services

3.3.1 Vulnerability index of pastoral systems

The vulnerability index (V) of pastoral systems to climate and land-use changes has been calculated as the sum of the *zud* index and the rangeland use index. The vulnerability of rangelands is higher when both *zud* and rangeland use intensity are higher. Hence,

$$V = \Delta N + \Delta S$$

where ΔN is the pasture use index and ΔS is the *zud* index, which incorporates previous summer drought conditions with current winter conditions (Natsagdorj & Sarantuya, 2004).

3.3.2 Integrated *zud* index

Natsagdorj *et al.* (2003) developed an integrated *zud* index based on the fact that *zud* occurred when winters were colder by 2–7°C and had mostly snow deeper than 15–20 cm. Also the biggest livestock loss happened during *zuds* with a previous summer drought. For example, 8 million livestock (one-third of the total livestock) were lost during the 1944–45 *zud*. A severe drought happened in 1944, and the temperature during November 1944 to April 1945 was colder by 5.9–11.4°C than the long-term average; snow depth was 15–28 cm, and cold storms were frequent. Thus, the integrated *zud* index (Natsagdorj & Sarantuya, 2004) is higher when the winter is colder and snowy, and the summer is drier and hot. This integrated *zud* index, which accounted for previous summer drought, had very good correlation with livestock numbers lost during a *zud*.

Summer drought index and winter indexes were calculated using Ped index-difference normalized temperature and normalized precipitation indexes (Ped, 1975) as:

$$S_{\text{summer}} = \sum_{i=1}^{n} \left(\frac{T - \overline{T}}{\sigma_T} \right) - \sum_{i=1}^{n} \left(\frac{R - \overline{R}}{\sigma_R} \right)$$

$$S_{\text{winter}} = \sum_{i=1}^{n} \left(\frac{T - \overline{T}}{\sigma_{\text{T}}} \right)_i - \sum_{i=1}^{n} \left(\frac{R - \overline{R}}{\sigma_R} \right)_i$$

where T_i and R_i are, respectively, the temperature and precipitation for particular months at the 'i' station; and σ_T and σ_R are the fluctuation of temperature and precipitation for particular months at the 'i' station, defined by the following formula:

$$\sigma = \sqrt{\frac{1}{n-1} \cdot \sum_{i=1}^{n} (x_i - \langle x \rangle)^2}$$

where x_i is the ith value of x and $<x>$ is the arithmetic average.

A drought is severe when S_{summer} has a high value (with high temperature and low precipitation), and a *zud* is severe when S_{winter} has a low value (with low temperature and high precipitation). The integrated *zud* index, which considers the previous summer drought condition, is defined as the difference between the drought and *zud* indexes (Natsagdorj & Sarantuya, 2004):

$$\Delta S = S_{\text{summer}} - S_{\text{winter}}$$

The integrated *zud* index (Natsagdorj & Sarantuya, 2004) has a high value (i.e. the *zud* is severe) when S_{summer} has a higher value (summer is drier and hot) and S_{winter} has a lower value (winter is colder and snowy).

We calculated the *zud* index for our study sites, making a slight modification in terms of winter and summer months. We assumed December, January, February, March and April as winter months because later onset of winter and snow storms in March and April. May, June and July were considered as summer or growing season months in the steppe, mountain steppe and forest steppe because rainfall during these months is critical for peak plant biomass, which happens early August. However, August was also considered a summer or growing season month in desert steppe and desert zones because rainfall in August still contributes to plant growth in short-grass steppe and desert steppe regions.

3.3.3 Rangeland use index

The rangeland use index was calculated using the formula:

$$\Delta N = \alpha \left(\frac{N - N_0}{N_0} \right)$$

where N is livestock density (sheep units/ha); N_0 is the carrying capacity (sheep units/ha) (Mongolian National Atlas, 1990; Tserendash, 2006); and α is the pasture management coefficient. Plant biomass data at study sites were obtained from the 'Gobi forage' project website (http://glews.tamu.edu/mongolia/).

3.3.4 Assessment of rangeland vulnerability to climate and land-use changes

Vulnerability of rangelands to climate and land-use changes has been assessed at *sum* scale between 1970-1990 and 1991-2003 (Chuluun & Altanbagana, 2005). Carrying capacity numbers were derived from the Mongolian National Atlas (1990). Vulnerability of rangelands to climate and land-use changes increased since the 1990s in all central and western parts of Mongolia (Figure 3.3), because of both increased *zud* and land-use intensity.

3.4 Coping scenarios

The key to sustainability lies in enhancing the resilience of communities (Walker & Salt, 2006). Resilience is the ability to absorb change and still retain basic function and structure. Interestingly, cooperatives based on traditional pastoral networks are re-emerging in response to recent drought and extreme winter conditions during 1999–2002 in Mongolia. A comparative study of pastoral communities showed that a cooperative based on a traditional network had lost fewer livestock compared to other communities during the recent climate disaster (Chuluun & Enh-Amgalan, 2003). Today the Mongolian rangelands are split between two pathways: predominantly private land ownership or a traditional land-use culture operated by traditional, resilient pastoral networks. It appears that enhancement of cooperatives based on traditional resilient pastoral networks will lead to a reduction in the future vulnerability of Mongolian pastoral systems.

The main elements for long-term sustainable development of pastoral animal husbandry are to use natural pastures and hay and to maintain ecosystem integrity. In Mongolia, almost half (151 out of 330 *sums*) of the county-level administrative units have no seasonal grazing lands due to the small land area and more homogeneous nature of landscapes occupied by these *sum*-level administrative units (Bazargur & Batbuyan, 2007). Because of this, pasture carrying capacity and forage biomass have decreased and been degraded. Development of regional rural policies that allow for greater flexibility of livestock movement following more traditional systems is re-emerging as a policy instrument. Support in government and among nomadic herders has led to a reorganization of regional government that encompasses territories of several ecological zones and restores culturally traditional landscapes similar to those existing in the early 1900s. This system provides greater flexibility of pastoral management, especially under high climate variability as experienced by herders in the Gobi region and other steppe areas where climate fluctuations are large and drought frequency is high.

Recent proposals for policy changes have suggested a modification to the major administrative boundaries to allow greater access to natural resources and seasonal grazing lands to better sustain the pastoral livelihood. The policy is designed to develop a settlement pattern that reduces the concentration of population around major civic centers and to promote usage of resources associated with rural areas of the country. These new administrative and territorial units have been proposed to enhance socio-economic optimality, environmental sustainability, and historical and cultural acceptability by citizens (Chuluun, 2005). Reforming and enlarging administrative and territorial units will give the opportunity to improve the adaptation of the pastoral sector and provide greater flexibility in managing livestock densities across a more diverse set of landscape types within a more comprehensive administrative unit. The overall result would be a greater utilization of the natural landscapes, which is now restricted by the smaller, fragmented administrative units of the *sums*. For this change to succeed, reinvestment in infrastructure to allow for longer movements

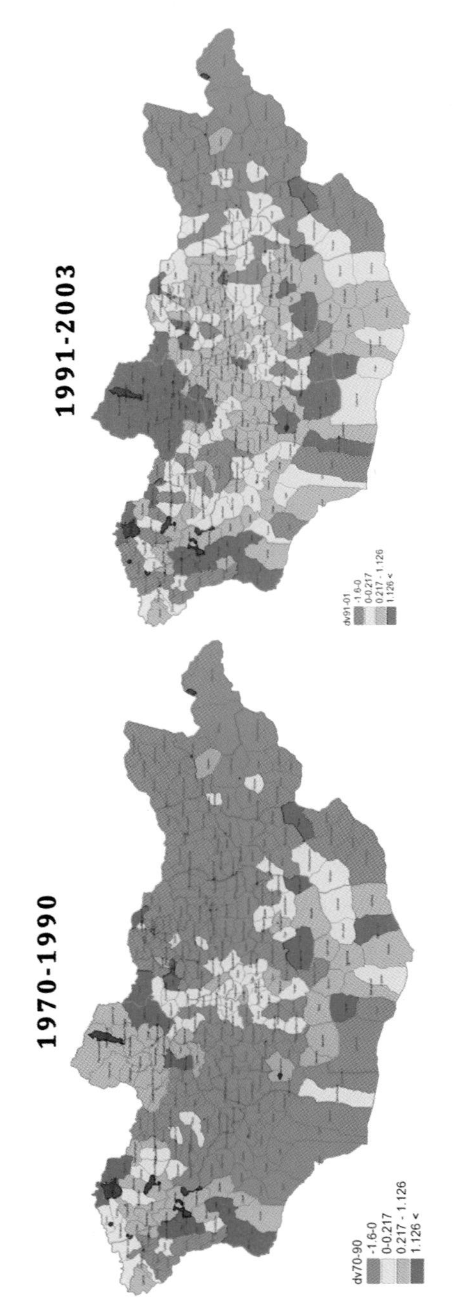

Figure 3.3 Vulnerability of rangelands to climate and land use changes

within these larger administrative units will be necessary, as well as the establishment of rules for access and allocation of seasonal pastures within these larger administrative units.

Adaptation options include the development of cultural landscape restoration, which incorporates community-based conservation and sustainable use of natural resources; the addition and protection of water points for additional pastureland; the agreement between neighboring *sums* for communal use of *otor* and reserve pastures; and the enlargement of administrative-territorial units, for instance, by combining several *sums* into one unit in order to restore cultural landscapes (Chuluun, 2005; Chuluun *et al.*, 2005b). For pastoral communities living in the riparian zones, diversification of the economy and intensification of the livestock industry through ecotourism and farming, the safeguarding of riparian ecosystems from degradation and desertification, and taking animals to *otor* pastureland during the summer period were suggested options. Protection of springs from degradation by livestock was critical for communities living in the mountain and forest steppe.

Enhancing collective actions among herders through strengthening the traditional customary arrangements is a key to achieving sustainable pastoral communities. These traditional networks based their livestock management and spatial sharing of key resources through consideration of common seasonal camping areas, water points or meadow areas. This system of social organization served a regulatory function for land-use management and a mechanism to provide safeguards against natural hazards. These community networks develop arrangements for regulating rangeland use, thus providing a good basis for future development of rangeland management cooperatives. The herders' interest in maximizing livestock in the current incentive structure is the primary challenge to building sustainable rangeland management (Enh-Amgalan, 2002). Productivity improvement and alternative income generation activities are crucial for changing the existing behavior and compensating potential income losses from restriction of animal numbers.

The development of modern cultural landscapes in Mongolian rangelands includes the strengthening of traditional pastoral networks with the introduction of modern technologies such as wireless communication, renewable energy resources, access to appropriate livestock breeding, use of good veterinary practices, and access to markets of more finished products. Further development of early-warning systems with use of integrated technologies, such as remote sensing and modeling and distribution through wireless technologies, can reduce risk in these very vulnerable but productive systems. Given the high literacy rate (97%) of the Mongolian people, there is a good probability that such innovative approaches to sustainable development of these cultural landscapes can be successfully implemented and lead to conservation of the critical cultural landscapes of the region.

Box 3.1 Improving resilience through community action and policies

Some steps towards closer cooperation include:

- Overcoming the constraint of inadequately sized (too small) households in terms of the number of members of the household.
- Better access to services – veterinary, breeding and social services.
- Strengthening of traditional mechanisms of coping with climate variability and extreme circumstances.
- Sustainable use of rangeland ecosystems, and decreasing the vulnerability of pastoral communities.

Lessons from the community strengthening activities in Mongolia include:

- Herders are eager to launch collective actions, ranging from simple gatherings especially among younger members to closer cooperation in production activities.
- Strengthening herding communities empowers initiatives for development thus encouraging equitable distribution of development resources.
- Because of the reduced requirements for establishment and lower transaction costs, herders prefer community-based networks.

Mongolia has about 190,000 herding families, of which about 67% own fewer than 100 livestock. Thus, the majority of the herders are poor and vulnerable to climate change. Increasing livestock numbers is difficult due to natural disasters such as *zud* and drought, which regulate numbers of livestock. A sustainable community-development approach, based on traditional pastoral networks, is indeed one of the most cost-effective adaptive strategies in the face of an uncertain climate and global changes.

Policy recommendations

Climate change adaptation policy must be based on a 'win-win' model. We could combine climate change adaptation strategy of Mongolia with the newly developing 'National program to combat desertification', following the Strategic goal No. 6, Environmental Policy, the 'National Development Comprehensive Policy based on the Millennium Development Goals'. *A restoration of cultural landscapes at multiple scales and strengthening of traditional pastoral networks with modern technologies will enhance socio-ecological resilience and reduce vulnerability of pastoral systems to climate change.*

3.5 Summary and conclusion

Global warming, a reduction of water and forage resources, an increase in goat numbers, human population changes, renewal energy and information communication technology increase may serve as critical slow variables driving pastoral social-ecological system dynamics. Key elements of pastoral social-ecological systems are water and its supporting riparian and forest ecosystems. Pastoral systems are very sensitive to any change in water resources due to global warming, including disappearance of water sources, reduction of water resources, delay in or early melting of snow, and lack of snow cover. *Protection of 'natural green walls' – the riparian ecosystems and forests – is more valuable for their ability to deliver water and water purification compared to building artificial 'green walls' – a recent development of creating vegetated walls to provide cooling of urban landscapes – which do not deliver any water service.*

The cumulative effect of climate change and overgrazing is much greater than their individual contributions because of their interaction with key ecosystem services. Vulnerability to climate change is amplified due to overgrazing of rangelands. Overgrazing of rangelands has become a large-scale problem of ecosystem degradation, not only near settlements and water sources. Thus ecosystem services have become degraded leading to desertification, water scarcity, increased dust events, and lack of forage, thus reducing carrying capacity. All of this leads to reduced well-being of herders.

Fragmentation of cultural landscapes in arid and semi-arid lands has increased vulnerability and reduced the capacities of pastoral systems to adapt to climate change. We observed a 'tragedy of commons' – the most environmental degradation in the most fragmented set of resources (Ostrom, 2008). There is some evidence of economic performance reduction with fragmentation of cultural landscapes. Social resilience based on traditional pastoral communities tends to be lost. Herders' groups not based on traditional pastoral communities may have a short life, but traditional pastoral communities, which have sustainably existed for centuries, are eroding.

3.5.1 A win-win model

Many projects are fragmented, aiming to achieve fragmented goals: only conservation or poverty reduction, etc. We need to reach a win-win situation both ecologically and socially. *The best transformation pathway is to strengthen the traditional pastoral community-cultural landscape system with renewable energy, wireless communication technology (further opening opportunities for distance learning and diagnosis), cultural and ecological tourism, and a developed industry based on livestock raw materials.* Traditional land-use strategies operating at larger spatial scales (cultural landscape) are in danger of being lost with contraction of the land-use scale near key resource areas in many regions. A reduction in the spatial scale of the cultural landscape has led to increased vulnerability of the coupled human-environmental systems.

Interestingly, cooperatives based on traditional pastoral networks are emerging in response to the complex climatic events of 1999–2002 in Mongolia. Mongolia has about 190,000 herding families, 67% of which own fewer than 100 livestock. Thus, most Mongolian herders are poor and vulnerable to climate change; moreover, they cannot increase their livestock numbers because natural climatic disasters such as *zud* and drought regulate livestock numbers. In this case, a sustainable community development approach based on traditional pastoral networks is indeed one of the most cost-effective adaptive strategies to deal with an uncertain climate and global changes.

A desirable future can be envisaged for the Mongolian rangelands in the new millennium with strengthening of traditional pastoral networks and the introduction of modern technologies such as wireless communication, the internet, renewable energy resources, etc. One can dream that the internet connection of pastoral communities with the outside world would give opportunities for distance learning and organization of ecotourism activities by the Mongolian herders directly. Given the high literacy rate (98%) of the Mongolian people, such a future might become reality given the high speed of information technology development and the positive aspects of globalization.

Acknowledgements

This research was supported by the Analysis of Integrated Assessment of Climate Change Project of START and funding to the Natural Resource Ecology Laboratory at Colorado State University on 'Northern Eurasian C-land use-climate interactions in the semi-arid regions' supported by NASA Project # NNG05GA33G. In addition, our research has been enhanced with funding provided by the Climate Development Knowledge Network run from the UK for our project entitled 'Climate Compatible Development in Dryland Systems of Mongolia and Surrounding Asian Systems'. This project was also support by the Mongolian Ministry of Environment and Green Development.

References

Batima, P., Natsagdorj, L., Gomboluudev, P., Erdenetsetseg, B. (2005) Observed climate change in Mongolia. *AIACC Working Paper No.* 12, 26 pp.

Bazargur, D., Batbuyan, B. (2007) Strategy of administrative-territorial division and socio-economic development of Mongolia. Ulaanbaatar. Admon printing. pp. 6 [in Mongolia].

Chuluun, T. (2005) A new administrative-territorial division of Mongolia as a mechanism to increase adaptive capacity to climate change. The 6th Open Meeting of the Human Dimensions of Global Environmental Change: 'Global environmental change, globalization and international security: new challenges for the 21st century', Bonn, Germany. Conference book, p. 436.

Chuluun, T., Altanbagana, M. (2005) Use of remote sensing information in administrative-territorial reform of Mongolia. The First National Conference on Remote Sensing and Geographic Information System Applications, Ulaanbaatar, pp. 71–6 [in Mongolian].

Chuluun, T., Enh-Amgalan, A. (2003) Tragedy of commons during transition to market economy and alternative future for the Mongolian rangelands. *African Journal of Range and Forage Science* **20**: 115.

Chuluun, T., Altanbagana, M., Sarantuya, G. (2005a) Vulnerability and adaptation assessment of the Mongolian pastoral systems to climate and land use changes. *Proceedings of the Department of Biology, School of Natural Sciences, Mongolian Education University* **4**: 182–9.

Chuluun, T., Nergui, P., Davaanyam, S., Altanbagana, M., Ariunmunkh, B. (2005b) Proposal on the new administrative-territorial division of Mongolia submitted to the President of Mongolia.

Ellis, J. and Chuluun, T. (1993) Cross-country survey of climate, ecology and land-use among Mongolian pastoralists. Report to Project on Policy Alternatives for Livestock Development (PALD) in Mongolia, Institute of Development Studies at the University of Sussex, UK.

Ellis, J., Price, K., Boone, R., Yu, F., Chuluun, T., Yu, M. (2002) Integrated assessment of climate change effects on vegetation in Mongolia and Inner Mongolia. In: Chuluun, T, Ojima, D. (eds), *Fundamental Issues Affecting Sustainability of the Mongolian Steppe*. Ulaanbaatar: Interpress Publishing and Printing, pp. 26–34.

Enh-Amgalan, A. (2002) Change and sustainability of pastoral land-use systems in Mongolia. In: Chuluun, T, Ojima, D. (eds), *Fundamental Issues Affecting Sustainability of the Mongolian Steppe*. Ulaanbaatar: Interpress Publishing and Printing, pp. 228–31.

Fernández-Giménez, M.E. (1999) Sustaining the steppes: A geographical history of pastoral land-use in Mongolia. *Geographical Review* **89**: 315–42.

Fernández-Giménez, M.E. (2006) Land use and land tenure in Mongolia: A brief history and current issues. In: Bedunah, D.J., McArthur, D.E., Fernández-Giménez, M. (eds), *Rangelands of Central Asia: Proceedings of the Conference on Tranformations, Issues, and Future Challenges*. USDA Forest Service Proceedings RMRS-P-39, pp. 30–6.

Janzen, J. (2005) Changing political regime and mobile livestock keeping in Mongolia. *Geography Research Forum* **25**: 62–82.

Khaulenbek, A. (2009) Desertification of Mongolia. In: *Proceedings of the first National Desertification Conference*, pp. 42–63.

Mearns, R. (1993) Territoriality and land tenure among Mongolian pastoralists: variation, continuity and change. *Nomadic Peoples* **33**: 73–103.

MNE (Ministry of Nature and Environment-Mongolia) (2001) *State of the Environment – Mongolia*. Klong Luang, Thailand: United Nations Environment Programme.

MNE (Ministry of Nature and Environment) (2006) *Mongolia: State of Environment, 2004-2005 Report*. Ulaanbaatar: MNE.

MNET (Ministry of Nature, Environment and Tourism), UNEP and UNDP (2009) MONGOLIA: Assessment Report on Climate Change 2009: 'MARCC 2009'. Ulaanbaatar: MNET. ISBN 978-99929-934-3-X.

Mongolian National Atlas (1990) Moscow: Mongolian Academy of Sciences.

Natsagdorj, D, Sarantuya, G. (2004) On the assessment and forecasting of winter disaster (atmospheric caused dzud) over Mongolia. In: *Proceedings of Sixth International Workshop on Climate Change in Arid and Semi-Arid Regions of Asia*, August 25-26, 2004, Ulaanbaatar, Mongolia, pp. 72–88.

Natsagdorj, L., Jugder, D., Chung, Y.S. (2003) Analysis of storms observed in Mongolia during 1937–1999. *Journal of the Atmospheric Environment* **37**: 1401–11.

NSO (National Statistical Office – Mongolia) (2000) *Mongolian Statistical Yearbook 2001*. Ulaanbaatar: National Statistical Office.

Ojima, D.S., Chuluun, T., Bolortsetseg, B., Tucker, C.J., Hicke, J. (2004) Eurasian land use impacts on rangeland productivity. In: DeFries, R., Asner, G.P. (eds), *Ecosystem Interactions with Land Use Change*. Geophysical Monograph Series vol. 153. Washington DC: American Geophysical Union, pp. 293–301.

Ostrom, E. (2008) Updating the design principles for robust resource institutions. Paper presented at Workshop in Political Theory and Policy Analysis, Indiana University Center.

Ped, D.A. (1975) On indicators of droughts and wet conditions. *Proceedings of the USSR Hydrometeorology Centre* **156**: 19–39 [in Russian].

Reading, R.P., Bedunah, D.J., Amgalanbaatar, S. (2006) Conserving biodiversity on Mongolian rangelands: Implications for protected area development and pastoral uses. In: Bedunah, D.J., McArthur, D.E., Fernández-Giménez, M. (eds), *Rangelands of Central Asia: Proceedings of the Conference on Transformations, Issues, and Future Challenges*. USDA Forest Service Proceedings RMRS-P-39, pp. 1–18.

Schmidt, S.M. (2006) Pastoral community organization, livelihoods and biodiversity conservation in Mongolia's Southern Gobi Region. In: Bedunah, D.J., McArthur, D.E., Fernández-Giménez, M. (eds), *Rangelands of Central Asia: Proceedings of the Conference on Transformations, Issues, and Future Challenges*. USDA Forest Service Proceedings RMRS-P-39, pp. 18–29.

Schmidt, S.M., Gansukh, G., Kamal, K., Swenson, K. (2002) Community organization – a key step towards sustainable livelihoods and co-management of natural resources in Mongolia. *Policy Matters* **10**: 71–4.

Sneath, D. (2003) Land use, the environment and development in post-socialist Mongolia. *Oxford Development Studies* **31**: 441–59.

Tserendash, S. (2006) Present conditions and strategies for management of Mongolian rangelands. Journal Erdem, **38**: 7–11 [in Mongolian].

UNDP (2005) *Economic and Ecological Vulnerabilities and Human Security in Mongolia*. Interpress Co., Ltd.

Walker, B., Salt, D. (2006) *Resilience Thinking: Sustaining Ecosystems and People in a Changing World*. Island Press.

4

Vulnerability of Pastoral Communities in Central Mongolia to Climate and Land-Use Changes

T. Chuluun[1], M. Altanbagana[1], S. Davaanyam[1], B. Tserenchunt[1] and D. Ojima[2]

[1] *Dryland Sustainability Institute and National University of Mongolia, Ulaanbaatar, Mongolia*

[2] *Natural Resource Ecology Laboratory, Colorado State University, Fort Collins, Colorado, USA*

4.1 Introduction

The pastoral community is more than the sum of its individual households. Community institutions, which develop through the interaction of individuals and subgroups within the community, are the principal emergent property of communities and vital in understanding land degradation at this scale. Environment, at this scale, is also more than a collection of individual phenomena. The emergent ecological properties of landscapes at the scale of pastoral communities are spatial pattern and diversity. Relative heterogeneity of patterns is an important factor for vulnerability to degradation.

Desertification or arid land degradation occurs as a result of a suite of processes in the coupled human-environmental system. Sustainable land use is possible only when the relevant changes and processes within community institutions and environment are spatially and temporally synchronous (Walker & Abel, 2002). Asynchronicity in these processes can lead to both social and environmental crises or collapse. The goal of management in drylands at the community scale is the facilitation of adaptive capacity of communities.

Vulnerability of Land Systems in Asia, First Edition. Edited by Ademola K. Braimoh and He Qing Huang.

Nomadic pastoral systems are dissipative structure-functions (Prigogine, 1989; Prigogine & Nicolis, 1977) immersed in arid ecosystems of great temporal and spatial heterogeneity (Chuluun, 2000). Historically, traditional pastoral networks emerged in drylands with scarce natural resources, subsequently evolving to increase human adaptive capacity in coping with climate variability and extreme climatic events such as drought and *zud*, a winter condition that can prove devastating for livestock. A large geographical landscape was critical in order to offset climate variability, as traditional pastoral networks used certain landscapes primarily for forage and water. There was thus a strong coupling between traditional pastoral groups and the landscapes they used. Traditional pastoral communities and their cultural landscapes, consisting of four seasonal land types in addition to reserve areas, *otor* pastures and haylands, provide prime examples of coupled social-ecological systems or human-environmental systems (Global Land Project, 2005: Science Plan and Implementation Strategy). These traditional pastoral social-ecological systems were sustainable for centuries.

The Mongolian cultural landscapes, however, were fragmented with the administrative-territorial division reform of the last century (Ojima & Chuluun, 2007). Now almost half of all *sums*, or subprovinces, lack one or two seasonal pastures. Interestingly, there wasn't much change in terms of cultural landscape use during the socialist period, although there were large changes in pastoral social-ecological systems during the socialist period between the late 1950s and 1990. More complex dynamic changes in pastoral social-ecological systems have occurred since 1990 in the transition to a market economy. The number of herders has more than doubled since the early 1990s as a result of the economic migration spurred by livestock privatisation. Traditional pastoral networks at the lowest level (*hot ail*) re-emerged and reorganised themselves during this period of time, as some younger and more inexperienced herders started to follow their parents or relatives who had more herding experience. These pastoralists continued to use traditional cultural landscapes under the leadership of experienced herders. Some new herders started to live near the settlements and water sources, causing overgrazing as a result of their low mobility. Due to a rise in the price of cashmere, goat numbers more than tripled from 5.1 million in 1990 to 18.3 million in 2007 after the transition to an open market economy (NSO, 2009).

In addition to economic and social factors, global warming is becoming a slow but critical variable, causing a reduction in water and food resources. Over the last 60 years, the surface air temperature in Mongolia has increased by 1.94°C, which, along with its socio-economic vulnerability, makes Mongolia one of the Earth's hot spots. Spring is also becoming increasingly dry as a result of warmer temperatures and decreased precipitation.

This research aims to investigate change and transformation of open pastoral social-ecological systems (Gallopin, 2006) and develop climate change adaptation options for pastoral communities with participation of herders, local and national governmental officers, and scientists (Vogel *et al.*, 2007). A social survey among herders on local climate change observation and its impact on pastoral systems was conducted, and participatory workshops held with pastoral communities and local policy-makers. A goal of this field survey was to study stakeholders' views on the current state of the social-ecological system and to identify policy solutions to reduce rangeland degradation. These workshops aimed to communicate the current and future risks of climate change, land-use changes and rangeland assessment techniques, as well as the socio-economic vulnerability of the herders to climate change.

A goal of this research is to study the vulnerability of pastoral communities to climate and land-use changes in different ecological zones, and the policy implications of reducing their vulnerability to climate change.

4.2 Study sites and methodology

Two out of six study sites for pastoral social-ecological systems were selected in the buffer zone of the Khustai Nuruu National Park, where wild horses known as *tahi* were reintroduced (Figure 4.1). These social-ecological systems in the buffer zone were selected so as to increase knowledge of the interaction between conservation and pastoral land systems, especially those in close proximity to the city of Ulaanbaatar. Four other sites were selected along ecological transects: forest steppe, mountain steppe, dry steppe and desert steppe (Figure 4.1). Prior to the socialist era, three of the sums along this gradient made up one administrative-territorial unit. One old herder from Sant sum said that his parents used to spend summer in the mountains of Khujist sum. This confirms that there was free pastoral movement between mountains and steppe within the old administrative unit, and old administrative-territorial divisions were primarily based on cultural landscape principles.

Hondiin Zaraa and Erdene-Ovoo, the herders' groups at Sant sum, and Ih Burd at Hujirt sum, were led or guided by the old experienced herders who had lived in these areas for generations. Thus, traditional indigenous knowledge was the basis for grazing management in these pastoral communities, and these groups followed their nomadic cultural legacy better than other herders' groups. Interestingly, the *zuds* of 1999–2002 prompted an increase in the formation of herders' groups due to several reasons such as legacy of cooperation and social learning, as well as government and donors' support. Generally, relatively poor herders tended to form herders' groups, exemplified by Batsumber and Santbayanbulag herders' groups along the Tuul and Orhon rivers. These groups were not led by an experienced herder, but by a former administrative worker or teacher. These group leaders

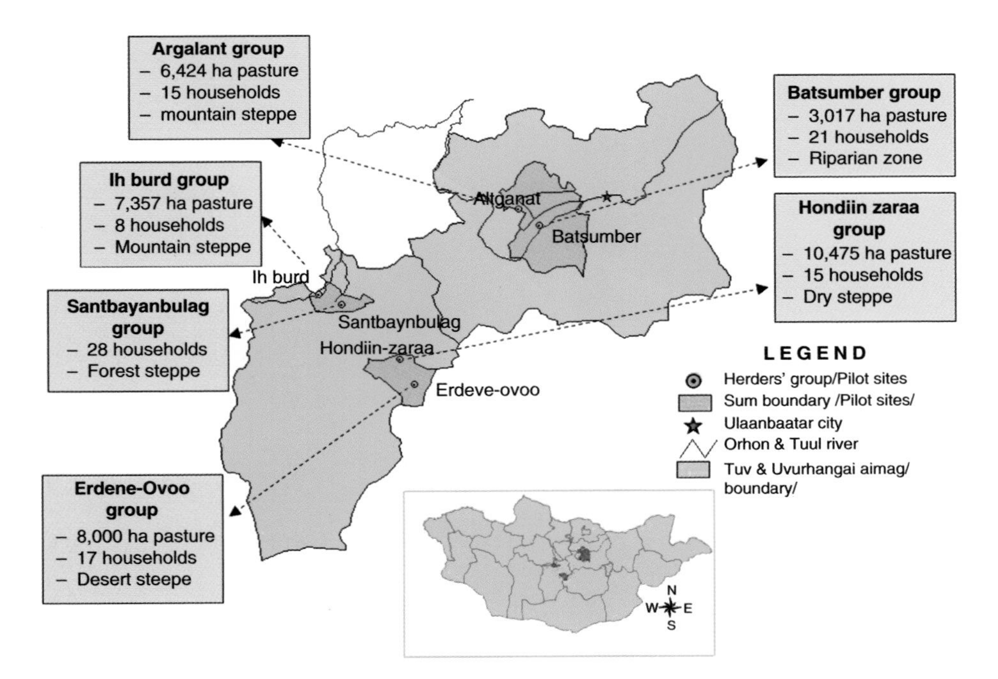

Figure 4.1 Map of pilot study sites

were intelligent people, quickly learning the advantage of cooperation for relatively poor households.

Migration from rural areas to the big cities of the central area started to increase in the mid-1990s. This was a result of environmental change, following summer droughts and intensifying after the 1999–2002 *zuds*. The herders who had lost their livestock during these extreme climatic events were forced to leave the area and can be regarded as environmental refugees. The Batsumber and Altganat herders are examples of the migration of herders from the rural to the central region of Mongolia. In addition, the Khustai Nuruu National Park probably attracted herders due to its beneficial buffer zone management programmes. Thus both the Batsumber and Altganat herders' groups have the shortest local ecological knowledge as only one household in each community was native, with the rest of the herders having migrated from the western Aimags. The Tuul river valley served as a market pathway for the transfer of animals from the western Aimags to Ulaanbaatar city, and it was kept free of grazing by local herders during the socialist era. State agricultural farms existed in the region north of the Hustai Nuruu National Park between the late 1950s and 1990, but the farmers have since moved out of the area. The fields, abandoned after the transition to a market economy in 1990, have still not recovered from severe soil erosion caused by the farming of these drylands.

The central region close to Ulaanbaatar, the capital city, and the Khustai Nuruu National Park was attracting people from remote areas of Mongolia. Thus, the majority of herders living in the buffer zones of the park are migrants, mainly from western Mongolia. They enjoy double economic benefits from being closer to the market of Ulaanbaatar and the support from the park (Table 4.1 shows their income level compared to herders from other

Table 4.1 Studied pastoral communities

Pastoral community name	Sum and aimag	Ecosystem type	Number of households	Livestock per capita: sheep units	Income per capita, US$	Cultural Landscape Use Index	Socio-economic vulnerability	Ecological vulnerability	Socio-ecological vulnerability
Batsumber	Altanbulag, Tov	Riparian/forest steppe	21	100	1,200	4/7	2.3	0.66	0.6
Altganat	Argalant, Tov	Forest steppe	15	181	1,877	5/7	2	0.46	0.31
Santbayanbulag	Hujirt, Ovorhangai	Riparian/forest steppe	8	41	574	4/7	3.3	0.47	0.75
Ihburd	Hujirt, Ovorhangai	Mountain steppe	8	49	618	5/7	3.2	0.4	0.65
Hondiin Zaraa	Sant, Ovorhangai	Dry steppe	15	83	827	6/7	3.4	0.13	0.47
Erdene-Ovoo	Sant, Ovorhangai	Desert steppe	17	79	972	6/7	3.5	0.14	0.51

regions). Interactions of the herders living in the park buffer zone are mutually beneficial. The herders assist in conservation of the park and they benefit from the park's assistance in building fences around springs, or in constructing wells. The herders are allowed to use park pasture during *zuds*. However, their impact on ecosystems outside the park is large, as overgrazing from their herds has led to ecosystem degradation. Herders living along the Tuul river developed a more sedentary lifestyle, moving only twice a year and covering only short, 2–3 km distances. This has greatly concerned the Altanbulag sum government, which has passed a regulation prohibiting grazing alongside the river between late June and late August. The herders, however, do not obey this regulation.

The social survey and examples of social-economic scenarios were done in the Tuin river basin area, located in Bayanhongor aimag. The Tuin river arises in the Khangai Mountains, flows through territory of Erdenetsogt, Olziit, Bayanhongor, Jinst and Bogd sums, and flows into the Orog lake. The questionnaire consisted of two parts, related to a dryland development paradigm (Reynolds *et al.*, 2007), rangeland degradation, and policy to reduce it. Diverse representation in terms of ecoregion, jobs and age – including herders, representatives of herders' groups, policy-makers, students and older people – was reflected in the survey. Eight people from the Erdenetsogt sum (forest-steppe), 14 people from the Olziit sum (primarily steppe), one person from Bayanhongor sum, four people from the Jinst sum (the Gobi), and five people from Bogd sum (the Gobi) participated in the survey. Seventy-five percent of the survey participants were male and 25% female; 59% were herders and 41% government workers; 75% were married and 25% unmarried; 3% had received elementary education, 59% primary, and 38% high (college) education; and participants were aged 17 to 68.

The following information was used for socio-economic vulnerability assessment of herders' groups (UNDP, 2005):

- GDP or total income;
- distance to the market;
- vulnerability to climatic disasters;
- diversity of production.

The vulnerability index of pastoral systems to climate and land-use changes has been calculated as the sum of the *zud* index and rangeland use index. Vulnerability of rangelands is higher when both *zud* and rangeland use intensity are higher. We refer to this index as an ecological vulnerability index (V):

$$V = \Delta N + \Delta S$$

where ΔN is the pasture use index (livestock density relative to carrying capacity) and ΔS is the *zud* index, incorporating previous summer drought (Natsagdorj & Sarantuya, 2004).

4.3 Research results

Some research findings are summarized in Table 4.1. Livestock per capita is well correlated with income per capita due to the fact that the herders' main income comes from livestock. The Batsumber and Altganat pastoral communities have the highest income. Four other communities, those of the forest steppe, the mountain steppe, the dry steppe and the desert steppe, live along the transect. In the herders' group, the livestock per household as well as the income per capita and overall richness of the cultural landscape has increased along this transect. In terms of cultural landscape, Batsumber and Santbayanbulag herders living in the riparian zones during the summer and fall lack three pasture types out of seven (cultural land use index = 4/7). Thus most ecosystem degradation was observed in the riparian zones where herders have become more sedentary. Cultural landscape is better conserved in Sant sum (6/7) as compared to Hujirt sum. As a consequence, the ecological condition in Hujirt sum is worse than in Sant sum, and it seems that this is already affecting the incomes of the herders (see Table 4.1). There are signs that this trend may continue unless proper measures are taken.

The socio-economic vulnerability of each community was calculated based on its proximity to the market, its income, the loss of animals during the 1999–2002 *zud* and the level of economic diversification. The pastoral communities of Tov aimag showed less socio-economic vulnerability due to higher income and shorter distance to the markets of Ulaanbaatar city.

The environment and poverty are interlinked, as was clear along the ecological transect in this study. The income per capita was greatest for communities living in the dry steppe and desert-steppe regions compared to the communities living in the mountain steppe and forest zones of the Ovorhangai aimag. Cultural landscape use was the best, and ecological vulnerability of the rangelands to climate and land-use changes was the lowest in desert-steppe and dry steppe. Communities living in the riparian zones had the highest socio-ecological vulnerability, despite one of these communities having less socio-economic vulnerability. For instance, the Batsumber community living in the riparian zone living close to Ulaanbaatar, had less socio-ecological vulnerability, although rangeland vulnerability was the highest due to overgrazing. The Altganat community had the least socio-ecological vulnerability among all pastoral communities mainly due to its having the least socio-economic vulnerability.

The herders were very sensitive to water availability during both the warm and cold seasons and there was also ecosystem degradation as a result of overgrazing around the wells and the few remaining springs that had not already shrunk due to the impact of climate change. The herders were also very sensitive to snow cover change. For instance, Nogoon Suuri, a herder of Hujirt sum, indicated that six springs had disappeared and that only a single watering point remained for 14 households with 3000 livestock. This sole remaining spring was prone to freezing, leaving these households without water in early December 2006. As there is usually snow on the ground during this time of year, the herders dispersed, moving away from their winter camps so that they could use the snow as a winter water source.

4.4 The results of a social survey related to the 'dryland development paradigm'

Sixty-six percent of the survey participants cited climatic disasters, 25% rangeland degradation, and the rest gold mining and human population growth as the main factors in defining the dynamics of social-ecological systems in the Tuin river basin area (Figure 4.2). As regards identifying a critical slow variable causing negative impact, 78% of people cited global warming (with its drying effects), 19% water shortage and 3% poverty (Figure 4.3). None of the respondents saw the brain drain as a critical threat to the region. Sixty-six percent of people responded that surface water shortage has already crossed threshold level and is leading to collapse of social-ecological systems in the Tuin river basin (Figure 4.4), while 34% cited a decrease in rangeland productivity. Interestingly, none regarded technology application as leading to development of the region. Forty-four percent of the

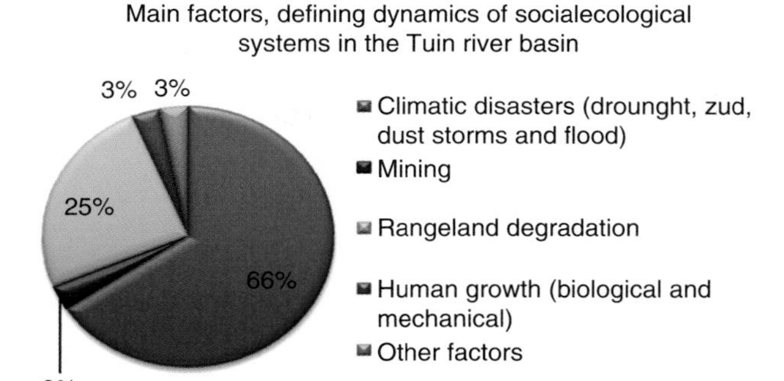

Figure 4.2 Main factors defining dynamics of social-ecological systems in the Tuin river basin

respondents answered that global regulation is the most important factor for sustainability of the social-ecological systems in the Tuin river basin (Figure 4.5), followed by 31% of respondents who cited Mongolia, 16% local government and 9% traditional communities. When asked to evaluate how effectively policy combines modern science and traditional knowledge, 59% of the survey participants said fair, 28% poor and 13% good (Figure 4.6).

The social survey conducted among the stakeholders in the Tuin river basin yielded interesting results for our understanding of coupled social-ecological systems in the Tuin river basin from the dryland development paradigm's point of view. The following statements are the main outputs of this study in the Tuin river basin:

1. Dynamics of social-ecological systems are defined primarily by climatic disaster events such as drought, *zud*, flood and dust storms.
2. Global warming is a critical determinant of social-ecological systems.
3. Surface water shortage has already crossed a threshold level and is leading to collapse of social-ecological systems.

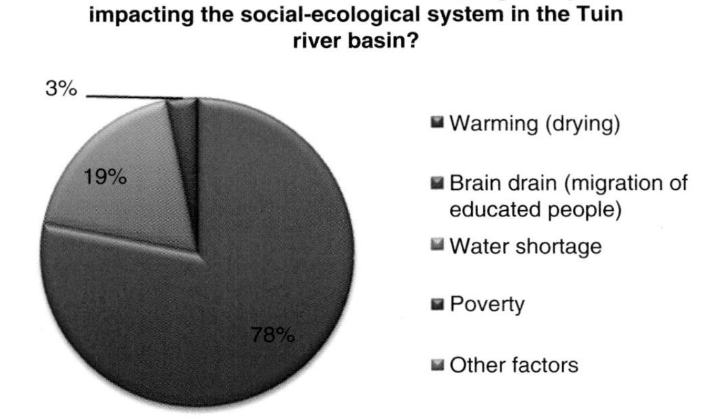

Figure 4.3 What is a critical slow variable, negatively impacting the social-ecological system in the Tuin river basin?

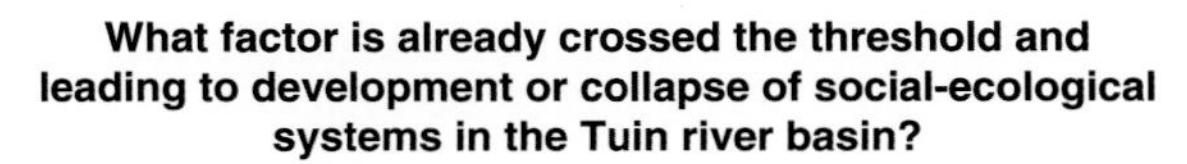

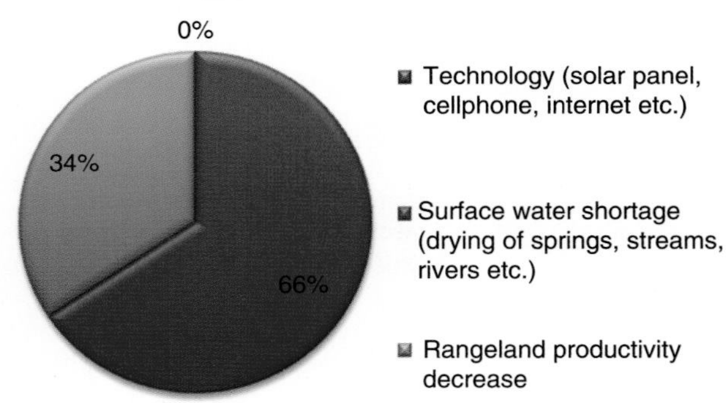

Figure 4.4 What factor has already crossed the threshold and is leading to development or collapse of social-ecological systems in the Tuin river basin?

4. Global (44%) and country (31%)-level regulations are more important than local government (16%) or community-level regulations (9%).
5. Level of policy, which combines up-to-date modern science and traditional knowledge, is fair.

Global warming impacts on surface water decrease. Nowadays only three rivers (Shargaljuut, Ortomt and Ovgon Jargalant) out of 99 original rivers and streams are still flowing into the Tuin river. The Tuin river is not reaching the Orog lake and the Orog lake has been dried out for several years. Climate change also impacts on increased (in frequency and intensity) climate disasters such as drought, *zud*, sand and dust storms, and floods. Thus it was logical to expect global regulation to be cited as having the highest importance for sustainability of social-ecological systems in the Tuin river basin. The survey showed

What level of regulation is the most important
for sustainability of the social-ecological
systems in the Tuin river?

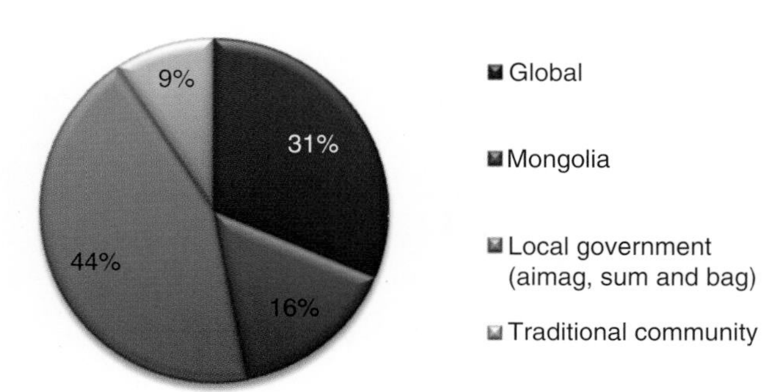

Figure 4.5 What level of regulation is the most important for sustainability of the social-ecological systems in the Tuin river?

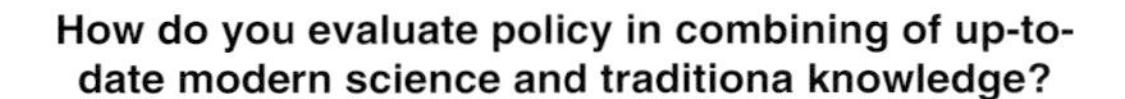

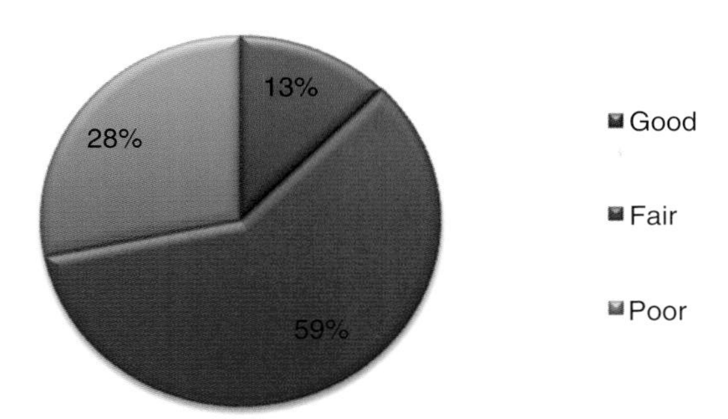

Figure 4.6 How do you evaluate policy in combining up-to-date modern science and traditional knowledge? Good, fair or poor?

a high level of environmental awareness, and a need for actions with knowledge based on modern science. However, actions without an up-to-date body of 'hybrid' environmental knowledge, which integrates local management and policy experience with science-based knowledge, may lead to maladaptation to climate change.

4.5 Pastoral social-ecological scenarios

Research findings and thoughts for future adaptation strategies for pastoral social-ecological systems can be summarized in a scenarios diagram (Figure 4.7).

Traditional system. Cooperation within traditional pastoral networks serves as a mechanism enhancing resilience to climatic disasters. Communal disaster relief mechanisms,

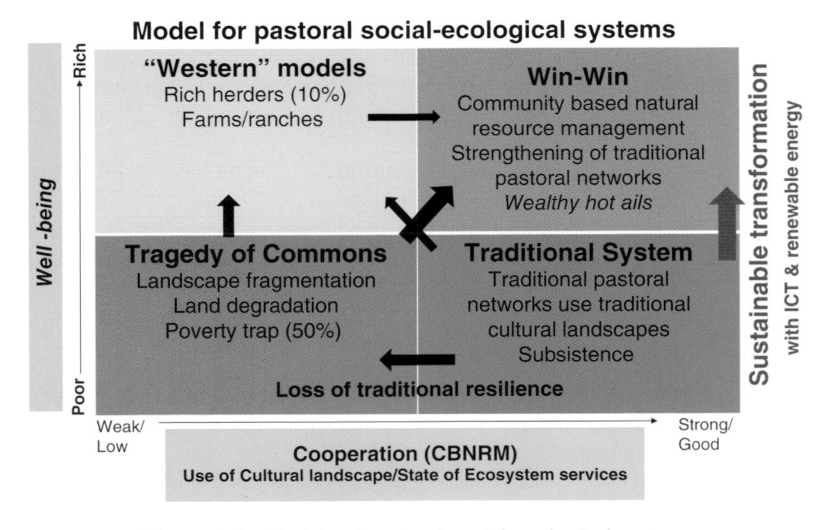

Figure 4.7 Models of pastoral social-ecological systems

assisting the most affected herders in many different ways, are in place. Traditional pastoral communities use cultural landscapes to cope with climate variability and climatic extremes. Due to proper management, rangeland ecosystems used for traditional grazing and ecosystem services are in good condition.

Tragedy of the commons. The rangelands are still State owned in Mongolia although livestock has been privatised. This has been the main reason for the increased overgrazing and ecosystem degradation near settlements and water sources under capitalism. Poor herders especially tend to become less mobile, living near towns, infrastructure and water sources, and this causes dryland fragmentation. Generally, herders do not cooperated and compete more for resources in this scenario. Many herders in this model lose their traditional resilience mechanisms to cope with climate variability and extremes, and potentially 50% of herders live in poverty. In this model there is deterioration of the social-ecological system with ecosystem degradation and increasing poverty.

'Western models'. Only 5–10% of herders became wealthier through the transition to a market economy. Generally, these rich herders don't cooperate with the larger pastoral community. They often take advantage of the current State ownership of pasture, often causing more damage to ecosystem services. Some herders have small communities and use traditional cultural landscapes. Thus, some of the traditional networks that use cultural landscapes in sustainable ways can be included in the win-win model, with social and ecological benefits. This group of herders needs to be encouraged through proper pasture and cultural landscape ownership mechanisms.

Win-win model. In the win-win scenario, the majority of herders must be transformed. The most desirable pathway for pastoral systems is direct transformation from a traditional system to a win-win state, strengthening traditional pastoral communities with modern technologies such as renewable energy and communication information technology. High levels of literacy among the Mongolian herders (98%) and the suitability of the nomadic culture in conjunction with wireless communication make such a sustainable transformation very attractive. There is a great opportunity to conserve natural, cultural and social capital in order to maintain the adaptive capacity and resilience of Mongolian pastoral social-ecological systems to climate change and globalization. Teaching sustainable farming techniques to herders living near settlements and water points would be another pathway to reach a win-win situation and escape the tragedy of the commons state. A reform of administrative-territorial divisions that restores cultural landscapes appears to be the best, most cost-effective adaptation option in order to promote the sustainability of coupled social-ecological systems with increased adaptive capacity and resilience to climate change at supra-pastoral community scales.

We present some case studies of different pastoral social-ecological system scenarios as studied in the Tuin river basin, Bayanhongor aimag.

Case study: Win-win model: Narmandah – young 'National champion' herder

Narmandah is 31 years old. He graduated from high school, married at age 17 and has three children aged between 6 and 14. The children study in Bayanhongor aimag center. The children come home in summer and assist their parents in herding. He lives in Nuramt Har of Olziit sum in Bayanhongor aimag (within the boundary area of dry steppe and desert-steppe). He lived with his grandfather from 6 to 18 years old, learning about traditional pastoral management. Thus he has 20 years' experience of herding. He keeps good contacts with his relatives, living

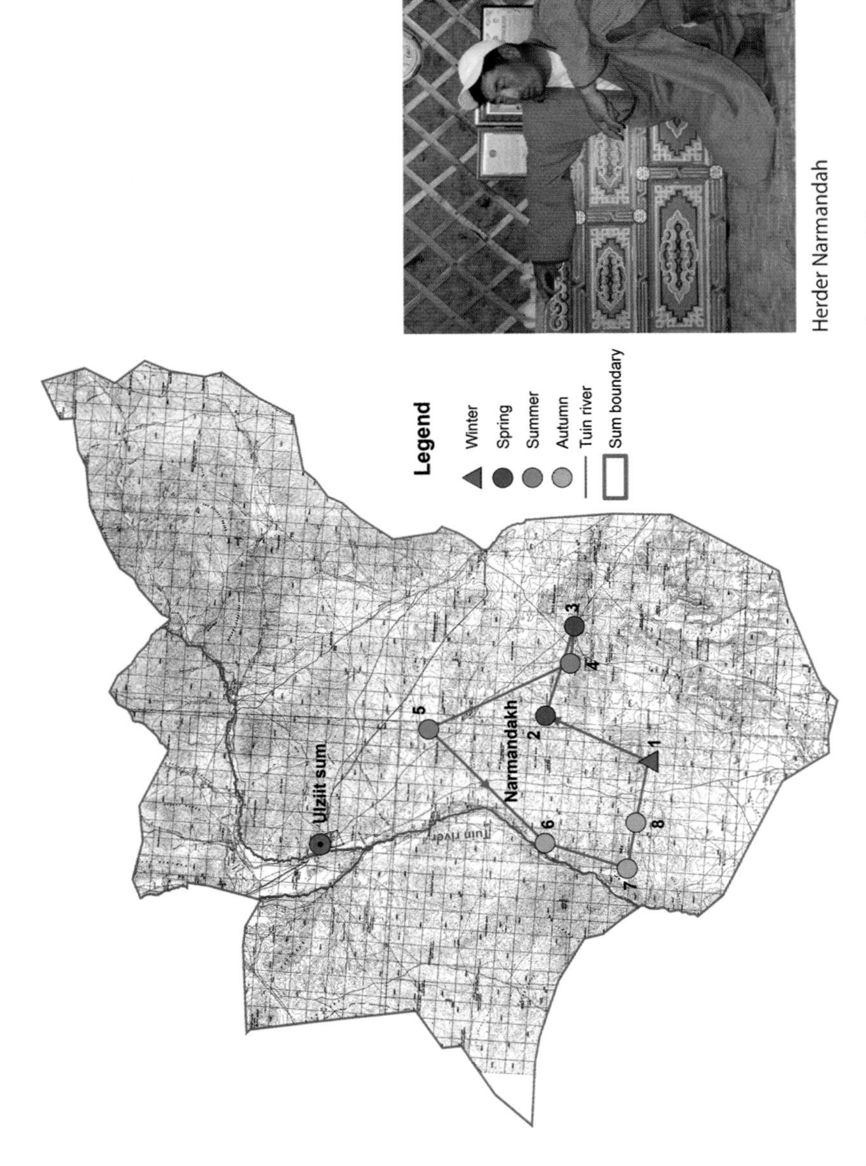

Figure 4.8 The annual movement pattern of Narmandah

Table 4.2 Moving season, duration and places

No.	Season	Duration	Name of place
1	Winter	12/12–2/20	Nuramt har
2	Spring	2/21–4/20	Dund hashaat
3	Spring	4/21–6/10	Nurangiin eh
4	Summer	6/10–7/10	Ulaadgain adag
5	Summer	7/11–8/20	Baruun hargana
6	Summer/Fall	8/21–9/20	Tsagaan ders
7	Fall	9/21–11/20	Gashuuny eh
8	Fall	11/21–12/11	Ovoot hanan

in the same area. However, he has to move far from them because of the large number of livestock and carrying capacity of pastures.

The household has over 70 horses, about 30 cattle, 250 sheep and over 2000 goats. His annual income is 24 million togrog. His wife, a hard-working woman, milks 180 goats every day, makes curd and cheese, and sells in Bayanhongor center. They make more income through sales of their dairy products in summer, instead of keeping the sun-dried products and selling through the fall and winter. They also produce Mongolian vodka every week, but never sell it.

Narmandah moves 8–10 times during a year, and can move 20–80 km for *otor* pasture (Figures 4.8 and 4.9; Table 4.2). He plans his destination and duration well because he has good knowledge about plants, specifically their quality, phenology and suitability for different kinds of livestock (Figures 4.8 and 4.9). For instance, he thinks that grazing in one area for longer than 1 month during the growing season is the main reason for overgrazing and sand movement. That's why he doesn't stay at *otor* longer than 1 month. Pasture that is not overgrazed can make a quick recovery, and this is good for the weight gain and productivity of the animals. Similarly, it is important to know which plants are suitable at which time of year in order to maximise productivity. For example, grazing of sheep and goats in fall pasture with budargana (*Salsola* sp. and *Reaumurea* sp.) and wild onion (*Allium* sp.), located in the desert-steppe zones, is critical for gaining weight and fat for winter survival.

Narmandah was left with 94 goats out of over 200 livestock after the 2002 *zud*. After this event Narmandah fully understood that disaster management policy based on both scientific and traditional knowledge is essential. His main policies for disaster management are:

- To keep the number of livestock constant (e.g., if you are expecting 500 kids, then sell the same number of goats).
- To loan money at an interest rate of 3–5% for people who need finance.
- To make *otor* movements often and fatten animals well in order to get more cashmere of better quality.
- To make sufficient hay to maintain animal weights through the winter.
- To clean winter shelters well in order to keep animals warmer.
- To sell animals at market before they lose weight in order to prevent from losses during a potential *zud*, following summer drought.

Narmandah spends the extra income from disaster management measures in buying housing, machinery and equipment, and also invests in education.

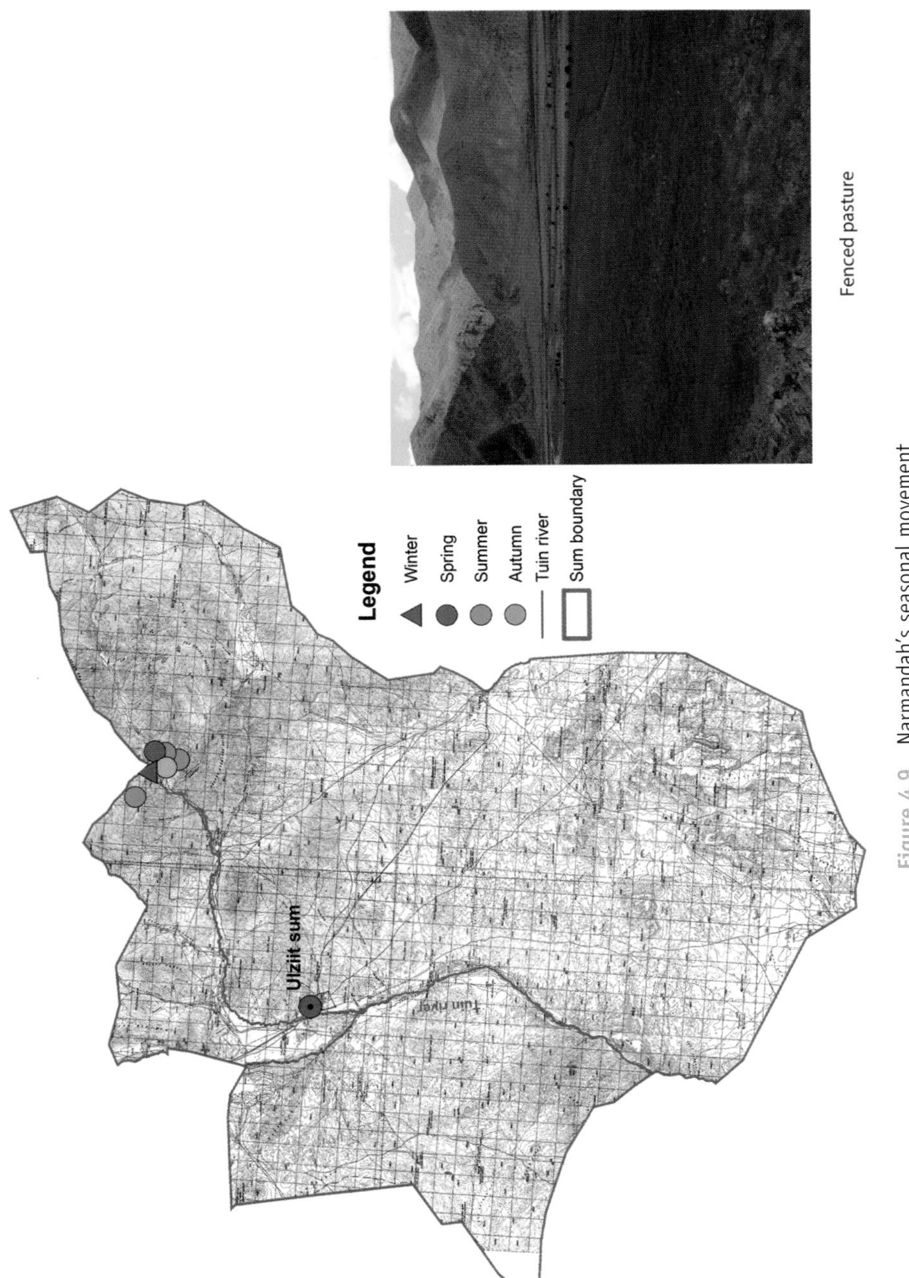

Figure 4.9 Narmandah's seasonal movement

Case study: Rich herder: Lhamhuu – a herder with fenced pasture

Lhamhuu lives at the confluence of three streams in the Shargaljuut river basin of Olziit sum (in the forest steppe zone). He is an elderly single man and lives with his nephew's family. He has a house located near his fenced pasture, and has owned 42,000 ha of fenced pasture since 1990. He doesn't face any pasture degradation or livestock loss during *zud* and droughts. He has also made a water channel from the river through his pasture, creating wetlands and protecting them. He can be considered a rich herder because of his fenced pasture. It is unusual to have fenced pasture, and according to current legislation, no herder can own such fenced pasture.

Fenced pastures have many benefits:

- The rangeland ecosystem inside the fences has a high diversity of plants, with over 50 plant species (*Carex* sp. and *Leymus chinensis* are dominant plant species, plus many annual plants and forbs). This plant diversity contributes to the weight gain of the livestock.
- Lhamhuu doesn't worry about drought or *zud* because his fenced pasture always has enough forage for his livestock, serving as the best disaster management method.
- The newborn animals gain weight faster in the fenced pasture.
- Lhamhuu also makes hay from his fenced pasture.
- He doesn't need to move long distances for *otor* pastures, saving the expense of *otor* movements and labor.
- Fenced pasture is best for the protection and recovery of rangeland ecosystems. There are even marmots inside the fences, which are absent in surrounding lands.

Lhamhuu has about 300 small ruminants, with an equal ratio of sheep and goats. He also has about 50 cattle and 10 or so horses, mainly as a winter food source. His strategy is to keep livestock numbers constant; however, he slaughters more animals if the summer is dry.

His annual income is 3 million togrog. He has everything he needs, such as a car, motorcycle, TV and solar panel. He is satisfied with his life. He pays 23,000 togrog as a pasture fee. His nephew lives with his wife and children. All his relatives of school age come to fix his fences and winter shelter during their summer vacation.

He thinks that improper and intensive use of water basins during the socialist era is one of the main reasons for water flow reduction and drying out. He protects wetlands with fencing, and uses surrounding mountain pastures seasonally. His fenced pasture has the following grazing schedule: sheep and goats in spring from March through 20 May 20, then cattle during May throiugh October.

Case study: Traditional pastoral system: 'Ogoomor Ortomt' – a one-river community

Herders of a traditional one-river community, living along the river Ortomt in Erdenetsogt sum, Bayanhongor aimag, formed an official herders' group 1 year ago. Twenty-six households self-organized the group, not artificially as happened in many parts of Mongolia to take advantage of an international project. This community lives on the southern slopes of the Khangai mountains. Yaks form a major proportion of the livestock because the herders live

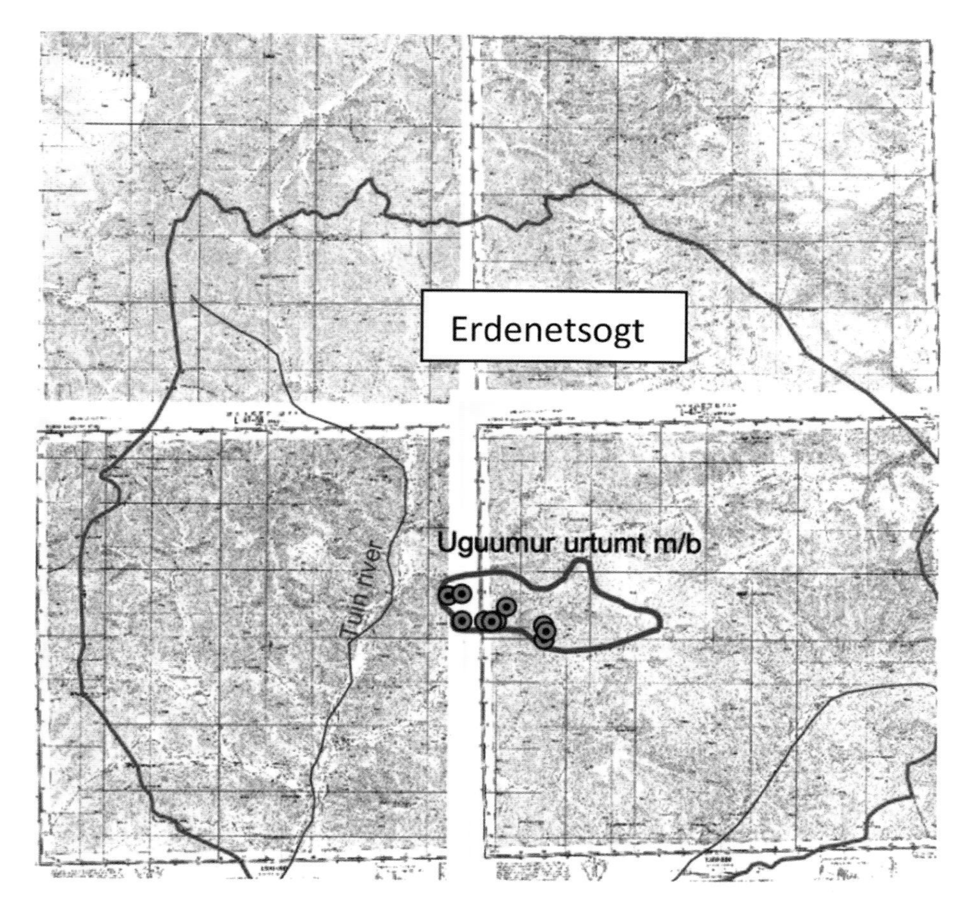

Figure 4.10 Boundary of pasture owned by the Ogoomor Ortomt community, Erdenetsogt sum, Bayanhongor aimag

at high altitudes. A major income source comes from the sale of yak milk and milk products. However, goat numbers have been increasing since 2000. Half of the herders' household members were aged under 40. A leader of the herders' group is Jadamba, 27 years old and a graduate from high school.

They have been doing the following activities:

- They have acquired ownership of their pasture from the sum government (Figure 4.10).
- After not grazing in 2008 plant biomass in the pasture was improved and it became possible to fatten their animals.
- Women have attended training courses and learned how to make sweet and fruit curds and cheese. Their income increased with the sale of these products in the aimag center or through salespeople.
- They observe traditional religious beliefs to protect nature, and clean the Ortomt river and river basin every month.

- They have a credit-savings pool, which is used for loans. Originally this fund was set up as an insurance to reduce the impact of disasters.

The herders' group is planning to find ways to sell yak meat (the Erdenetsogt sum government was going to export 2000 yaks to Russia), especially to increase their income through the sale of dried yak meat. This community can attain the 'win-win model' if they continue to protect pasture, water and nature, and improve their living standards.

Case study: Tragedy of the commons

One family was introduced as an example of this model by local government officers. This family lives in the southern part of Ovoot Mountain in Olziit sum. This household doesn't move seasonally, but lives in one place all year around. They survive by milking a few animals. The household head died many years ago, thus a woman was head of the family and living with her three daughters and one son. She used to be the state champion in milking, and family life was once very good. However, during privatization she acquired few livestock and her children were not able to raise animals, hence they became poor. Having become elderly, the woman now survives on pensions. The family has lost contact with relatives, who could help them. They are not only poor, but they have degraded their surrounding lands, digging for gold and making many holes.

About 50% of herders own less than 100 animals, which can be considered poor or vulnerable to becoming poor. Many of these households are in the poverty trap because they are only subsisting and don't have the opportunity to expand their livestock herds. If they degrade ecological services and don't cooperate in order to improve management of natural resources, they end up in the tragedy of the commons scenario.

4.6 Policy-related social survey

Precipitation decrease (cited by 50% of respondents) was the number one factor impacting on rangeland degradation (Figure 4.11). Human factors such as an increase in goat

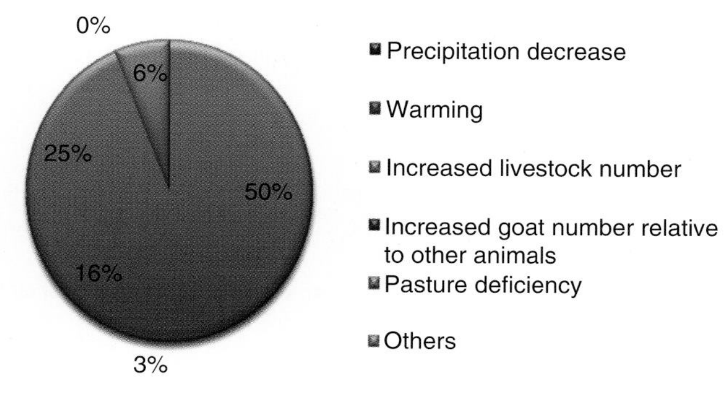

Figure 4.11 Factors impacting on rangeland degradation

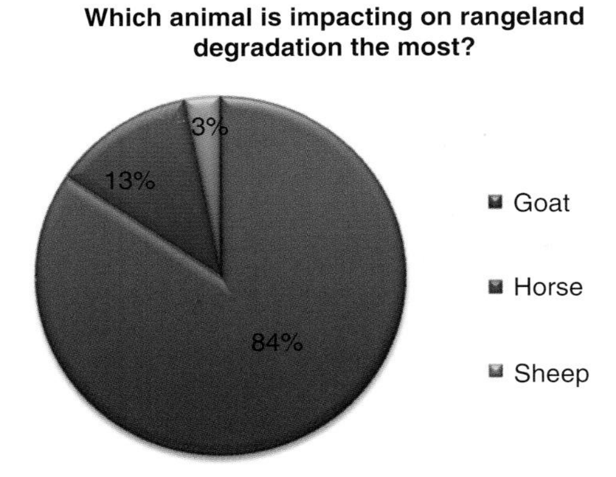

Figure 4.12 Which animal is impacting on rangeland degradation the most?

numbers relative to other animals (25%) and an increase in total livestock numbers (16%) were given less prominence. Others cited warming, disappearance of traditional pasture management, increased gold mining, people's short-sighted view of nature protection, and lack of *otor* movements by herders' groups as other factors for rangeland degradation. Eighty-four percent of people supported a leading role in rangeland degradation of goats relative to other animals (only 13% named horses, and 3% sheep) (Figure 4.12). No one listed 'not enough pasture' as a factor for its degradation.

When asked what the government should do to reduce rangeland degradation, 34% of participants responded that the government should implement a pasture use tax for animals in order to reduce their numbers, and a higher tax for goats to reflect their role in rangeland degradation (Figure 4.13). Thirty-one percent wanted to give pasture ownership to herders' groups; 19% wanted to restore degraded rangelands; 10% to give pasture ownership to herders; and 6% to limit the number of goats per household.

A majority (75%) responded 'yes' to the question of whether pasture ownership is the way to stop rangeland degradation (Figure 4.14). When asked if pasture use by herders' groups in a collective and rotational way can benefit rangeland productivity and

Figure 4.13 What should the government do in order to reduce rangeland degradation?

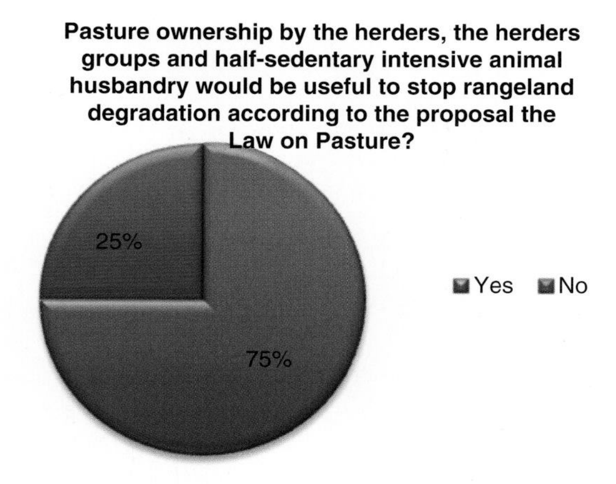

Figure 4.14 Pasture ownership by herders, herders' groups and half-sedentary intensive animal husbandry would be useful to stop rangeland degradation according to the proposed Law on Pasture – yes or no?

nutrient quality, 60% of the participants were positive, and 34% negative (Figure 4.15). Others responded that it would conserve rangeland productivity and quality at a certain level, cited the need to reduce the number of herders, that managers of herders' groups should undergo short-term training, that pasture specialists were needed or that nomadic pastoralism should be developed.

The social survey among the stakeholders in the Tuin river basin yielded interesting results for our understanding of rangeland degradation and policies needed to

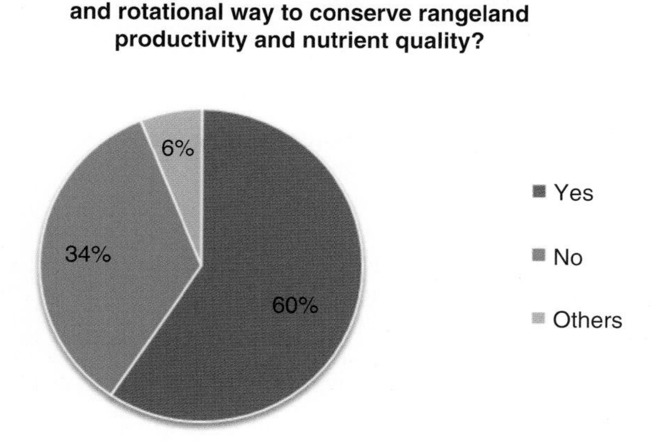

Figure 4.15 Can pasture use by herders' groups in a collective and rotational way conserve rangeland productivity and nutrient quality – yes, no or others?

reduce it. The following statements are the main outputs of this study in the Tuin river basin:

1. Meteorological factors, particular decreased precipitation (50%), were pre-eminent in impacting on rangeland degradation. Human factors such as goat numbers increasing (25%) and total livestock number increasing (16%) were secondary for rangeland degradation.
2. The role of goats was greatest (84%) relative to other animals in rangeland degradation, according to stakeholders' views.
3. A majority (34%) of the participants responded that the government should implement a pasture use tax for animals in order to reduce their numbers, with a higher tax for goats to reflect their contribution to rangeland degradation. Other government measures suggested were to give pasture ownership to herders' groups (31%), and to restore degraded rangelands (19%).
4. A majority (75%) responded that pasture ownership is the way to stop rangeland degradation.
5. Sixty percent of the participants agreed that pasture use by herders' groups in a collective and rotational way could benefit rangeland productivity and nutrient quality, while 34% responded negatively.

People living along the Tuin river view meteorological changes, particularly a decrease in precipitation, as a dominant factor for rangeland degradation. However, land use was another important aspect of land degradation. The goat was without doubt the prime culprit relative to other domestic animals in terms of causing damage to rangelands. Taxing animal ownership in order to reduce their numbers, especially goats, and transferring pastures into the collective ownership of herders' groups were the highest-priority measures for the government. Pasture ownership was the way to reduce rangeland degradation, and most survey participants thought that pasture use by herders' groups in a rotational manner would conserve rangeland productivity and nutrient quality for livestock.

There were many other suggestions, such as:

- Enact laws for pasture resting, and protection of winter and spring pastures.
- Develop a livestock raw materials processing industry.
- Support herders for four seasonal and *otor* movements, making regulations for *otor* movements and investing in herders' mobility.
- Improve livestock quality and build wells in order to improve pasture use.
- Improve ecological education and invest in nature protection.

4.7 Discussion

Livestock density exceeds carrying capacity in central Mongolia. The herders' economic well-being has generally improved with increased livestock numbers. Due to a rise in the price of cashmere, goat numbers have tripled since 1990. Water and forage resources are becoming depleted due to climate change in central Mongolia, and the depletion has been amplified due to increasing land-use intensity. Herders complain that goats exacerbate ecosystem degradation because they dig out the roots of young plants in the spring. Plant species' composition is shifting, with a decreasing number of edible plant species. Plant biomass may have already been reduced in non-linear fashion due to the interaction between climate change and overgrazing in central Mongolia. Spring drying trends have delayed the onset of plant growth in the Gobi-dry steppe boundary area (Ellis *et al.*, 2002) and plant biomass decreased in central Mongolia during the 1990s, primarily due to climate change (Ojima *et al.*, 2004). Livestock numbers have increased since the early 1990s causing increased impact of overgrazing, with a reduction of dominant plant species known as Mongolian grasses by the herders. Observations of grassland ecosystem conditions in Inner Mongolia (China) and in Ovorhangai Aimag made in 2002 and 2007 indicate that central Mongolia may be headed towards ecosystem degradation and desertification problems of the type already experienced in Inner Mongolia.

In the central Mongolian study sites, the disappearance of small streams, lakes and springs was also observed. Decreases in snowfall, increased tree cutting, the melting of permafrost, intensifying drying trends, destruction of riparian shrubs and swamps, and overgrazing all interacted in a non-linear way, resulting in the disappearance of water sources. Regional climate may be affected due to the albedo change that comes with land and snow cover changes. Last summer, severe floods in Hujirt sum territory due to a combination of heavy rain and drought were observed. Riparian ecosystems appear to have keystone value in coupled pastoral social-ecological systems. The collapse of these critical ecosystems' ability to provide water would greatly impact the pastoral community, as water is the most valuable resource for both people and animals in drylands.

The complexities of coupled social-ecological systems increased with Mongolia's transition to a market economy and there are three general categories of herders and communities that are affected to different degrees. A wealthy class of herders with more than 500 livestock per household is emerging, making up only about 5% of herder households. A middle class with 200–500 livestock per household now makes up almost 20% of all herder households, and this group of herders has more choices to increase their resilience and adaptive capacity. Herders with fewer than 200 animals per household will gain advantage by joining formal herders' groups such as NGOs or informal traditional networks, as well-organised cooperation will give opportunities for economic, social, ecological, technological and cultural benefits. More than half of herders are considered poor, with fewer than 100 animals per household. These poorer herders typically live near the cities and along the rivers, and the link between environmental degradation and poverty is notable among this group. Some would benefit from retraining and the institution of sustainable farming systems with the introduction of productive livestock breeds and the diversification of their economy to include pigs, chickens and vegetables as sources of income.

Climate change adaptation options for cultural landscape restoration suggested in participatory community workshops included the introduction of community-based conservation and sustainable use of natural resources, the addition and protection of water points for additional pastureland, agreement between neighboring sums for communal use of *otor* and reserve pastures, and the enlargement of administrative-territorial units, for instance, by combining several sums into one unit in order to restore cultural landscapes. For pastoral communities living in the riparian zones, diversification of the economy and intensification of the livestock industry through ecotourism and farming, the safeguarding of riparian ecosystems from degradation and desertification, and taking animals to *otor* pastureland during the summer period were suggested options. Protection of springs from degradation by livestock was critical for communities living in the mountain and forest steppe.

4.8 Conclusion

Pastoral land systems in central Mongolia are becoming very vulnerable to climate change, which has become a critical slow variable. Water and forage availability is changing due to global warming. Land-use changes, especially since Mongolia's transition to a market economy in 1990, have become critical factors in the vulnerability of pastoral social-ecological systems. The traditional coping mechanisms enhancing the resilience of pastoral communities in the face of climate variability will be lost in Mongolia as in the surrounding countries of central Asia, China and Russia unless alternative development agendas are followed. Using the existing cultural landscape at community and cross-administrative boundary scales in Mongolia appears to be the most cost-effective resilience option for climate change adaptation in pastoral communities. Many international projects on pastoral development, poverty reduction or nature conservation in Mongolia only consider parts of the problem. More holistic approaches are needed to achieve win-win scenarios, strengthening both ecological and social resilience. Strengthening traditional pastoral networks with modern technologies to enhance social well-being as well as development of a legal framework for cultural landscapes at community and administrative unit scales for ecosystem service conservation are required to promote sustainability in pastoral social-ecological systems. Finally, any policy for reduction of vulnerability of pastoral communities must differ in different ecological zones.

We need to develop a 'Strategy for the commons' to avoid a 'Tragedy of the commons' (Hardin, 1968). A detailed analysis shows that a tragedy of the commons scenario may seem attractive if herders are poor and have lost social or ecological resilience. Thus actions that are not based on an up-to-date body of 'hybrid' environmental knowledge, that integrates local management and policy experience with science-based knowledge, may become maladaptation to the critical slow driver that is climate change.

Acknowledgements

The studies were supported by the grants ACCCA and APN. The project 'Policy Framework for Adaptation Strategies of the Mongolian Rangelands to Climate Change at Multiple Scales' (PARCC) is supported by a grant from Advancing Capacity in Support of Climate Change Adaptation (ACCCA), managed by UNITAR and START and funded by the European Commission EuropeAid Cooperation Office, the UK Department of Environment and Rural Affairs, and the Netherlands Climate Change Support Programme. The project 'Dryland Development Paradigm (DDP) Application for Pastoral Systems in the Southern Khangai Mountains, the Most Vulnerable to Climate and Land Use Changes in Mongolia' (CBA2009-12NMY-Togtohyn) was supported by the APN on global environmental change. Both projects were endorsed by the Global Land Project.

References

Chuluun, T. (2000) Climate variability, nomadic society and turbulent history: A Mongolian case study. Update-IHDP. *Newsletter of the International Human Dimensions Program on Global Environmental Change* **1**: 10–12.

Ellis, J., Price, K., Boone, R., Yu, F., Chuluun, T., Yu, M. (2002) Integrated assessment of climate change effects on vegetation in Mongolia and Inner Mongolia. In: Chuluun, T., Ojima, D. (eds), *Fundamental Issues Affecting Sustainability of the Mongolian Steppe*. Interpress Publishing.

Gallopin, G.C. (2006) Linkages between vulnerability, resilience, and adaptive capacity. *Global Environmental Change* **16**: 293–303.

Global Land Project (2005) Science Plan and Implementation Strategy. IGBP Report No. 53/IHDP Report No. 19. Stockholm: IGBP Secretariat, 64 pp.

Hardin, G. (1968) The tragedy of the commons. *Science* **162**: 1243–8. Available at: http://www.sciencemag.org/cgi/reprint/162/3859/1243.pdf

Natsagdorj, D., Sarantuya, G. (2004) On the assessment and forecasting of winter disaster (atmospheric caused dzud) over Mongolia. In: Sixth International Workshop Proceedings on Climate Change in Arid and Semi-Arid Regions of Asia, August 25–26, 2004, Ulaanbaatar, Mongolia, pp. 72–88.

NSO (National Statistical Office – Mongolia) (2009) *Mongolian Statistical Yearbook 2008*. Ulaanbaatar: National Statistical Office.

Ojima, D., Chuluun, T. (2007) Policy changes in Mongolia: implications for land use and landscapes. In: Galvin, K.A., Reid, R.S., Behnke, R.H., Hobbs, N.T. (eds), *Fragmentation in Semi-arid and Arid Landscapes: Consequences for Human and Natural Systems*. Dordrecht: Springer, pp. 179–93.

Ojima, D.S., Chuluun, T., Bolortsetseg, B., Tucker, C.J., Hicke, J. (2004) Eurasian land use impacts on rangeland productivity. In: DeFries, R., Asner, G.P. (eds), *Ecosystem Interactions with Land Use Change*. Geophysical Monograph Series, Vol. 153. Washington DC: American Geophysical Union, pp. 293–301.

Prigogine, I. (1989) The Behavior of Matter under Nonequilibrium Conditions: Fundamental Aspects and Applications: Progress Report for Period August 15, 1989 - April 14, 1990. Center for Studies in Statistical Mechanics at the University of Texas-Austin, United States Department of Energy-Office of Energy Research.

Prigogine, I., Nicolis, G. (1977) *Self-Organization in Non-Equilibrium Systems*. John Wiley & Sons, Inc.

Reynolds, J.F., Smith, M.S., Lambin, E.F. *et al.* (2007) Global desertification: building a science for dryland development. *Science* **316**: 847–51; doi:10.1126/science.1131634.

UNDP (2005) *Economic and Ecological Vulnerabilities and Human Security in Mongolia*. United Nations Development Programme.

Vogel, C., Moser, S.C., Kasperson, R.E., Dabelko, G.D. (2007) Linking vulnerability, adaptation, and resilience to practice: Pathways, players, and partnerships, *Global Environmental Change* **17**: 349–64.

Walker, B., Abel, N. (2002) Resilient rangelands – adaptation in complex systems. In: Gunderson, L.H., Holling, C.S. (eds), *Panarchy: Understanding Transformations in Human and Natural Systems*. Washington DC: Island Press, pp. 293–313.

5 Vulnerability Assessment Diagram: A Case Study on Drought in Middle Inner Mongolia, China

Xiaoqian Liu[1], He Yin[1] and Ademola K. Braimoh[2]
[1]*College of Urban and Environmental Sciences, Peking University, Beijing, China*
[2]*The World Bank, Washington, DC, USA*

5.1 Introduction

Global environmental change is a critical challenge to socioeconomic development and human wellbeing. As such, 'vulnerability' has become a major element of developmental activities in the last few years (IPCC, 2001). The term 'vulnerability' derives from the Latin *vulnerare*, 'to wound', and *vulnus*, 'a wound', 'the potential for damage to' (Collins English Dictionary, 1979; Kelly & Adger, 2000). Although the discipline has undergone significant expansion, especially in its academic dimensions, it never strays too far from this original etymology (Birkmann, 2007; Kates *et al.*, 1985). Now, vulnerability has become a health-related, socioeconomic measure for the comprehensive state of a regional ecosystem. Recently, these social and biophysical aspects have combined to examine the vulnerability of 'human-environment' (Turner *et al.*, 2003a) or 'socio-ecological' (Eakin & Luers, 2006) systems or social-'economical' systems (e.g. Yohe & Tol, 2002).

Although vulnerability assessment studies have been widely carried out, scientists find that many theoretical issues, including a 'conflicting conceptual framework, and inadequate evaluation model', have become major constraints to further development of vulnerability research (Luers *et al.*, 2003). Earlier vulnerability assessments were mainly focused on countries, but variability among locations calls for a more focused approach especially on adaptive capacity management at the local level. Besides, some studies have addressed the

Vulnerability of Land Systems in Asia, First Edition. Edited by Ademola K. Braimoh and He Qing Huang.

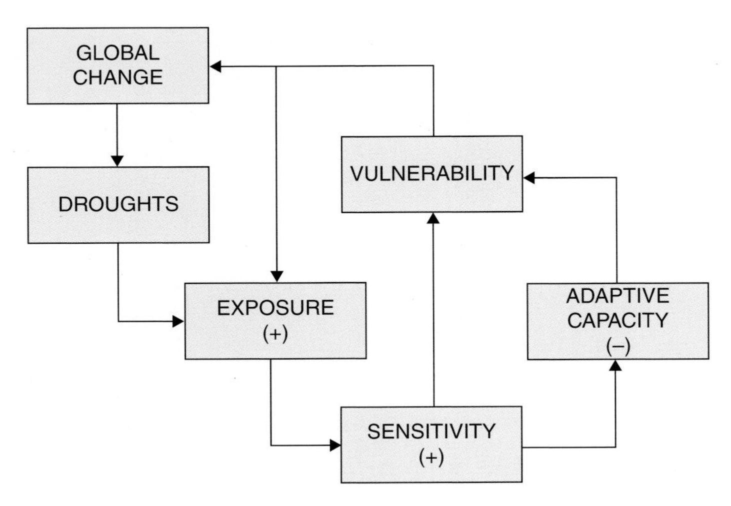

Figure 5.1 A framework for analyzing vulnerability to drought (modified from Turner *et al.*, 2003b)

question of developing indicators to measure vulnerability (Winograd, 2007). Most studies on vulnerability have focused on the extent of exposure to certain hazards and sensitivity to climate change, with little done to examine the comprehensive interaction of exposure and sensitivity, as well as the adaptive capacity from a social-economical perspective on vulnerability reduction.

Therefore, an integrated assessment is required to address the theoretical divergence in vulnerability assessment (Clark *et al.*, 2000; Liu *et al.*, 2009; McLaughlin & Dietz, 2008; Turner *et al.*, 2003b). The central challenge in integration includes evaluating the components of vulnerability and estimating a vulnerability index relevant for policy and decision-making (Figure 5.1).

5.2 An integrated diagram for vulnerability assessment: the VSD model

From its birth, 'vulnerability assessment' has not been a purely quantitative method, but rather a combination of quantitative and qualitative concepts to describe and manage frangibility (McCarthy *et al.*, 2001). However, questions about the great divergence in theory and practical assessment of vulnerability are emerging as a central focus of debates, which has hampered broader academic communication and the further application and referencing of existing works. Depicting a comprehensive and integrative vulnerability assessment framework and resolving the great divergence becomes one of the top issues. In reviewing the most influential theoretical and practical works on vulnerability assessment, we found three essential concepts (see also Brooks, 2003; Adger, 2006; Adger *et al.*, 2005; Luers *et al.*, 2003).

Exposure (E) measures the extent, duration or frequency of a stress on a system. Type and severity of exposure varies with regions and focus of discussion. Sensitivity (S) is the degree to which a system is affected by a stress or perturbation, either positively or negatively. It is an inherent attribute of the human-environment system existing prior to perturbation, and is influenced by both ecological and socioeconomic conditions. Adaptive capacity (A) is the ability of the system to cope with actual or expected stress. It includes the ability of the system to initiate measures to prevent future damage, and/or to extend the range of conditions to which it is adapted (Smit & Wandel, 2006; Walker & Abel, 2002). It is a function of several factors including income, education, information, skills, infrastructural access and management capabilities (McCarthy *et al.*, 2001; Tol & Yohe, 2007).

Polsky and his colleagues were inspired by the success of PPS (Project for Public Spaces) on integrated assessment of increasing environment vulnerability, and urged use of a clear three-dimensional framework to consolidate the divergence (Polsky *et al.*, 2007). They point out that the major source of divergence originated in constructing the evaluation framework and indicator selection when scientists are sharing the same basic data. Thus, they devised a Vulnerability Scoping Diagram (VSD Model), which could be used to manage data in a unified concept and consolidated analytical process. The diagram, shown in Figure 5.2, describes the evaluation specification process in the form of spheres. When exposure and other factors are not yet specified, a general framework is composed of exposure, sensitivity and adaptive capacity. The process of specific evaluation entails four progressive steps. Firstly, the central vulnerability is defined. Then, the three dimensions are defined, followed by an indicator layer to explain the various dimensions; a final layer provides specific quantitative or qualitative parameters/factors. Conceptual functions are given as an equation of three dimensions based upon the definition of agricultural vulnerability from Luers' study (Luers *et al.*, 2003). In summary, the VSD model shows a progressive specification process of vulnerability assessment in the form of spheres, and is a practical step towards organizing data and increasing compatibility of research (Olga & Donald, 2002).

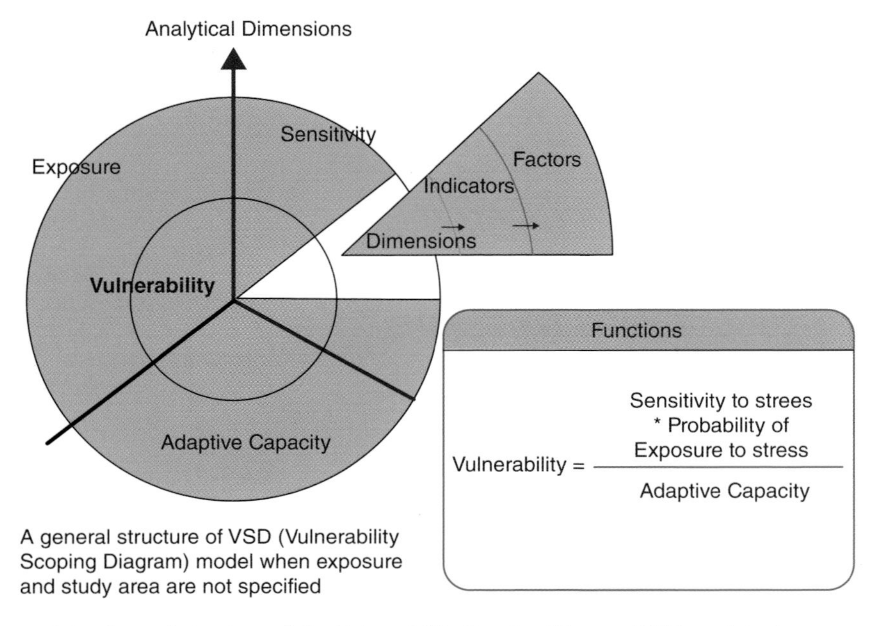

Figure 5.2 General structure of the Vulnerability Scoping Diagram (VSD) model when exposure and study area are not specified

5.3 Case study using the VSD model

5.3.1 The study area

In order to show how the diagram works and to demonstrate its application, we processed a case study of vulnerability assessment in Middle Inner Mongolia, China, using the VSD model (see also Liu *et al.*, 2013). Based upon the diagram, three main components of vulnerability – exposure, sensitivity and adaptive capacity – were described. Indicators were further combined to a vulnerability index in a geographical information system to highlight the differential effects of the components on overall vulnerability. The study demonstrates that vulnerability reducing actions that increase adaptive capacity are crucial in the mitigation of vulnerability at the local level. Social and environmental variables were acquired for this study (Table 5.1). To render them suitable for spatial analysis, we created indicators from the original variables using the county as the unit of analysis.

Middle Inner Mongolia (MIM) is located between latitudes 38°N and 49°N and stretches between longitudes 107°E and 119°E in the north of China. Desertification caused by

Table 5.1 Description of data used in the study

Type and date	Indicators	Source
Vegetation (1998–2008)	Standard deviation of Normalized Difference Vegetation Index (NDVI)	Spot/vegetation images from Institute of Technology in Belgium Flemish (VITO) (1 × 1 km) http://www.vito.be/VITO/EN/HomepageAdmin/Home
Climate (1960–2008)	Coefficient of variance of precipitation, coefficient of variance of temperature, standardized precipitation index	Daily maximum/minimum/average precipitation and temperature data of meteorological stations from National Meteorically bureau of China http://cdc.bjmb.gov.cn/
Digital Elevation Model (2000)	Elevation	US Geological Survey GTOPO30 DEM (1 × 1 km) http://www.usgs.gov/pubprod/data.html#data
Socioeconomic data (2006)	Per Capita Cultivated Area, Physicians per 1000 Persons, Ratio between agriculture and industrial output, Technologists per 1000 Persons, Per Capita Savings Deposit, Per Capita Business Volume of Post and Telecom Service (access to information), Population Density, Per Capita GDP	Statistical Yearbook of China (2007) and Statistical Yearbook of Counties and Cities in Inner Mongolia Autonomous Province (Inner Mongolia Statistical Yearbook, 2007)

natural and anthropogenic factors has dominated changes of landscape. We roughly divided Inner Mongolia Autonomous region into three parts and focused on the middle part for its geographical homogeneity (Figure 5.3).

As a typical inland plateau, most of the area is covered by grassland or desert, occupying the farming-grazing transitional zone in China, and is very sensitive to climate fluctuation and anthropogenic impact. The grassland forms a continuum through the southern mountains across the Inner Mongolia Plateau, to the south of the Ordos Plateau and the Loess Plateau. Cultivation of grain is restricted to a few areas suitable for agriculture, and also consumes a lot of underground water, which further exacerbates hydrologic scarcity in the semi-arid ecosystem (Luo & Xue, 1995). MIM has a temperate continental monsoon climate, with low and unevenly distributed rainfall, wind, and extremes of cold and heat. Annual solar radiation increases from northeast to southwest, precipitation decreases from northeast to southwest. Mean annual precipitation ranges from less than 50 mm to more than 450 mm, showing a strong northeast-west gradient (Niu, 2000; Yin, 2009).

Desertification of land caused by drought is one of the biggest challenges for sustainable development and environmental management. In order to balance ecological fragility with land-use requirements for socioeconomic development, the government has embarked on several environmental programs through regulations and technical improvement (Leng, 1994). The programs include grassland restoration or prohibition of grazing in degraded areas, implementation of scientific methods of ecosystem services conservation and enhancement of land productivity.

5.3.2 Vulnerability profile at the county level

Drought is the main risk in the study area and was therefore chosen as the exposure variable. We used the standardized precipitation index (SPI) as a simple indicator of exposure. SPI has been used in much research related to drought hazard in the past (e.g. McKee *et al.*, 1993; Wu, *et al.*, 2007; see also Chapter 2). In this study, SPI was calculated from a time series of annual precipitation data for 50 years from 1958 to 2008 using the equation:

$$\mathrm{SPI} = \frac{P_x - \bar{P}}{P_{\mathrm{std}}} \tag{5.1}$$

where P_x is the average precipitation for county x for the time series, $\bar{P}$ is the average for all the counties and P_{std} the standard deviation for all the counties. Negative values obtained from this equation indicate precipitation deficits (drought events), while positive values represent wet conditions. SPI values for MIM ranged from −2.5 for Erenhot county to 1.5 for Horqin Left county, with over 40% of the counties having values less than 0. To enable a direct comparison with other components of vulnerability (sensitivity and adaptive capacity), the SPI was rescaled to a range between 0 and 1 using the maximum standardization method.

To measure sensitivity we performed a principal components analysis[1] on the coefficients of variation of temperature, precipitation, vegetation index and elevation (OuYang *et al.*, 2000). The coefficient of variation of climate measures climatic variability whereas that of the normalized difference vegetation index measures the dependence of

[1] Apart from helping us to deal with the problem of data redundancy, principal components analysis also provides an objective basis for weighting the indices, which is crucial for the overall vulnerability index.

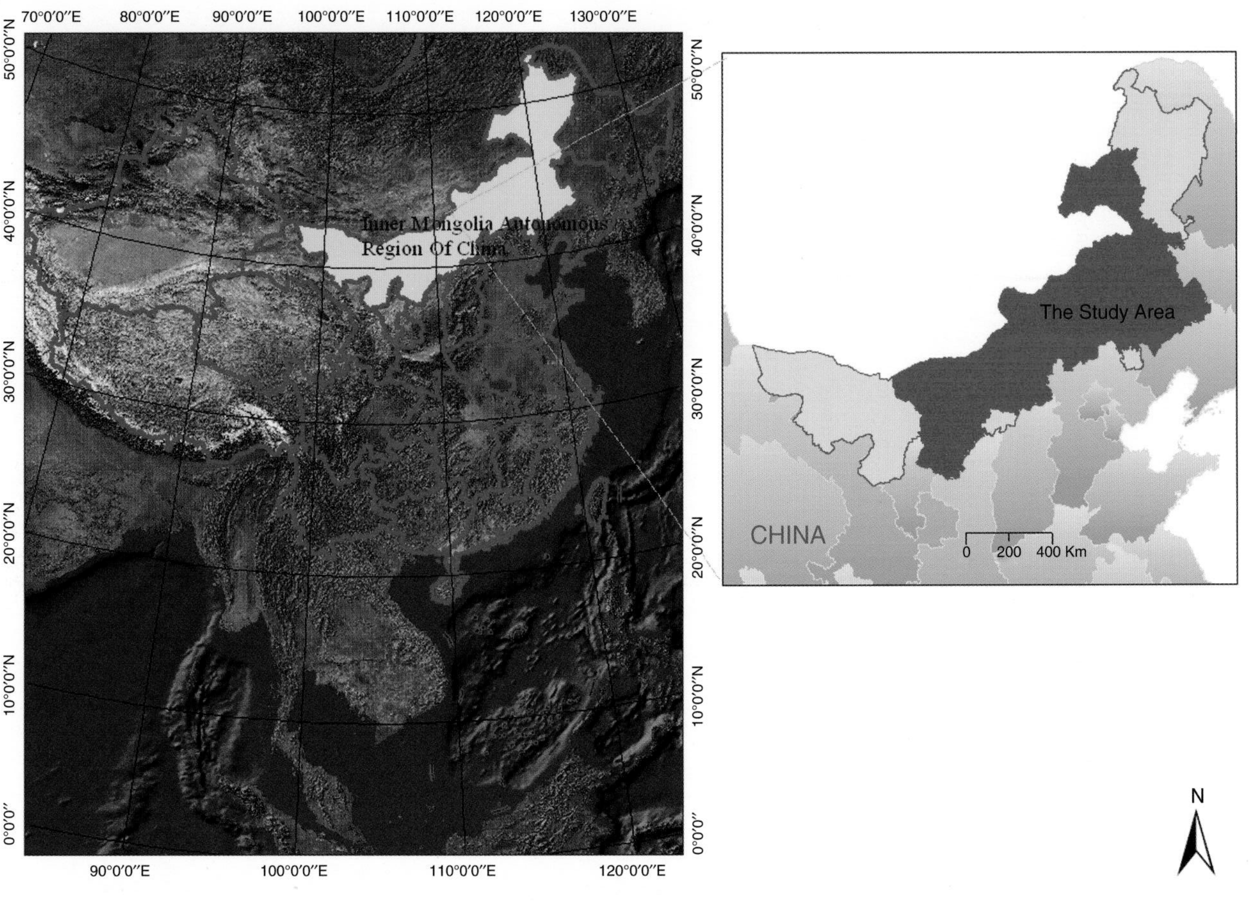

Figure 5.3 Geographical location of the study area

vegetation on climate. Elevation potentially influences the other variables across the landscape (Li *et al.*, 2006).

To measure adaptive capacity, we also performed a principal components analysis on standardized indices of adaptive capacity parameters, including 'Per Capita Cultivated Area', 'Ratio of Agriculture and Industrial Output' and 'Population Density' to measure the population and economic production conditions; 'Per Capita Saving Deposit', 'Per Capita GDP' and 'Per Capita Business Volume of Post and Telecom Service' to measure the income level; also, parameters such as 'Physicians per 1000 Persons' and 'Technologists per 1000 Persons' could indicate the level of information access. The overall vulnerability index (V) was calculated using the equation:

$$V = \frac{|-E + S|}{A} \tag{5.2}$$

in which E indicates the level of exposure, S indicates the sensitivity and A indicates the value of the Adaptive Capacity Index. We assume that E, S and A are positive values. Equation 5.2 ensures that the most vulnerable county is the one with the lowest adaptive capacity, and the highest exposure and sensitivity to drought. Similar results on calculating V can be found in other studies (e.g. O'Brien *et al.*, 2004; Polsky *et al.*, 2007).

5.4 Results and discussion

5.4.1 Relative impact of the components on the vulnerability index

Maps of three vulnerability dimensions are given in Figure 5.4: (a) exposure index; (b) sensitivity index; and (c) adaptive capacity index. The overall vulnerability index is given in Figure 5.4d; this shows that the counties with the highest vulnerability index form a line through the mid-zone of the region. Close examination of Figure 5.4, the correlation analysis in Table 5.2 and the cumulative distribution functions provided in Figure 5.5, provide multiple insights into the relative impact of the components on the estimated vulnerability index.

The linear correlations between vulnerability and its components reflect the human-environment conditions in MIM. Hydrologic scarcity and poor socioeconomic conditions markedly influence the differential vulnerability of the counties to drought. The high correlation between exposure and sensitivity (Table 5.2) suggests that the counties that are most sensitive to drought are also the most exposed. On the other hand, correlations between adaptive capacity and the other vulnerability components are much lower, signifying limited ability of most of the counties to cope with the adverse consequences of drought. Adaptive capacity mitigates or reduces the variability in exposure and sensitivity. The rescaled median adaptive capacity of 0.22 (Table 5.3) suggests that adaptive capacity is generally low across the counties. This is a result of the underdeveloped socioeconomics of MIM. The highest adaptive capacity is for Ordos region in the southwest of MIM, which encompasses the highest industrial production and per capita GDP emanating from its relatively rich natural resources and energy industry (Figure 5.4c). Medium adaptive capacity is associated mostly with counties in the middle of MIM. These are predominantly urbanized areas with higher levels of social infrastructure compared to rural areas.

The spread of the components measured by the inter-quartile range (IQR; Table 5.3) is in the order: sensitivity > exposure > adaptive capacity. The lowest value of IQR for adaptive capacity is another clear manifestation of the uneven economic development in

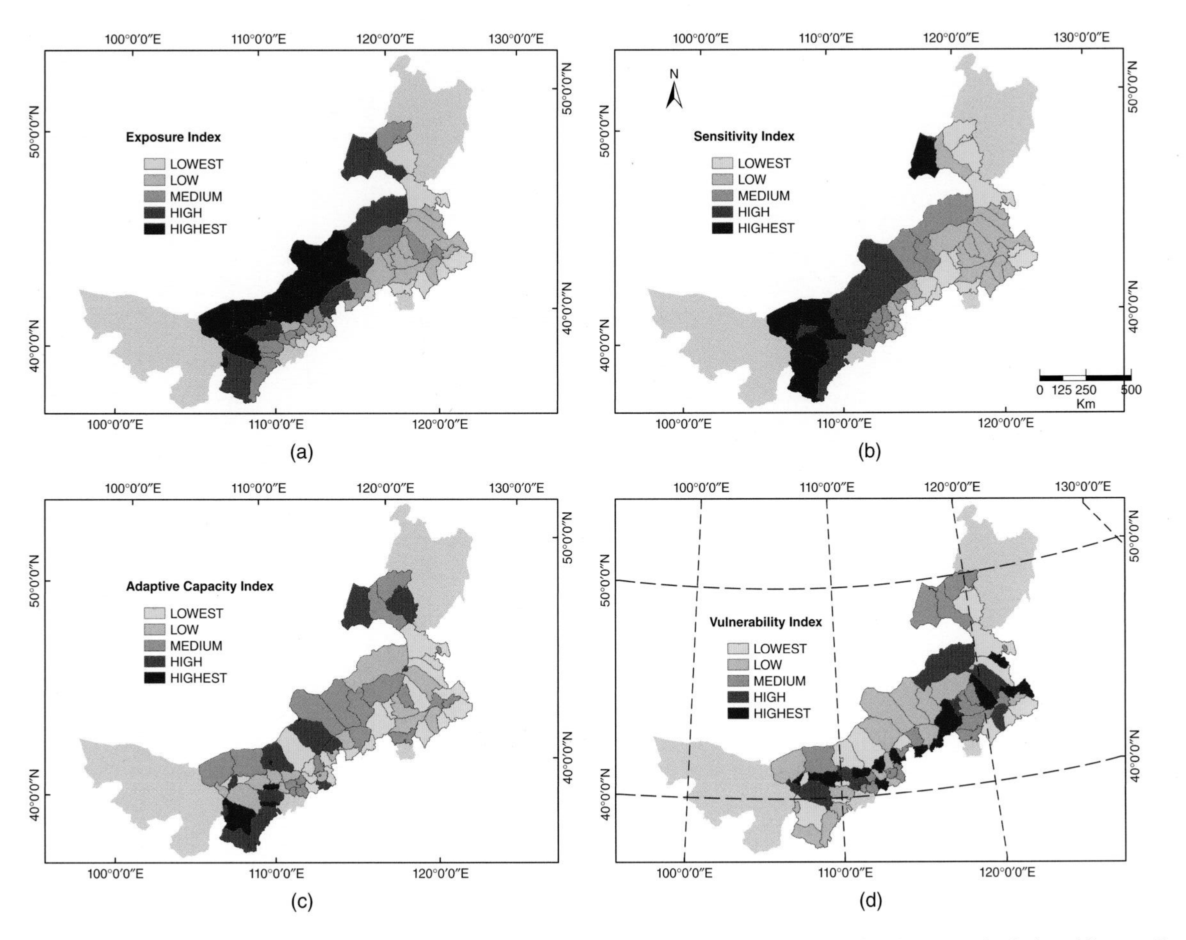

Figure 5.4 Maps of vulnerability indices in the study area: (a) exposure index; (b) sensitivity index; (c) adaptive capacity index; (d) overall vulnerability index

Table 5.2 Intercorrelation of the vulnerability indices

	Exposure	Sensitivity	Adaptive capacity	Vulnerability
Exposure	1.00			
Sensitivity	0.80	1.00		
Adaptive capacity	0.33	0.45	1.00	
Vulnerability	0.90	0.86	0.04	1.00

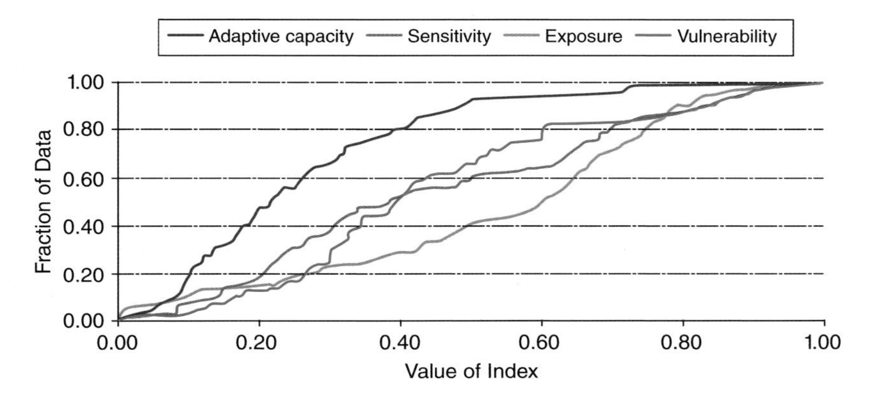

Figure 5.5 Cumulative distribution functions for vulnerability indices in the study area

MIM. Ordos region is experiencing unprecedented development with its abundant reserves of mineral and energy resources. Its industrial production dominates the regional economy and over the past few years there has been a huge government investment in ecological restoration through a program returning farmland to grassland (Song & Zhang, 2007). This restoration program markedly enhances adaptive capacity and reduces vulnerability.

5.4.2 Model calibration

Vulnerability is a measure of the potential for damage to a threatened system, and socioeconomic losses incurred in repairing the damage can indicate the severity of damage (O'Brien *et al.*, 2004; Tol & Yohe 2007). Social assistance funds have proved to be important for creating projects to support sustainable development, especially at a community-based level.

Table 5.3 Percentiles of the rescaled vulnerability indices

	Percentiles			
	25th	Median	75th	Interquartile range
Exposure	0.35	0.61	0.73	0.38
Sensitivity	0.22	0.39	0.68	0.46
Adaptive capacity	0.12	0.22	0.36	0.24
Vulnerability	0.30	0.40	0.60	0.30

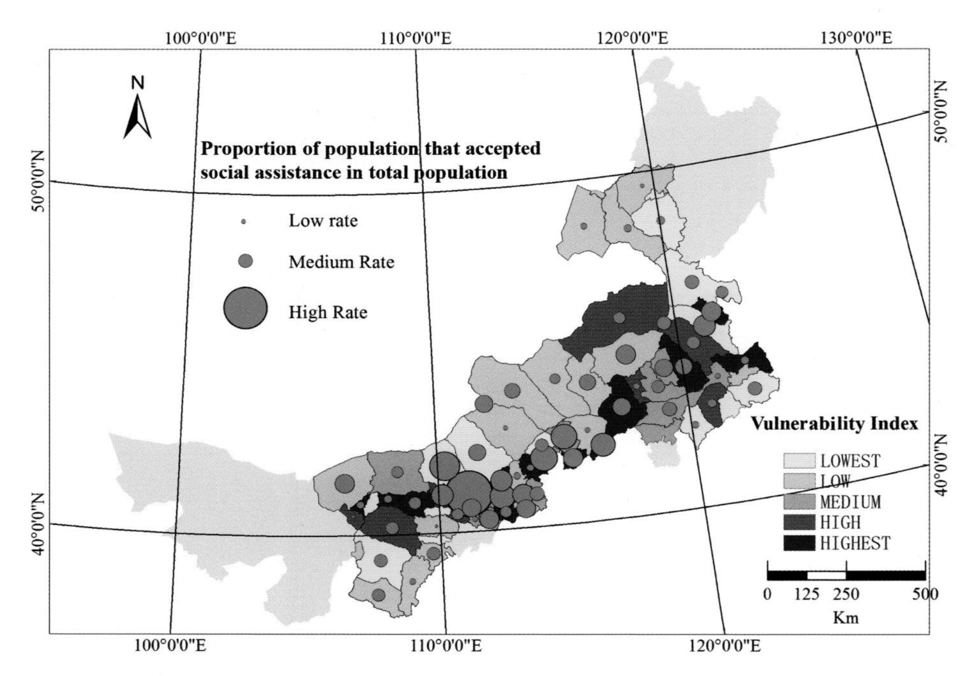

Figure 5.6 Proportions of the population receiving social assistance in the study area

Facing challenges from increasing vulnerability, central and local governments increasingly highlight the importance of social welfare in balancing economic growth and social stability (Leung, 2006). Technical and social assistance were provided when vulnerable communities suffered welfare loss. By using statistical data from the Ministry of Civil Affairs of China for 2008, which was the second year of social economic data used in the above vulnerability evaluation, we could evaluate the significance of social assistance upon our assessment. Two sets of statistical data at the county level were applied in the model calibration: annual social assistance investment and annual social assistance population.

The proportion of the social-assisted population in 2008, normalized by the total population in each county, is shown in Figure 5.6 superimposed on the vulnerability evaluation results. A dot map of the distribution of social assistance investment (in units of 10,000 yuan) depicted a similar trend in the proportion of people who accepted social assistance (Figure 5.7). By comparing with the integrated vulnerability levels, we could easily find that the most vulnerable areas well matched those with the highest social assistance data. Thus, our evaluation diagram can provide general practical guidance in directing environmental management and social welfare programs at the county level.

5.5 Conclusion

Vulnerability induced by climate change or natural disaster is increasingly attracting attention in both environmental science and geography (Adger, 2006; Brooks, 2003). Much has been achieved in both theory and practice in different research fields, including biophysical geography, human ecology, political economics, and constructivist and political ecology

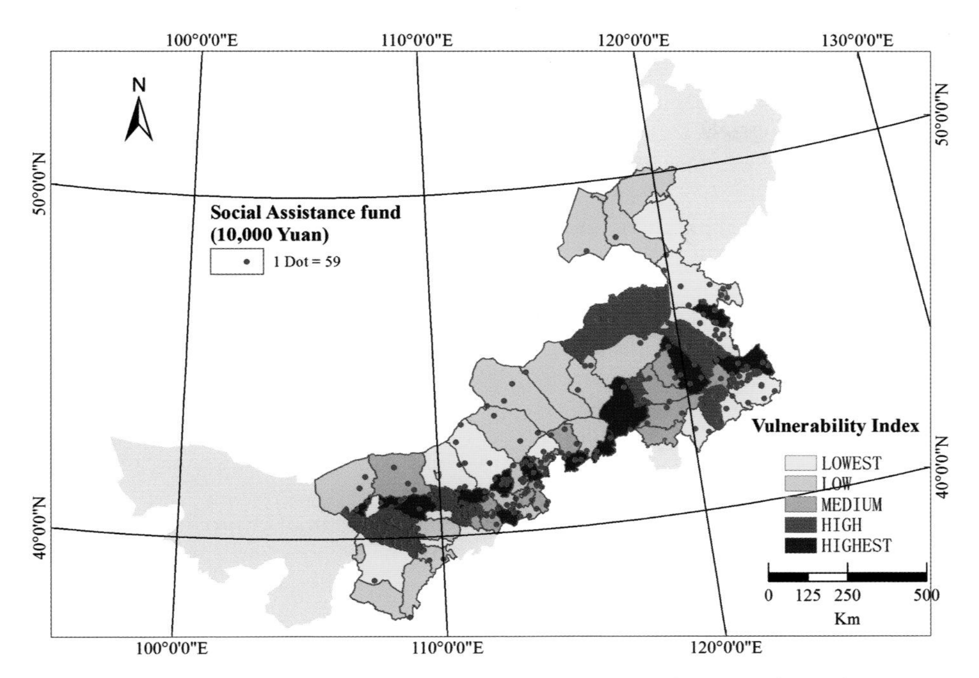

Figure 5.7 Dot map of distribution of the social assistance fund over the study area

(McLaughlin & Dietz, 2008). These researches provide scientific bases for improvements in the adaptability and resilience of systems, helping to enhance human development and prevent environmental degradation. Future research must focus on developing a clearly defined, standardized and scientific evaluation process, and this trend should be reflected in domestic vulnerability studies. The VSD model is an example of constructing a standardized and retroactive diagram for vulnerability assessment in a clear progressive way.

Our case study demonstrates how vulnerability can be assessed from its components using relevant indicators in specific areas subject to rapid natural and socioeconomic changes. Even though the vulnerability literature is replete with the need to assess vulnerability using exposure, sensitivity and adaptive capacity indices, the relationships between the indices are hardly emphasized. The interactions of the vulnerability indices can guide government and local communities in vulnerability-reducing actions. It should be pointed out that this study is not about identifying the causes of drought, which are multi-factorial, but rather identifying areas susceptible to drought given their location characteristics and socioeconomic attributes. The fact that drought hazard is very sensitive to the timescale of analysis calls for repeated monitoring to reduce uncertainty in vulnerability analysis. Furthermore, changes in land-use policy and ecological protection could have a long-lasting and delayed impact on adaptive capacity in facing natural hazards, thereby increasing the need for long-term observation.

The technique outlined in this chapter is a first step in elucidating vulnerability to drought in Middle Inner Mongolia of China. In future, we hope to employ integrated models to fully model and quantify the interactions between the vulnerability components at various scales and the implication for sustainable land management.

References

Adger, W.N. (2006) Vulnerability. *Global Environmental Change* **163**: 268–81.

Adger, W.N., Arnell, N.W., Tompkins, E.L. (2005) Successful adaptation to climate change: perspectives across scales. *Global Environmental Change* Part A, **15**: 77–86.

Birkmann, J. (2007) Risk and vulnerability indicators at different scales: Applicability, usefulness and policy implications. *Environmental Hazards* **7**: 20–31.

Brooks, N. (2003) *Vulnerability, Risk and Adaptation – a Framework*. Norwich: Tyndall Centre for Climate Change Research, University of East Anglia.

Clark, W.C., Jager, J., Corell, R., *et al.* (2000) Assessing vulnerability to global environmental risks: Report of the Workshop on Vulnerability to Global Environmental Change: Challenges for Research, Assessment and Decision Making, 22–25 May, Airlie House, Warrenton, Virginia. Research and Assessment Systems for Sustainability Program Discussion Paper 2000-12. Cambridge, MA: Environmental and Natural Resources Program, Belfer Center for Science and International Affairs BCSIA, Kennedy School of Government, Harvard University.

Eakin, H., Luers, A.L. (2006) Assessing the vulnerability of social-environmental systems. *Annual Review of Environment and Resources* **31**: 365–94.

Collins Dictionary of the English Language (1979) Hanks, P. (ed.), 1st edn. London: William Collins & Sons.

Inner Mongolia Statistical Yearbook (2007) Beijing: China Statistics Press.

IPCC (2001) *Impacts, Adaptation, and Vulnerability. The Contribution of Working Group II to the Third Scientific Assessment of the Intergovernmental Panel on Climate Change.* Cambridge: Cambridge University Press.

Kates, R.W., Ausubel, J.H., Berbarian, M. (1985) *Climate Impacts Assessment: Studies of the Interaction of Climate and Society*. New York: John Wiley & Sons, Inc.

Kelly, P.M., Adger, W.N. (2000) Theory and practice in assessing vulnerability to climate change and facilitating adaptation. *Climatic Change* **47**: 325–52.

Leng, S.Y. (1994) Human impacts on environmental degradation of the Ordos. *Chinese Journal of Arid Land Resources and Environment* **8**: 44–52.

Leung, J.C.B. (2006) The emergence of social assistance in China. *International Journal of Social Welfare* **15**: 188–98.

Li, A., Wang, A., Liang, S., Zhou, W. (2006) Eco-environmental vulnerability evaluation in mountainous regions using remote sensing and GIS – A case study in the upper reaches of Minjiang River, China. *Ecological Modelling* **192**: 175–87.

Liu, X.Q., Wang, Y.L., Peng, J. (2009) Progress in vulnerability analysis of coupled human–environment systems. *Advances in Earth Science* **24**: 917–27 [in Chinese].

Liu, X.Q., Wang, Y.L., Peng, J. *et al.* (2013) Assessing vulnerability to drought based on exposure, sensitivity and adaptive capacity: A case study in middle Inner Mongolia of China. *Chinese Geographical Science* **23**: 13–25.

Luers, A.L., Lobell, D.B., Sklar, L.S., Addams, C.L., Matson, P.A. (2003) A method for quantifying vulnerability, applied to the agricultural system of the Yaqui Valley, Mexico. *Global Environmental Change* **134**: 255–67.

Luo, C.P., Xue, J.Y. (1995) Ecologically vulnerable characteristics of the farming-pastoral zigzag zone in northern China. *Chinese Journal of Arid Land Resources and Environment* **91**: 1–7.

McCarthy, J.J., Canziani, O.F., Leary, N.A., *et al.* (2001) *Climate Change: Impacts, Adaptation and Vulnerability*. Cambridge: Cambridge University Press.

McKee, T.N., Doesken, J., Kleist, J. (1993) The relationship of drought frequency and duration to time scales. Eighth Conference on Applied Climatology, American Meteorological Society, Anaheim, CA, pp. 174–84.

McLaughlin, P., Dietz, T. (2008) Structure, agency and environment: Toward an integrated perspective on vulnerability. *Global Environmental Change* **18**: 99–111.

Niu, J.M. (2000) Relationship between main vegetation types and climatic factors in Inner Mongolia. *Chinese Journal of Applied Ecology* **11**: 47–52.

O'Brien, K., Leichenko, R., Kelkar, U. *et al.* (2004) Mapping vulnerability to multiple stressors: climate change and globalization in India. *Global Environmental Change Part A* **14**: 303–13.

Olga, V.W., Donald, A.W. (2002) Assessing vulnerability to agricultural drought: a Nebraska case study. *Natural Hazards* **25**: 37–58.

OuYang, Z.Y., Wang, X.K., Miao, H. *et al.* (2000) China's Eco-environmental sensitivity and its spatial heterogeneity. *Acta Ecological Sinica* **20**: 9–12.

Polsky, C., Neff, R., Yarnal, B. (2007) Building comparable global change vulnerability assessments: The vulnerability scoping diagram. *Global Environmental Change* **17**: 472–85.

Smit, B., Wandel, J. (2006) Adaptation, adaptive capacity and vulnerability. *Global Environmental Change* **16**: 282–92.

Song, N.P., Zhang, F.R. (2007) The changing process and mechanism of the farming-grazing transitional land use pattern in Ordos. *Acta Geographica Sinica* **62**: 1300–08.

Tol, R.S.J., Yohe, G.W. (2007) The weakest link hypothesis for adaptive capacity: An empirical test. *Global Environmental Change* **17**: 218–27.

Turner, B.L., Matson, P.A., McCarthy, J.J., *et al.* (2003a) Illustrating the coupled human-environment system for vulnerability analysis: Three case studies. *Proceedings of the National Academy of Sciences of the U. S. A.* **100**: 8080–5.

Turner, B.L., Kasperson, R.E., Matson, P.A., *et al.* (2003b) A framework for vulnerability analysis in sustainability science. *Proceedings of the National Academy of Sciences of the U. S. A.* **100**: 8074–9.

Walker, B., Abel, N. (2002) Resilient rangelands – adaptation in complex systems. In: Gunderson, L., Holling, C. (eds), *Panarchy: Understanding Transformations in Human and Natural Systems*. London: Island Press, pp. 293–361.

Winograd, M. (2007) Sustainability and vulnerability indicators for decision making: lessons learned from Honduras. *International Journal of Sustainable Development* **1** (1): 93–105.

Wu, H., Svoboda, M., Hayes, M., *et al.* (2007) Appropriate application of the Standardized Precipitation Index in arid locations and dry seasons. *International Journal of Climatology* **27**: 65–79.

Yin, H. (2009) Desertification assessment in Inner Mongolia based on vegetation dynamics. Master's thesis, Peking University.

Yohe, G., Tol, R.S.J. (2002) Indicators for social and economic coping capacity – moving toward a working definition of adaptive capacity. *Global Environmental Change* **12**: 25–40.

6 Vulnerability of Agriculture to Climate Change in Arid Regions: a Case Study of Western Rajasthan, India

R.B. Singh and Ajay Kumar
Department of Geography, Delhi School of Economics, Delhi, India

6.1 Introduction

Today the existence of mankind faces a serious threat in the form of global climate change. Vulnerability to this threat differs with income level, state of the environment and the relationship of the local community with the natural resources. The most vulnerable groups include the poor and tribal peoples, who are totally and directly dependent on natural resources. Their vulnerability increases with depletion of these resources. In this pressing situation the scientific community is still debating the rate of change. However, all agree that the climate is changing, and politicians and planners are trying to formulate mitigation strategies.

Despite efforts to abate the human causes, human-driven climate change will continue for decades and longer (IPCC, 2001a). Population growth and economic growth are likely to intensify uses of and pressure on natural resource systems. Global climate change, which has already impacted natural resource systems across the Earth, is adding to the pressures and is expected to substantially disrupt many of these systems and the goods and services that they provide (Fischlin & Midgley, 2007; MEA, 2005). Therefore climate change mitigation steps by themselves are not enough; there is also a need to strengthen the adaptive capacity of people and regions. But prior to this a vulnerability study of every

Vulnerability of Land Systems in Asia, First Edition. Edited by Ademola K. Braimoh and He Qing Huang.

region, sector and community has to be carried out to assess the sensitivity and capacity of each ecosystem.

When we talk about vulnerability to climate change, the developing countries stand on the frontline. India, as a developing country, scores high on the vulnerability scale. The factors that make India most vulnerable are its fast-growing population, depleting natural resources, high dependence of the majority of the population on climate-sensitive sectors like agriculture and forestry, and a lack of infrastructure, resulting in a large number of poor people, malnourished children and mothers, a deprived marginal class and a large unskilled labor population (Bohle *et al.*, 1994; Downing, 2003; O'Brein & Leichenko, 2000). The arid regions of India are already very poor in natural resources, and these are highly susceptible to loss through climate change. The problem is further intensified by unsustainable agricultural practices, which lead to water losses, soil degradation and permanent loss of the land as a resource. The solution to this problem lies in traditional techniques and local knowledge, which contribute in building the adaptive capacity of the system in a sustainable manner to cope with climate change.

The objective of this chapter is to find out the extent of climate variability in Western Rajasthan and to assess the vulnerability of agriculture in response to climate change. The chapter will also try to prove that the availability of water resources in arid regions is the most important factor in determining vulnerability to climate change.

6.2 Climate change scenarios: global, national and local levels

Climate change is not a new phenomenon; the Earth's climate has been in a state of change ever since the formation of the planet. The evidence shows that the climate of Earth swings between global cooling and global warming. Then the question arises, what is the difference between climate change past and present? The answer is the role of anthropogenic activities. In the past climate change was totally induced by nature, either due to periodic solar changes or by Earth's tectonic activity. But current climate change is more human induced in nature. Also there is a difference in the rate of climate change; at present the change is occurring at a far greater pace compared to rates in the past. Human activities like burning of fossil fuels, emissions from industries, paddy culture, etc. are accelerating the process of climate change (Kumar & Singh, 2003). The ability of humans to predict future conditions based on the present situation is also responsible for creating panic about climate change. However, this ability is also useful in identifying the most vulnerable sectors and communities, so that a better strategy against this threat can be adopted.

An increasing body of evidence suggests that the global average surface temperature has increased over the 20th century by about 0.6 ± 0.2°C (IPCC, 2001a). According to the World Meteorological Organization (WMO), the decade 1998–2007 was the warmest on record. The global mean surface temperature for 2007 is estimated at 0.41°C/0.74°F above the 1961–1990 annual average of 14.00°C/57.20°F. The WMO states that among other remarkable global climatic events recorded in 2007, a record low extent of Arctic sea ice was observed, which led to the first recorded opening of the Canadian Northwest Passage.

India, being a developing country in which most people live in a poor state of economy, society and environment, falls into the most vulnerable group (Drunen *et al.*, 2006; Shukla *et al.*, 2003). India has diverse climatic conditions, which makes it susceptible to different types of climatic hazards like floods, cyclones, droughts, landslides, avalanches, etc. (Blaikie *et al.*, 1994; Dash, 2007). Besides these hazards, the variability of the Indian monsoon and the reliance on agriculture of the majority of the population adds to the vulnerability factor. A small rise in temperature can cause a huge fall in crop production (Sen & Singh, 2002).

Observations over India show that the mean annual surface air temperature has increased by 0.4°C in the last 100 years (Hingane & Rupa Kumar, 1985). Subsequently, Rupa Kumar *et al.* (2006) indicated that the warming trend over India is about 0.57°C per 100 years. The increasing warming and the changing rainfall pattern over the Indian region are expected to adversely affect the human population, infrastructure and ecosystems. It might also lead to high instances of water-borne diseases such as malaria, loss of soil fertility and declines in agricultural productivity. Preliminary studies undertaken in India indicate an increased occurrence of extreme weather events such as floods, droughts and cyclones. The impacts of extreme climate have so far fallen most heavily on the poor, suggesting that the impacts of future changes in climate extremes would also fall disproportionately on the poor (IPCC, 2001b).

Considering a range of equilibrium climate change scenarios that project a temperature rise of 2.5 to 4.9°C for India, Kumar and Parikh (2001) estimated that: (a) without considering the yield losses due to carbon dioxide fertilization effects for rice and wheat, the yield losses vary between 32 and 40%, and 41 and 52%, respectively; (b) GDP would drop by between 1.8 and 3.4%. Their study also showed that even with carbon fertilization effects, losses would be in the same direction but somewhat smaller. Using an alternative methodology, Kumar and Parikh (2001) showed that even with farm-level adaptations, the impacts of climate change on Indian agriculture would remain significant. They estimated that with a temperature change of +2°C and an accompanying precipitation change of +7%, farm-level total net revenue would fall by 9%, whereas with a temperature increase of +3.5°C and precipitation change of +15%, the fall in farm-level total net revenue would be nearly 25%.

The predicted impact of climate change on arid regions is that their climate is likely to become more extreme, with few exceptions; they are projected to become hotter but not significantly wetter. Temperature increase is a major threat to organisms that exist near their heat tolerance limit. The impacts on water balance, hydrology and vegetation are also predicted to be severe. Desertification is more likely to become irreversible if the environment become drier and the soil become further degraded through erosion and compaction.

In semi-arid and arid environments, rainfall is short-lived and often very intense. Because soils tend to be thin, much of the rainfall runs directly off the surface, only to infiltrate deeper soils down slope or along river beds. However, in semi-arid and arid areas, where groundwater recharge occurs after flood events, changes in the frequency and magnitude of rainfall events will alter the number of recharge events (IPCC, 2001a; Ribot, 1996). There are other problems associated with some water bodies in the region, especially in areas of high human density. In these areas, habitat degradation often is important, causing many semi-enclosed water bodies to become eutrophic (Kharin, 1995; Ramesh & Yadava, 2005). Dryland salinization also is having an impact on water quality in some countries where groundwater is contaminated with salt; in other countries (e.g., Oman and United Arab Emirates), sea water has intruded into freshwater aquifers (IPCC, 2001b).

In an area dominated by arid and semi-arid lands, water is a very limited resource. Droughts, desertification and water shortages are permanent features of life. Rapid development is threatening some water supplies through salinization and pollution, and increasing standards of living and expanding population are increasing demand (Sainath, 2002). Water is a scarce resource, and will continue to be in the future. Projections of changes in runoff and water supply under climate change scenarios vary (Sharma & Bharat, 2009). Some countries are developing programmes to conserve and reuse water or to achieve more efficient irrigation (Sathaye & Najam, 2007). Some countries are vulnerable to reductions in runoff. Hence there is a grave need to develop a sustainable and efficient strategy for combating climate change (Yohe & Lasco, 2007).

6.3 Study area

Rajasthan alone accounts for about 61.9% of the total arid area of India, spread over 12 districts, namely Barmer, Bikaner, Churu, Ganganagar, Hanumangarh, Jaisalmer, Jalore, Jhunjhunu, Jodhpur, Nagaur, Pali and Sikar in the western Rajasthan. The hot arid region of western Rajasthan, part of the Thar desert, is highly prone to wind erosion and represents a fragile ecosystem that has resulted from the continued effect of various natural processes such as low and erratic rainfall, intense heat, high evaporation, low relative humidity, poor edaphic conditions, high biotic pressure, high wind speed, etc. Agricultural productivity in the region remains limited due to an unconducive environment, limited choice of crops and aberrant weather conditions. About 38% of the total population of the state, with a density of 84 persons per sq km, area live in the arid region. This makes it one of the most densely populated deserts of the world (Figure 6.1).

6.4 Research methodology

The secondary data sources include historical data for 30 years from 1961 to 1990 of WMO standard in respect of rainfall, maximum temperature, minimum temperature and rainy days from the Indian Institute of Tropical Meteorology (IITM), Pune. For construction of a vulnerability index, information related to population density, rate of population growth, level of literacy, land holding size, and area under irrigation was collected from the census.

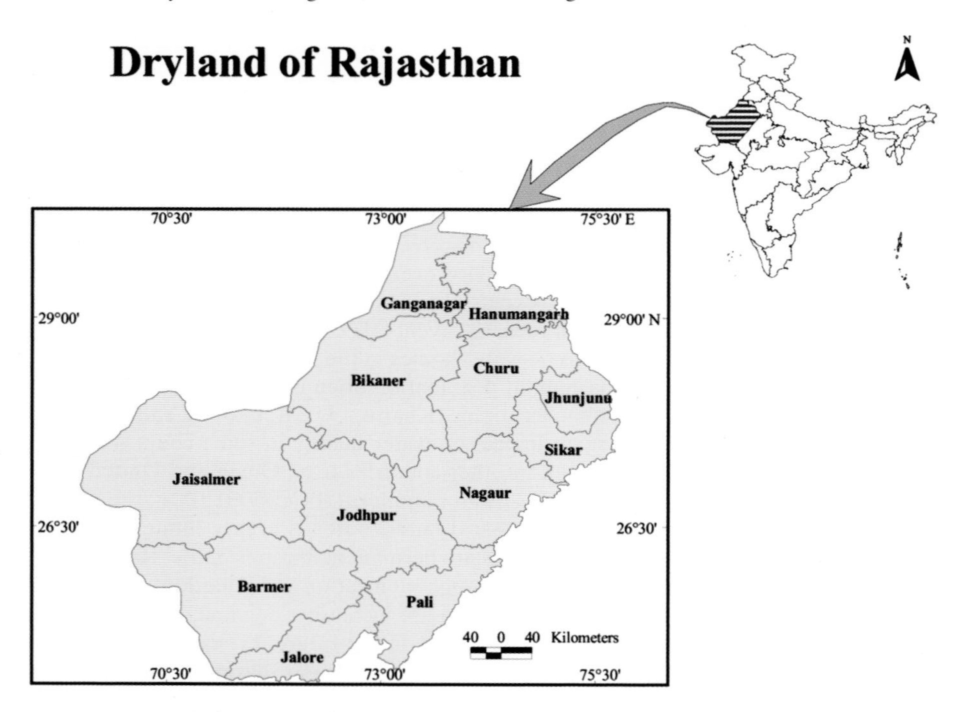

Figure 6.1 Location map of the study area

Information related to forest cover and water resources/water bodies was extracted from the Survey of India Toposheet. Data on rates of food production was taken from the District Agriculture Statistics Handbook. Information about groundwater position was collected from the Central Ground Water Commission. Data related to poverty, per capita income and income source was obtained from the website of the Government of Rajasthan.

The temporal and spatial analysis of data was done by using the Moving average (Five years) technique. This tool is used to find out the trend of temperature and rainfall variation over the years. The vulnerability index used in this study is based on the index developed by scientists (T.N. Balasubramanian, A. Arivudai Nambi and Diya Paul) working at M.S. Swaminathan Research Foundation, Chennai, on the 'Vulnerability Assessment and Enhancing the Adaptive Capacity' project under the Climate Change Programme. The index is developed as a statistical tool to assess the existing vulnerability of a region to climate change in terms of the current level of degraded resources. A few carefully chosen indicators representing the physical, biological, social and socio-economic scenarios of the region are used for this purpose.

The index uses 15 indicators that are sensitive to climate. Each indicator was assigned an optimum value reflecting the average value of the indicator. Values below or above the chosen optimum mark signify the level of sensitivity expressed in terms of severe vulnerability, moderate vulnerability and slight vulnerability scales.

These 15 indicators are further subclassified into three groups. The first group (Group 1) consists of six indicators that would suffer severe impact in a region on account of climate change:

- population density;
- forest cover;
- water resources/water bodies;
- rate of food production;
- area under irrigation;
- level of literacy.

If the value of an indicator rates as severe (S) according to the assigned scale, 100 points are awarded. Similarly, for moderate (M) and lesser (L) categories, 50 and 25 points are awarded, respectively.

The second group (Group 2) consists of three indicators:

- rate of population growth;
- groundwater status;
- percent of people below the poverty line.

These indicators are considered to have lesser or moderate impact as a result of climate change as compared to the indicators in Group 1. For these indicators in the second category, 75 points are assigned for severe vulnerability, 38 points for moderate vulnerability and 19 points for the lesser vulnerability categories.

The third group (Group 3) consists of six indicators:

- per capita income;
- income source;
- frequency of seasonal dry spells;
- pest and disease outbreaks;
- land holding size;
- intensity of soil degradation.

These are considered to suffer relatively less impact from climate change in comparison with the chosen indicators in Groups 1 and 2. The severe vulnerability category in Group 3 carries 50 points, while moderate and lesser categories carry 25 and 12.5 points, respectively.

Thus, the points awarded to each indicator in the three groups based on values obtained from the collected data serve as the 'Main Factor Effect'. Alternatively points assigned to severe, moderate and lower categories for the indicator given in the first, second and third groups could also be derived based on the expected damage to the system. An interaction table for 2025 (15 × 15 × 9) interaction points is produced based on the assumptions elaborated already relating to interactions between the main factors in Group 1 and Groups 1, 2 and 3; between Group 2 and Groups 1, 2 and 3; and between Group 3 and Groups 1, 2 and 3 as nine combinations of s × s; s × m; s × l; m × s; m × m; m × l; l × s; l × m; and l × l (s = severe, m = moderate, l = low). Based on these interactions between different factors the weightage score is derived for particular regions and finally the vulnerability scale of a region/area is calculated. The results from these tools and techniques are represented using various diagrams, mapped using computer-aided cartography, and interpreted.

6.5 Results and discussions

6.5.1 Climate variability

Climate is a sum of meteorological elements in a given region over an extended period of time, the standard period being 30 years. In a normal run, climate shows only small or negligible variations, which are induced by natural factors. If these variations are amplified or persist for a longer period of time, the condition is referred to as climate change. In the present scenario, anthropogenic factors are responsible for amplifying climate variations to greater extent. In arid regions the in situ anthropogenic activities are not concentrated enough to cause any climate variation, nor are natural factors responsible. The climatic variations in arid regions are externally induced. All meteorological phenomena are inherently linked with each other across regions. Therefore, any change in one or more factors in a nearby region can bring variations in the climate of another region. The temperature data of western Rajasthan recorded at 12 meteorological stations, one from each district, show that there was an increase of 0.45°C in average temperature over the 30 years from 1976 to 2006 (Figure 6.2). A very high variability is observed in the rainfall pattern (Figure 6.3).

6.5.2 Vulnerability assessment

At present the world needs effective mitigation measures to cope with climate change along with an adaptive strategy. However, there is also a prior need to identify the hierarchy of vulnerability so that priority can be given to the most vulnerable people and sectors. The population of western Rajasthan is mainly dependent on agriculture for its livelihood and therefore it is very important to assess the impact of climate change on the agriculture sector. Agriculture is dependent on many factors, which differ not only in importance but also in their individual vulnerability to climate change.

6.5.2.1 Population density Demographic factors such as population growth and density are quite determinant. A higher density implies that more people are dependent on available resources for their livelihood. Humans can alter the type and degree of their impact on their environment by, for instance, increasing the productivity of land through

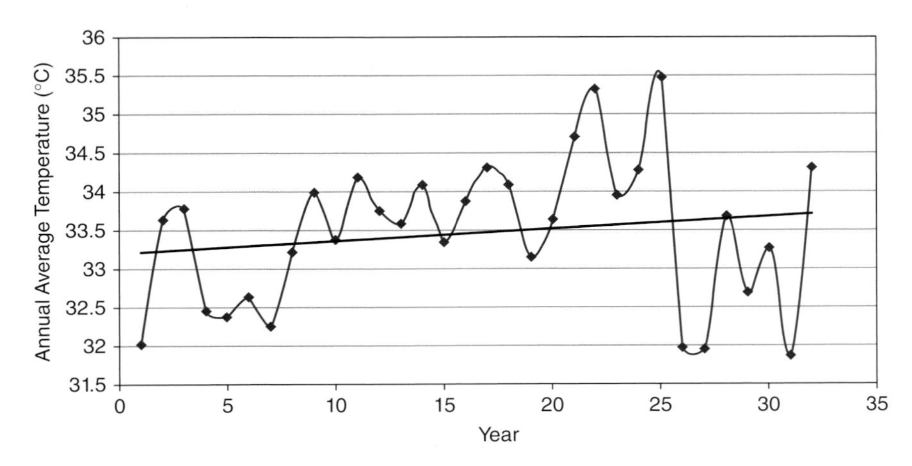

Figure 6.2 Rise in temperature over the 30 years from 1976 to 2006

more intensive farming techniques, or leaving a defined local area; and, of course, humans may also irreversibly decrease the productivity of their environment or increase consumption (Figure 6.4).

The study area has a meager capability to sustain human life because of its hostile natural environment and other socio-cultural factors; but still the region has the highest population density of any arid region at the same latitude. The western part of the region has a very low population density, but the population density of the eastern part is quite high, which makes it comparatively more vulnerable compared to the western part.

6.5.2.2 Literacy levels Literacy plays an important role in the decision-making process of farmers. On average, literacy is above 50% except for some parts like the extreme west and southeast. Better education is reflected in the type of farm practices adopted by

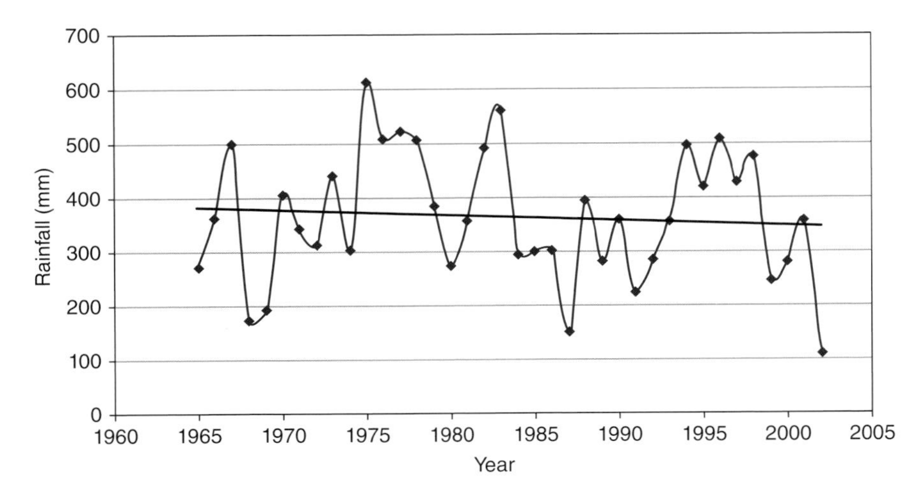

Figure 6.3 Variation in the rainfall pattern in western Rajasthan

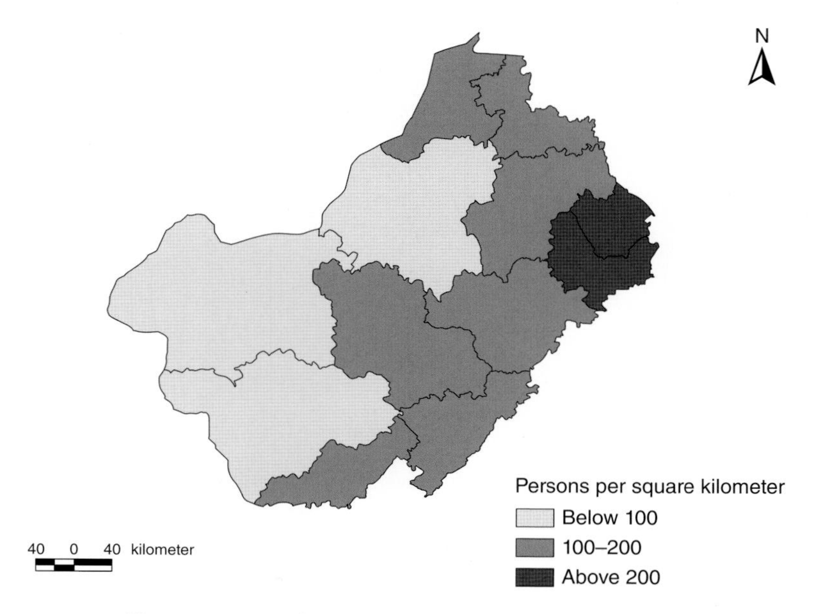

Figure 6.4 Population distribution in western Rajasthan

farmers. The importance of literacy can be seen, for example, in the application of nitrogen fertilizers to crops. Heavy applications of nitrogen fertilizer can contribute to global warming. A literate farmer is able to apply nitrogen fertilizer in optimum quantities, thereby reducing the emission of nitrous oxide, which is a harmful greenhouse gas (Figure 6.5).

6.5.2.3 Groundwater Most of the region lies in the critical zone for groundwater. The northeast and southeast sections are the only exceptions – these are classed as over-exploited zones owing to the fact that they contain highly irrigated agricultural land. These areas have the highest proportions of cultivated land and thus require large irrigation facilities. The other exception is the northern section, which lies entirely in a safe zone due to a very good network of canals. They not only prevent the groundwater from being over used but also help in groundwater recharge (Figure 6.6).

6.5.2.4 Food production Food is a primary requirement for all living organisms. It is very important to match the rate of growth of food production with the population growth rate. If the rate of increase in food production fails to match the rate of population growth then there is a probability of mass deaths due to starvation. A proper food supply and adequate buffer stocks enhance capability to fight climate change. It is evident that the growth rates of food production in the eastern and southern parts of the study region are higher than in the western part; the reason for such a pattern is the availability of water and land resources. The western part of the region is devoid of both, and any growth in food production in such an arid region is credited to the Indira Gandhi Canal Project (Figure 6.7).

6.5.2.5 Land degradation In India the state of Rajasthan has the largest area suffering land degradation, and the main agent is wind. As the area has very low drainage

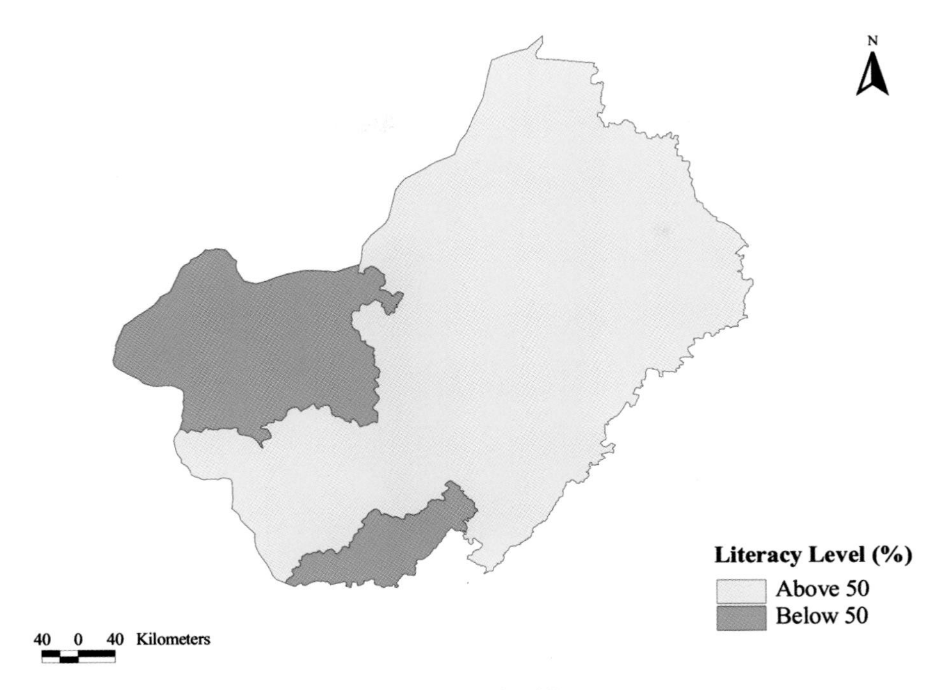

Figure 6.5 Levels of literacy

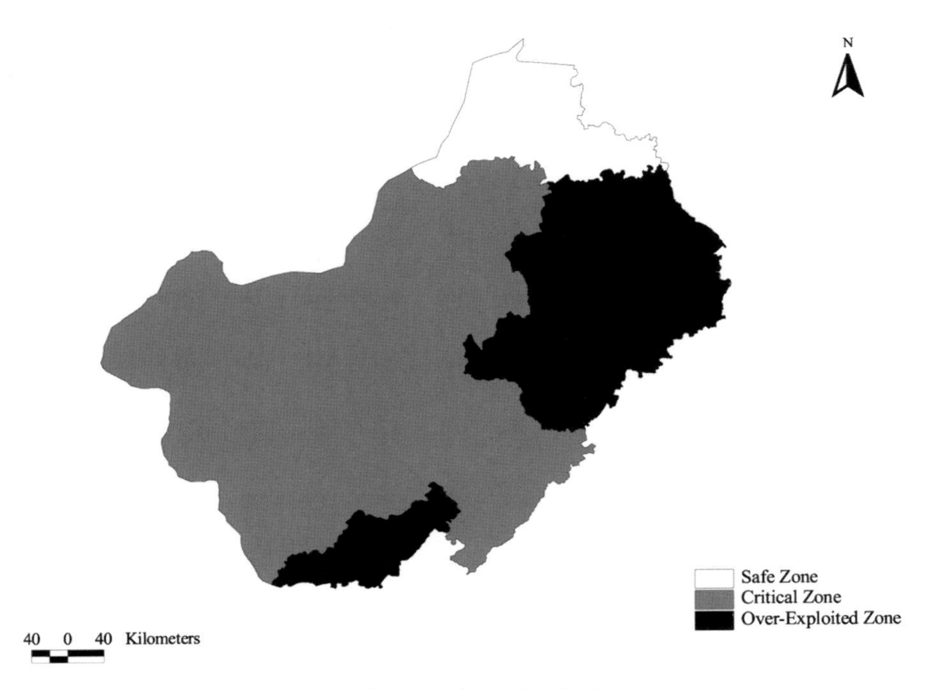

Figure 6.6 Groundwater status in the region

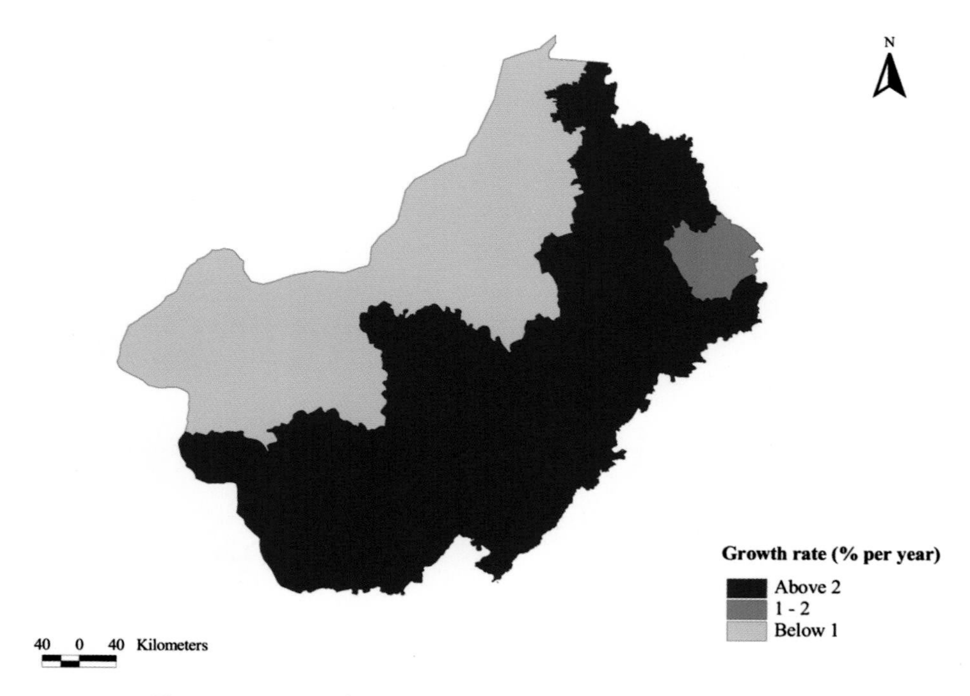

Figure 6.7 Rate of growth in food production in western Rajasthan

intensity and lacks any perennial streams, the area degraded by water is much less. An exception to this scenario is Jalore district where land degradation is caused by water running down from the hilly tracts during the rainy season (Figure 6.8).

6.5.2.6 Income Too much dependence on agriculture has harmful effects on the fertility of the land. Off-farm employment helps in reducing pressure on land resources. People in the western and middle parts of this region source gain income from both on-farm and off-farm activities, mainly because either less fertile land is available for pursuing agricultural activities or literacy levels are sufficiently high to make them competent to diversify their economic activities (Figure 6.9).

6.5.3 Vulnerability

The region has been divided into three zones depicting various levels of vulnerability.

6.5.3.1 High vulnerability The extreme western part of Rajasthan, comprising the state of Jaisalmer, is highly vulnerable to climate change. The extremely hot, dry weather of this region makes this district inherently vulnerable. Thus, it is certain that here natural factors have played a big role in determining its level of vulnerability. But, additionally, human factors have enhanced its vulnerability. The problem of land degradation is quite prevalent in this district, and this can be well correlated with low literacy levels of its population. Education affects natural resource management in several ways. Educated farmers are more likely to adopt new technologies to improve their land. Land degradation in

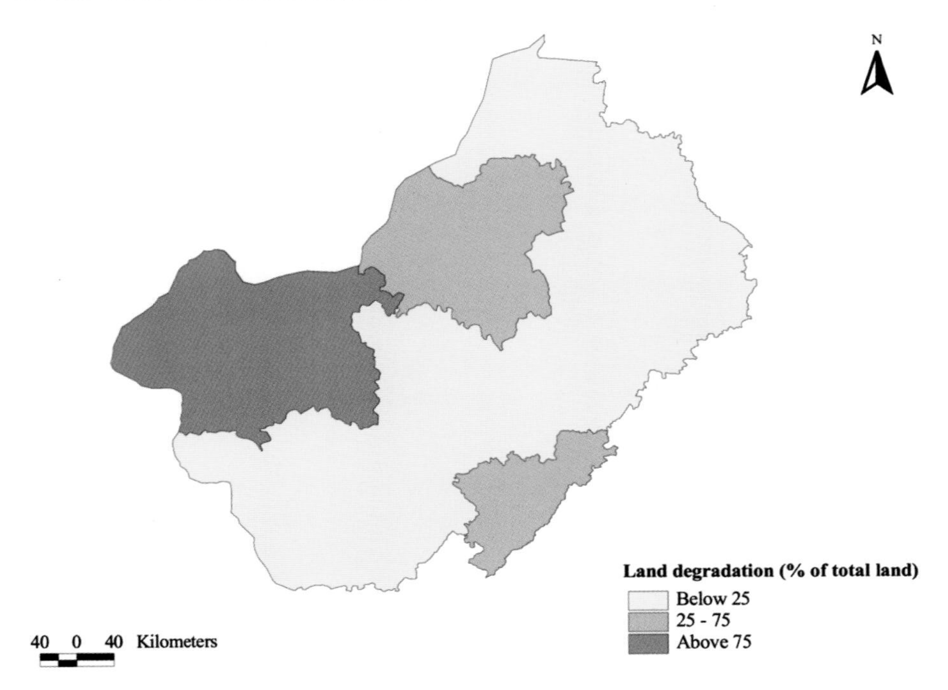

Figure 6.8 Land degradation status map

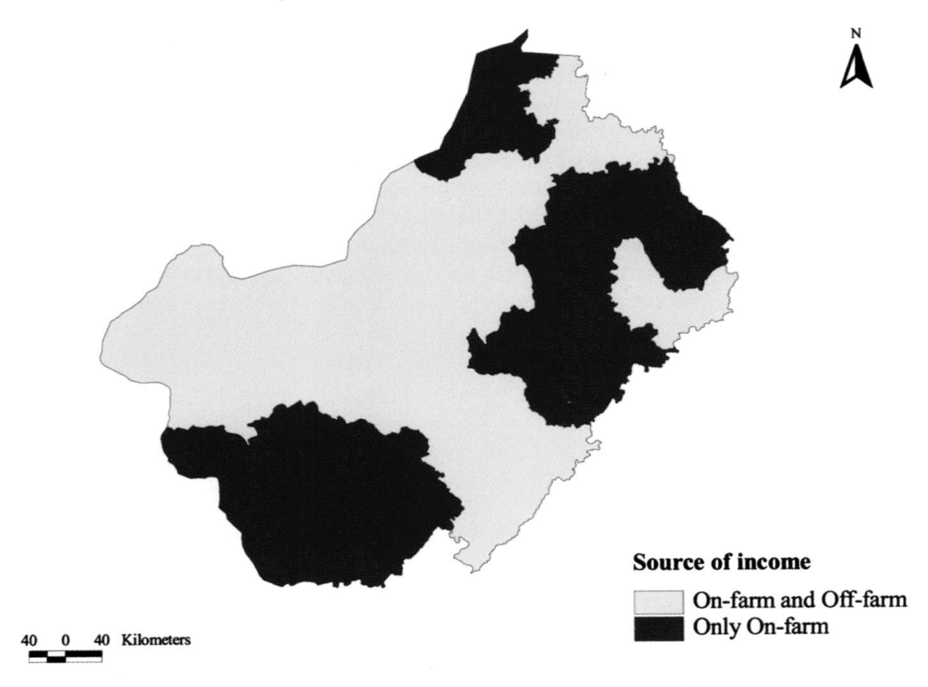

Figure 6.9 Income from on-farm and off-farm activities

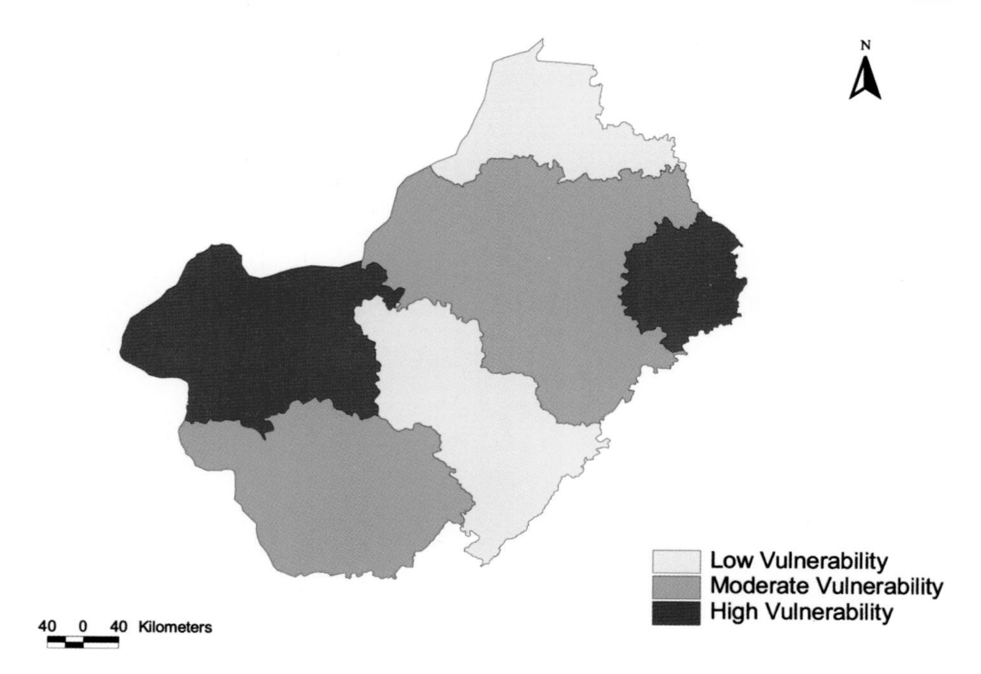

Figure 6.10 Climate change vulnerability map for western Rajasthan

drylands is a very complex process – not the outcome of a single factor but rather an emergent property of interacting human and biophysical factors. In the context of drylands, degradation is closely related to desertification. According to the United Nations Environment Programme (UNEP) desertification is defined as land degradation in arid, semi-arid and dry sub-humid areas that is caused by human and climatic factors.

In the northeast, the districts of Jhunjunu and Sikar also are highly vulnerable. Human-induced factors have mainly contributed to its high level of vulnerability. Here the population factor is relevant to the level of direct pressure on resources. By and large employment is on farms, which further raises the dependency of its dense population on its land resources, making it susceptible to land degradation (Figure 6.10).

6.5.3.2 Moderate vulnerability This applies mainly to the districts in the central zone of the region: Barmer, Bikaner, Churu, Nagaur and Jalore. All have moderate values for all factors, with some exceptions. The major determining factors of vulnerability, i.e. population density and groundwater, are at safe levels except in Jalore, which lies in an over-exploited zone. The area is less dependent on on-farm activities and the extent of land degradation is also low. Literacy is above 50% in this area, except for Jalore. The performance of Jalore district is good on all other measures, which qualified it for the moderate vulnerability zone.

6.5.3.3 Low vulnerability The middle section of the region including Jodhpur and Pali appears to be in a low-vulnerability zone. This can be attributed to higher literacy levels and less dependence on farm employment. People have diversified their economic activities. Moreover, the problem of land degradation is not serious enough to make it more vulnerable to climate change. This section has witnessed a comparatively higher growth rate in food production, indicating the adoption of better farm management and conservation techniques by its population.

It is also apparent that Ganganagar and Hanumangarh districts in the north show signs of low vulnerability. Being located in the safe zone for groundwater means its people might have resorted to better water resource management. High literacy levels reflect that people are aware of the latest techniques of conserving and managing resources and are more likely to adopt them in order to improve the fertility of the land and their overall development.

6.6 Conclusion

The available natural resource base and physical factors make the entire Indian arid region highly vulnerable to climate change, but its low population density reduces that vulnerability to some extent. The creation of the Indira Gandhi Canal and introduction of modern tools and farming techniques mean that the region is experiencing an increase in population density, which in turn may lead to an increase in vulnerability of region. The state's literacy level has very good scope for improvement. Literacy gives people knowledge about climate change and enables them to innovate and adopt new tools and techniques to cope with climate change. Less dependence on on-farm practices is a positive pointer for the region, because when the agriculture sector suffers from climate change other sectors may not be affected to the same extent.

The growth of agriculture has a positive record in the region, with most of the area having a growth rate in food production of more than 2% per year, which makes it less vulnerable in the context of food supply. Water has been a most important factor in determining the condition of the agriculture sector, and western Rajasthan suffers from a water deficit. The importance of water is shown by the fact that the over-exploited groundwater zones fall into the very highly vulnerability zone. Along with groundwater, surface water is also present in meager quantities in this region, although the western part has reduced its vulnerability with the building of the Indira Gandhi Canal, which provides water for irrigation and drinking purposes. Land degradation is mainly due to wind erosion, but is very dominant only in the western part.

When combined, all the factors make the region moderately vulnerable as a whole with some highly vulnerable and slightly vulnerable zones.

References

Blaikie, P., Cannon, T., Davis, I., Wisner, B. (1994) *At Risk: Natural Hazards, People's Vulnerability, and Disasters.* London: Routledge.

Bohle, H., Downing, T., Watts, M. (1994) Climate change and social vulnerability. *Global Environmental Change* **4**: 37–48.

Dash, S.K. (2007) *Climate Change: An Indian Perspective*. Ahemdabad: Environment Education.

Downing, T.E. (2003) Assessing vulnerability for climate adaptation. Technical paper 3: Oxford: Stockholm Environment Institute.

Drunen, M. van, Lasage, R., Dorland, C. (2006) *Climate Change in Developing Countries*. Wallingford: CABI.

Fischlin, A., Midgley, G.F. (2007) Ecosystems, their properties, goods and services. In: Parry, M.L., Canziani, O.F., Palutikof, J.P., van der Linden, P.J., Hanson, C.E., (eds), *Contribution of Working Group II to the Fourth Assessment Report of the Intergovernmental Panel on Climate Change*. Cambridge: Cambridge University Press, pp. 211–72.

Hingane, L.S., Rupa Kumar, K. (1985) Long-term trends of surface air temperature in India. *International Journal of Climatology* **5**: 521–8.

IPCC (Intergovernmental Panel on Climate Change) (2001a) The regional climate information – evaluation and projections. In: *The Scientific Basis, Third Assessment Report of the IPCC*. Cambridge: Cambridge University Press [online].

IPCC (Intergovernmental Panel on Climate Change) (2001b) In: McCarthy, J., Canziani, O., Leary, N., Dokken, D., White, K. (eds), *Climate Change 2001: Impacts, Adaptation, and Vulnerability*. Cambridge: Cambridge University Press, pp. 725–800.

Kavi Kumar, K.S., Parikh, J. (2001) Socio-economic impacts of climate change on Indian agriculture. *International Review for Environmental Strategies* **2**: 237–93.

Kharin, N.G. (1995). Change in biodiversity in ecosystems of central Asia under the impacts of desertification. In: *Biological Diversity in the Dry lands of the World*. Washington, DC: United Nations, pp. 23–32.

Kumar, B., Singh, R.B. (2003) *Urban Development and Anthropogenic Climate Change*. New Delhi: Manak Publications.

MEA (2005) *Ecosystems and Human Well-being: Synthesis. Millennium Ecosystem Assessment*. Washington, DC: Island Press.

O'Brien, K., Leichenko, R. (2000) Double exposure: Assessing the impacts of climate change within the context of globalization. *Global Environmental Change* **10**: 221–32.

Ramesh, R., Yadava, M.G. (2005) Climate and water resources of India. *Current Science* **89**: 818–24.

Ribot, J.C. (ed.) (1996) *Climate Variability, Climate Change and Social Vulnerability in the Semi Arid Tropics*. Cambridge: Cambridge University Press.

Rupa Kumar, K., Sahai, A.K., Krishna Kumar, K., *et al.* (2006) High-resolution climate change scenarios for India for the 21st century. *Current Science* **90**: 334–45.

Sainath, S. (2002) *Everybody Loves a Good Drought: Stories from India's Poorest Districts*. New Delhi: Penguin.

Sathaye, J., Najam, A. (2007) Climate change 2007: sustainable development and mitigation. In: Metz, B., Davidson, O.R., Bosch, P.R., Dave, R., Meyer L.A. (eds), *Contribution of Working Group III to the Fourth Assessment Report of the Intergovernmental Panel on Climate Change*. Cambridge: Cambridge University Press [online].

Sen Roy, S., Singh, R.B. (2002) *Climate Variability, Extreme Events and Agriculture Productivity in Mountain Regions*. New Delhi: Oxford and IBH Publications.

Sharma, D., Bharat, A. (2009) Conceptualizing risk assessment framework for impacts of climate change on water resources. *Current Science* **96**: 1044–52.

Shukla, P.R., Sharma, S.K., Ravindranath, N.H., Garg, A., Bhattacharya, S. (eds) (2003) *Climate Change and India: Vulnerability Assessment and Adaptation*. Hyderabad: University Press.

Yohe, G.W., Lasco, R.D. (2007) Perspectives on climate change and sustainability. In: Parry, M.L., Canziani, O.F., Palutikof, J.P., van der Linden P.J., Hanson, C.E. (eds), *Contribution of Working Group II to the Fourth Assessment Report of the Intergovernmental Panel on Climate Change*. Cambridge: Cambridge University Press.

7 Dendrogeomorphological and Sedimentological Analysis of Debris Flow Hazards in the Northern Zailiiskiy Alatau, Tien Shan Mountains, Kazakhstan

Vanessa Winchester[1], David G. Passmore[2], Stephan Harrison[3], Alaric Rae[4], Igor Severskiy[5] and Nina V. Pimankina[5]

[1]*School of Geography and the Environment, University of Oxford, Oxford, UK*
[2]*School of Geography, Politics and Sociology, Newcastle University, Newcastle upon Tyne, UK*
[3]*Department of Geography, University of Exeter, Cornwall Campus, Penryn, Cornwall, UK*
[4]*Department of Geography, Coventry University, Coventry, UK*
[5]*Institute of Geography, Academy of Sciences, Almaty, Republic of Kazakhstan*

7.1 Introduction

Vulnerability, in terms of landscape processes in the wide, steep-sloped valleys of the northern Zailiyskiy Alatau, Tien Shan Mountains, is related to catastrophic change caused by earthquakes or debris flows. In the last 300 years in the region there have been over 130

Vulnerability of Land Systems in Asia, First Edition. Edited by Ademola K. Braimoh and He Qing Huang.

earthquakes above 6 on the Richter scale, with four of these measuring 8–8.5 on the scale (Severskiy & Blagoveshchenskiy, 1983; Severskiy *et al.*, 2000), the last two of these, in 1887 and 1911 destroyed, on each occasion, a large part of the city of Almatay. Further, in 1921 there was a debris flow that also caused widespread damage and loss of life. The population of the city today stands at around two million and there is an expanding leisure industry. At the foot of the Malaya Almatinka valley (M. Almatinka) an Olympic-standard ice-skating rink has been built and a fashionable ski resort is sited on the slopes above. In the Bolshaya Almatinka valley (B. Almatinka), there are many villas, small farms and holiday homes on the valley floor. Two large earth dams have been built at the foot of each valley to protect the city and its amenities from debris flows and flooding.

A debris flows is a mass movement of unconsolidated, water-saturated sediment that flows down-slope propelled by gravity. Flows range from gently flowing sand and water slurries to violently surging bouldery masses; they include events described as debris slides, mudflows, mudslides, earth flows and lahars. Many of these events may be regarded as debris flows (Iverson, 1997), and it is this term that is used below without further distinction. Apart from earthquakes, debris flows can be triggered by freeze/thaw processes, especially where there has been an increase in altitude of the permafrost level (Marchenkoa *et al.*, 2007), also by intense snowmelt, seasonally intense rainstorms, lake outbursts or a combination of these.

Many moraine-dammed lakes are particularly susceptible to outbursts. Moraines are composed of loose, poorly sorted sediment (possibly containing ice cores or interstitial ice), often with surrounding steep slopes prone to snow, ice avalanches and rock falls whose activity can generate waves overtopping the dam. The dam can also be overtopped or burst by sudden drainage of an upstream river or ice-dammed lake (jökulhlaup). Melting of moraine ice cores and piping are other possible failure mechanisms (Clague & Evans, 2000). In the Zailiyskiy Alatau, since the Little Ice Age (LIA) maximum 150 years ago, when many of the moraines were formed, there has been a widespread trend towards glacier retreat and an increase in potentially hazardous geomorphic activity. There is thus a pressing need to better understand the long-term pattern and frequency of debris flows in susceptible valley settings to enable future trend forecasting (VanDine & Bovis, 2002).

Concern over the frequency and magnitude of hazardous geomorphological processes and the implications of declining ice cover for long-term water resources in the region, has resulted in establishment by the International Centre of Geoecology of the Mountain Countries in Arid Regions (ICGM, Institute of Geography, Almaty) and the Institute of Permafrost, Russian Academy of Sciences (e.g. Severskiy & Blagoveschenskiy, 1983; Severskiy & Severskiy, 1990) of two mountain research stations at the head of the Almatinka valleys. These stations have maintained unusually long and detailed records of recent historic environmental changes, with these including glacier fluctuations from the late 19th century to the present, a wide range of climate parameters, and the frequency and character of avalanche and rock-glacier activity over the past 40 years (Passmore *et al.*, 2008).

Despite these records and much research in the B. Almatinka study area, attempts at developing resource management frameworks have been hindered by a lack of attention to the rates and patterns of slope and valley floor geomorphic processes in currently non-glaciated lower-valley areas – areas exhibiting numerous avalanche debris cones, debris flow landforms, alluvial fans and cones, and high-magnitude flood deposits attesting to frequent, and occasionally large-scale, geomorphological events, with many of these predating the 20th century (Harrison *et al.*, 2004). There has also been little attempt at analyzing documented geomorphological and climatological data with a view to assessing the response of valley side and valley floor instability in relation to 20th century climate changes. It is anticipated that climate fluctuations spanning the last 150 years of climate warming will promote widespread and large-scale environmental change, including geomorphic activity. The purpose of the present study is to address these shortcomings, with the overall aim of assisting environmental policy makers:

- by establishing the spatio-temporal patterns of sediment assemblages resulting from debris flows in the B. Almatinka River drainage basin;
- by expanding 20th century records of climate, glacier and geomorphological events through analysis of patterns of valley-side and valley-floor instability over the past 400 years using dendrogeomorphological methods and archival and meteorological records provided by the Institute of Geography, Almaty.

7.2 Study area

The Tien Shan mountains of Central Asia stretch over 2000 km west to east and many of the mountains over 3000 m a.s.l. are heavily glaciated. The mountains form a barrier playing a critical role in the atmospheric circulation of the region. Snow melt and water relations are governed by a continental climate strongly influenced in winter by Siberian anticyclonic circulation and in summer by frontal cyclonic development. In the Northern Tien Shan in summer, the northern jet stream merges with the subtropical jet stream producing an influx of cold moist air masses from the northwest and heavy summer precipitation (Aizen *et al.*, 1995). Warming temperatures and increasing precipitation over the last 150 years have led to an increase in melting that, coupled with intense summer storms during the season of maximum snow and glacier melt, has frequently led to catastrophic moraine-dam breaks and debris flow events.

Almaty, standing on the alluvial fans of the M. Almatinka and B. Almatinka rivers, is principally supplied with water from small glaciers at the head of the B. Almatinka valley. This research focused on two contrasting study areas in the valley's drainage basin (Figure 7.1).

- Kumbelsu Creek is a 1 km reach of a major eastern drainage tributary joining the B. Almatinka River, 2 km downstream from a weather station. The study reach lies 1 km upstream of the confluence and occupies a narrow (50–80 m), deeply entrenched valley floor that features several terraced depositional sequences (Figure 7.2). A walk was made along the margins of the main B. Almatinka River downstream of the Kumbelsu creek confluence, centered on latitude 43°06′07″N and longitude 76°55′35″E.
- The second study area is at the base of the Ozernaya valley (43°02′N 76°59′E) at the head of the B. Almatinka valley above the lake. The area extends over 1.5 km between the lakeshore at 2510 m a.s.l. and the Ozernaya River's confluence with three steep, east valley-side tributary gullies at approximately 2650 m a.s.l. (around 250 m below the tree line). Here the Ozernaya valley floor is between 200 and 500 m wide. The present Ozernaya River channel occupies a deeply incised (up to 6 m deep) trench that is confined to the west side of the valley floor by debris flows and coarse flood deposits largely derived from the eastern gully systems.

7.3 Methods and materials

7.3.1 Geomorphology and sedimentology

A geomorphological survey of valley side and valley floor landform assemblages in each of the study areas was undertaken to (i) identify the distribution and relative age of

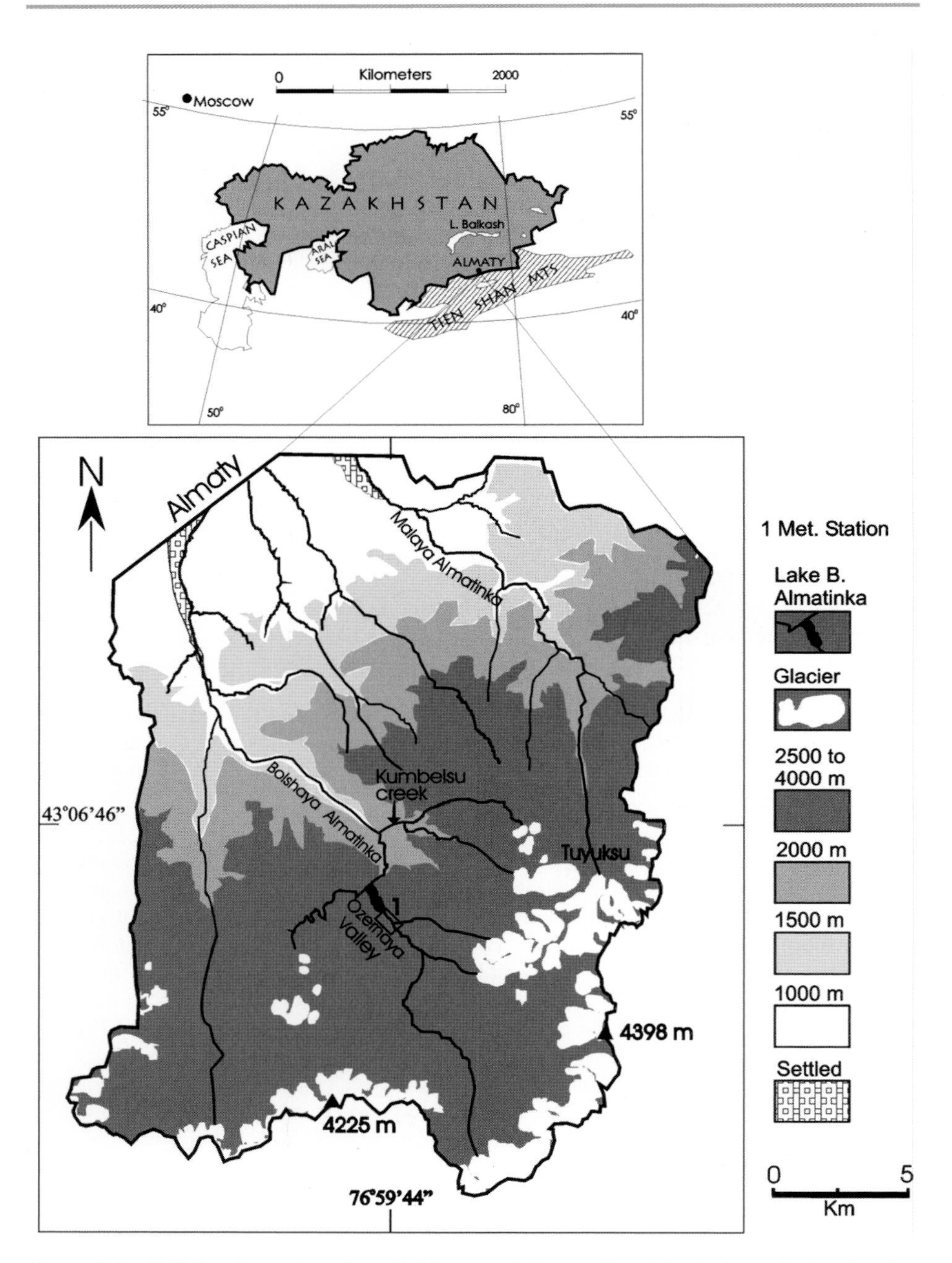

Figure 7.1 Relief and location of the Bolshaya and Malaya Almatinka drainage basins showing Ozernaya River and Kumbelsu Creek study areas

Figure 7.2 View of Kumbelsu Creek. (a) Looking downstream showing sediments exposed following 1977 debris flow event. (b) Looking upstream; terraces T1, T3 and T4 can be seen on the right

mass-wasting and fluvial landform assemblages, and (ii) to identify localities prone to hazardous geomorphic activity over timescales considerably longer than existing documentary records. A log was kept of sedimentary sequences deposited on poorly vegetated boulder and cobble berms (a narrow ledge or shelf typically at the top or bottom of a slope) and of valley-side exposures and terraces created by mass-wasting events, to assist in interpretation of depositional environments associated with large-scale floods and debris flows.

7.3.2 Archive datasets

Records of 19th and 20th century large-scale debris flow events in the northern Zailiyskiy Alatau have been compiled by the Institute of Geography, Almaty (Gorbuhov & Severskiy, 2001). These records include written accounts of events impacting Almaty and the settlements of Kaskelen (24 km west of Almaty), Talgar and Issyk (respectively 27 and 44 km east of Almaty). Over 450 individual events have been documented in the region between 1841 and the present; detailed and regular field-based records associated with the establishment of high-elevation meteorological stations are confined to the middle and latter part of the 20th century. Archive sources used in this study include monthly precipitation records (from 1938 to 1998) from two meteorological stations: one in the Ozernaya valley above B. Almatinka Lake at 2516 m a.s.l. (records from 1977 to 1988 only) and one in the M. Almatinka valley at Mynzhilki at 3017 m a.s.l. (Passmore *et al.*, 2008).

7.3.3 Dendrogeomorphology

Annually formed tree rings can date landscape events to the year of occurrence and be used to analyze rates of environmental change (Alestalo, 1971; Shroder, 1980; Fantucci & Sorriso-Valvo, 1999; Stefanini & Schweingruber, 2000; Bollschweiler *et al.*, 2007). Dendrogeomorphological techniques differ from those used in dendrochronology in that they depend on careful field observations as well as the conventional cross-dating used in standard dendrochronological work. The study's first dating requirement was to construct a tree-ring reference series to show non-geomorphological effects on tree growth (mainly climatic) in the study area; and the second was to compare the reference series with tree growth affected by other events.

Cores and cross-sections were taken from a total of 48 Schrenk's spruce (*Picea schrenkiana* Fish. & C.A. Mey; subsp. *tienshanica* Rupr.). This spruce, growing up to 50 m tall, has well-defined annual ring structures. The sampled trees were growing on eight depositional assemblages (S1 to S8) defined by differences in sediment types and vegetation cover, on the S9 delta fan (Figure 7.3). The reference series was chosen from trees that showed a high degree of cross-correlation (see below). Tree cores and sections, particularly of damaged stems and branches were also collected from Kumbelsu Creek and the lower B. Almatinka valley, and the height of flood damage on trees above the creek and river beds was noted.

Details of each tree and its surroundings were recorded: stem circumference, stem-tilt angle, crown deformation and damage, and burial or exposure of roots. Also noted were local landscape features: slope angle and aspect and the competitive status of each tree (the number of sides a crown is touched by other crowns and distances to the three nearest trees) and estimated tree height (whether it was dominant, the same height, or below the canopy). Where possible, two cores were taken from each stem on upslope and downslope sides. Differing core heights were noted to allow calculation of maximum and minimum estimates for the delay period before seedling establishment (see below). Where a core did not hit a tree's centre (pith), a clear acetate sheet scribed with 5 mm concentric circles was fitted over the earliest rings and their widths used to estimate the likely number of missing rings.

7.3.4 Cross-dating, reference series identification and skeleton plotting

Standard dendrochronological methods were used to prepare the cores; ring widths were measured from bark to pith using the LINTAB measuring system (Rinn, 2003). Matching

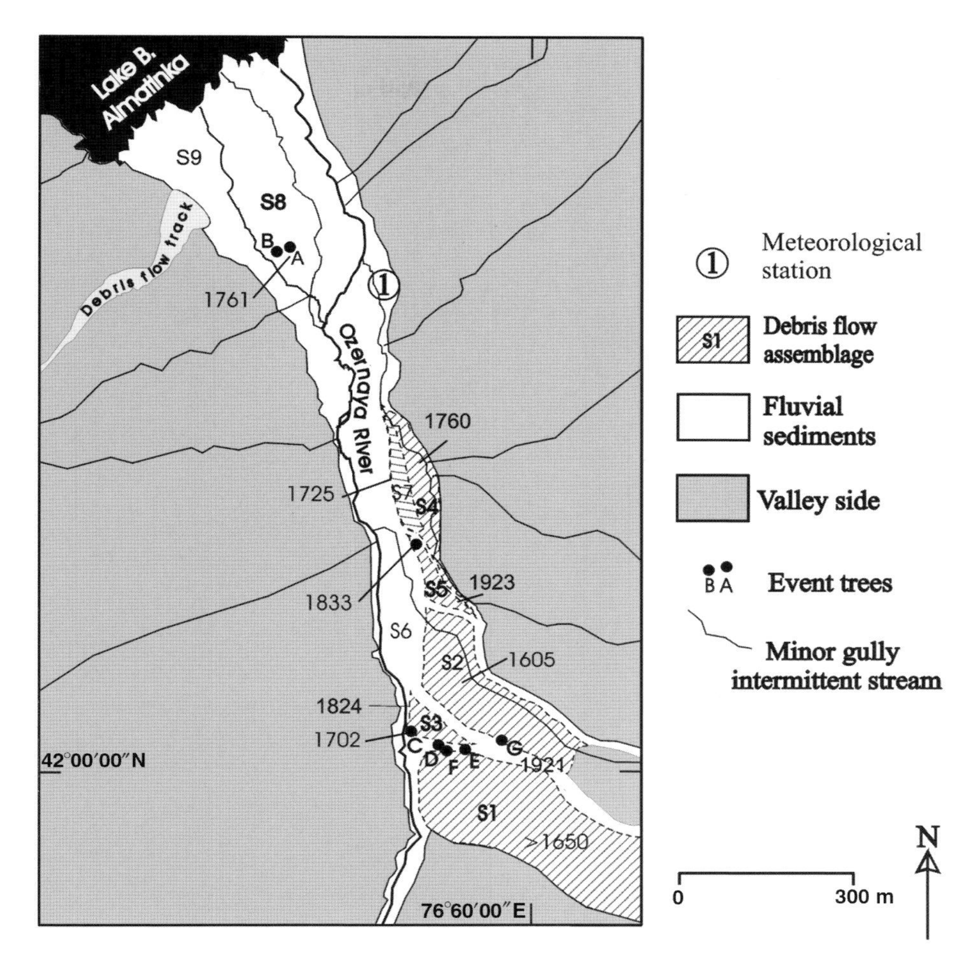

Figure 7.3 Geomorphological map of the Ozernaya valley showing major landform-sediment assemblages and selected trees used for debris flow event analyses. Also shown are maximum dates for vegetation clearance

ring-width patterns were cross-dated employing a range of statistical tests from the TSAP-Win program (Rinn, 2003). Each matching core pair was averaged and similarly cross-dated with all the other averaged pairs to form a reference series. Trees that did not cross-date either showed visible signs of damage or grew in positions, on the floodplain, beside or in gullies created by geomorphic events.

- All trees younger than 60 years were removed from the reference series (Table 7.1) to eliminate bias in the growth curve produced by younger tree growth (Esper & Gärtner, 2001).
- After elimination, the matching series contained 10 trees (110 to 340 years old). And, although the two oldest trees were included in the series for possible future reference, analysis using skeleton plotting (see below) was truncated at 1770 due to the requirement that sample size should contain at least three trees (Esper & Gärtner, 2001).

Table 7.1 Summary of chronology of landform-sediment assemblages identified in the Ozernaya Valley supplying minimum and maximum dates (ecesis minus 7 years and 33 years respectively) for events and a basis for the mean curve. Tree age is calculated from a combination of annual ring counts plus estimated years below core height plus estimated missing rings (where core did not reach pith center). Main events noted in text are highlighted

Unit – ID	Ring numbers	Est. years below core height	Missing rings (est.)	Tree age/date	Minus 7 years ecesis: event date	Minus 33 years ecesis: event date
S2						
B5	339	17	6	1638	**1631**	**1605**
C3	255	9	4	1732	1725	1699
B4	210	25	24	1741	1734	1708
C2	231	18	50?	1751	1744	1718
S7						
E4	223	19	–	1758	**1751**	**1725**
E5	181	20	25?	1774	1767	1741
S4						
E1	189	18	–	1793	**1786**	**1760**
E2	167	19	14	1800	1793	1767
E3	192	6	–	1802	1795	1769
S5						
C4	118	11	5	1866	**1859**	**1833**

- The matching reference series, derived from the averaged raw-ring-width data, was fitted with an 11-year moving average. The TSAP-Win moving average routine uses a sliding iteration with a non-parametric weighted kernel and the statistical properties of a spline (Gasser & Müller, 1984; Rinn, 1988, 2003). The 11-year value was chosen empirically as best emphasizing decadal variability while removing inter-annual variability and signal noise.
- Standardization about a mean, a normal dendrochronological procedure, eliminates variations in tree growth fluctuations due to ageing and averages the combined tree-ring series into a mean chronology. This was used with caution in the analysis since standardization of series removes differences in growth between individual trees due to geomorphic events and also removes growth effects due to climate change: a particularly noticeable feature here where growth, instead of diminishing as would be expected with age, showed a marked increase in recent years.
- Seven trees (A to G, Figure 7.3) not cross-dating with the reference series were visually checked and cross-dated with each other to identify ring-width patterns associated with mass movement episodes of tree damage, tilting, sediment accumulation or degradation.
- The impact of geomorphological events on tree growth was assessed using skeleton plotting, a technique focusing on identification of pointer years signaling abrupt growth changes (Schweingruber *et al.*, 1990). Here, pointer years suggesting geomorphological events are defined by the occurrence of extra narrow ring widths in two or more individual trees in a series. Schweingruber *et al.* (1990) stipulated that a pointer year should be recorded by 80% of the sample number at any one location on a plot for it to be acceptable. However, following the requirement specified by Esper and Gärtner (2001) of a minimum sample of three trees and the small sample size available, the present study has adopted a 50% criterion.

7.3.5 Seedling establishment, growth rates below coring height and earthquakes

Cores only provide ring counts above the core: minimum dates for geomorphic events were derived from a combination of (i) core ring counts, (ii) estimates of the potential delay period before seedling establishment (ecesis) on freshly exposed surfaces and (iii) numbers of years' growth below core heights (Winchester & Harrison, 2000; Gutsell & Johnson, 2002; Winchester *et al.*, 2007).

The length of delay before ecesis was provided by the most recently deposited sediments in the Ozernaya Valley and a time horizon created by a lake outburst from the head of Kumbelsu Creek in 1977. The outburst stripped vegetation from terraces up to 5 m above the creek bed allowing the calculation of an average delay period before ecesis: this was obtained from ring counts from cross-sections cut from the bases of 11 young spruces (<140 cm tall) on the 5 m terrace in Kumbelsu Creek. However, in the colder conditions of the mid-19th century when the glaciers were considerably advanced (Solomina *et al.*, 2004), seedling establishment could have taken rather longer at the 500 m higher altitude of the Ozernaya study area. Thus, Kumbelsu Creek provided a minimum ecesis value; a maximum value was tentatively inferred for the Ozernaya valley based on the growth pattern and age, respectively, of trees C and D (see sections 7.4 and 7.5). The same 11 trees from Kumbelsu Creek and small trees in the Ozernaya valley also provided a mean height/age relationship for growth below coring height to add to core ring counts (Table 7.2).

Table 7.2 Tree height/age correlations from Kumbelsu Creek and Ozernaya valley. The mean value is used as the basis for calculating seedling growth below core height. A delay of 7 years before seedling establishment is deduced from the difference between the age of the oldest tree (height 154 cm) on the middle terrace (T3) and the 1977 date of the last recorded debris flow that cleared the terrace of vegetation

Tree ht (cm)	Age (years)	Mean growth (cm/year)	Site
			Kumbelsu Creek
24	7	3.4	Terrace 4
28	7	4.0	Terrace 4
33	6	5.5	Terrace 4
41	7	5.8	Terrace 4 with rockslide partly covering stem
107	12	8.9	Terrace 4
53	9	5.9	Terrace 3 outer edge
59	8	7.4	Terrace 3 outer edge
126	14	9.0	Middle terrace 3
154	16	9.6	Middle terrace 3
			Ozernaya Valley
130	22	5.9	Outwash plain: wind-seared tree, defoliated down valley
135	18	7.5	Outwash plain: healthy tree
140	16	8.7	Gully west side of Lake B. Almatinka. Tree damaged by rock fall
Mean growth: 6.8 cm/year			

The data were also examined to see if local earthquake events could be separated from those caused by climatic or other factors (Jacoby, 1997). In addition, consideration was given to the possibility of spruce budworm (*Choristoneura fumiferana*) infestations that could have been responsible for, or contributed to, cyclical growth oscillations (Weber & Schweingruber, 1995; Schweingruber 1996). However, the lack of synchronicity between the reference series and tree pointer years suggests that infestations were not involved.

7.4 Results

7.4.1 Growth rate and establishment periods

An average growth rate of 6.8 cm/year (Table 7.2) was derived from the height/age relationship of the trees beside Kumbelsu Creek and Ozernaya valley together with an establishment period from a 7-year minimum (obtained as described above) to a provisional maximum of 33 years in the Ozernaya valley. The 33-year maximum was derived from the difference between the 1860 age of tree D growing in the middle of the gully separating the S1 and S3 assemblages and tree C growing ~60 m lower at the mouth of the gully, with severely depressed ring growth in 1827 suggesting the occurrence of a major debris flow in the gully (Figure 7.3 and Table 7.3). Accordingly, age ranges for the major stripping of

Table 7.3 Summary of oldest non-cross-dating tree ages in the Ozernaya Valley. Sections and cores taken from stream terraces in Kumbelsu Creek illustrate declining magnitude of events. Ecesis estimates supply maximum and minimum dates for geomorphological events. Dates mentioned in text are highlighted

Unit – ID	Ring numbers	Est. years below core ht	Missing rings (est.)	Tree age/date	Minus 7 years ecesis: event date	Minus 33 years ecesis: event date
Ozernaya Valley						
S3						
C	243	17	5	1735	**1728**	**1702**
D	122	10	8	1860	1853	**1827**
E	103	8	7	1882	1875	1849
F	122	9	3	1866	1859	1833
G*	117	11	5	1867	1860	1834
S8						
A	188	18		1794	**1787**	**1761**
B	180	16	4	1800	1793	1767
Kumbelsu Creek						
Top terrace T1	47	13	7	1928	**1921**	–
Terrace T2	33	15	Damage 1977	1952	1945	–
Terrace T3	16	0 – cut at root collar	Damage 1977	1984	1977	–
Lowest terrace T4	7	0 – cut at root collar	–	1993	**1986**	–

Tree G* stands on north side of main gully flanking S2 sediment assemblage (not marked on Figure 7.3).

vegetation from land surfaces in the Ozernaya study area are based on tree age plus an establishment period from a 7-year minimum to a maximum of 33 years.

7.4.2 Kumbelsu Creek

A total of four terraces were identified lying 9.2 m (T1), 7–8 m (T2), 4–5 m (T3) and 2 m (T4), respectively, above the current channel bed (Figure 7.2). The maximum tree age for terrace T1 is 1921, while terrace T2 dating to 1977 is cut through earlier deposits, with these revealed on the opposite valley side below the T1 terrace level in an exposure showing stacked sequences of massive, crudely stratified boulders and finer deposits consistent with earlier debris flows and transitional flow events (Blair & McPherson, 1994). The T3 terrace dates to 1977 and T4 to 1986. A preliminary survey of the lower reaches of the B. Almatinka riverbanks below the Kumbelsu Creek confluence revealed terraces with similarly diminishing age and height relationships above the riverbed (Table 7.3).

7.4.3 Sedimentology, geomorphology and dating of fluvial terraces, fans and debris flows in the Ozernaya valley

Nine large-scale landform-sediment assemblages were identified in the Ozernaya study area. Assemblages S1–S5 are spread across the eastern lower valley sides and floor of the valley where they form crudely lobate deposits flanking steep tributary channels draining the eastern valley-side (Figures 7.3 and 7.4). These landforms are broadly defined as composite suites of debris flow and transitional-flow deposits that have been locally reworked by stream flows, with soil development extending to approximately 20–40 cm deep (Figure 7.4a,b). Boulder sizes on the older deposits vary (maximum 3 m^3), while areas affected by more recent flows, S6–S9, generally have smaller boulder sizes and little-to-no vegetation cover. Although the assemblages themselves may have been formed at an early period, their surrounding major gullies have remained active as evidenced by the lack of plant development on the floors and lower sides of the two gullies defining S2. Continuing low-level river flooding (possibly on a seasonal basis) is also highlighted by discontinuous grass cover, small shrubs and occasional small trees on the S6 terrace extending from the most southerly debris flow assemblage (S1) down valley for some 600 m to meet the youngest surface, S9, covered only by occasional grasses and annual plants forming the contemporary active floodplain of the Ozernaya River (Figures 7.3 and 7.4).

The position of the river lying against the western valley-side cut down through the S6 deposits in a 6 m-deep channel, and the character of the exposed deposits suggest that the S6 terrace has largely been formed from multiple debris flow events issuing from the flanking S2 gullies descending the M. Almatinka divide to the east (Figure 7.3). The passage of events across the terrace is marked by the highly channeled topography with flanking boulder and cobble banks, pebble berms, bars and splays (Figure 7.4a,b). Multiple events are also marked by large difference in tree ages on S5, with at least one tree surviving from the first half of the 19th century at the lower end of the assemblage, whereas the trees above it date to the 1930s and 40s (Table 7.1). Finally, the much lower angled assemblages S7–S9 are interpreted as fluvial fan-delta deposits formed from sediments derived from reworking of up-valley deposits in the higher reaches of the Ozernaya (Figures 7.3 and 7.5).

On the basis of dendrochronological dating (cf. Alestalo, 1971) the oldest debris flow unit (S1) could have been deposited sometime before the mid-17th century. Estimated dates for

(a)

(b)

Figure 7.4 Views, looking north towards the lake, showing (a) a partially vegetated boulder berm and (b) superficial boulder berm deposits, with S5 in the foreground and S4 deposits in the background

development of later debris flow assemblages have minimum and maximum ages, respectively, of 1605–31 (S2), 1702–28 (S3), 1760–86 (S4) and 1833–59 (S5), 1725–51 (S7) and 1761–87 (S8) (Figure 7.3 and Tables 7.1 and 7.3). The initial development of debris flows in the valley is thus dated to an early period of the LIA (1605–31) by the S1 and S2 assemblages at the head of the study area where they stand on either side of the main channel descending from the eastern divide. In the mid-Ozernaya study area, a residual, fragmentary deposit, S7, remaining from the first half of the 18th century, is flanked by the larger S4

Figure 7.5 View, southwards from the lakeshore, of the Ozernaya valley study area. Trees A (to the left) and B stand on the S8 delta-fan surface. Very small, young trees can be seen behind on the S6 surface, the S9 surface is in the foreground

assemblage deposited between 1760 and 1786. That this was a time of considerable activity is illustrated by the closely similar dating of the two lone trees on the S8 upper floodplain (Figure 7.5 and Table 7.3).

7.4.4 Debris flow events in the Ozernaya Valley and archival records

The pointer-year dates shown in the skeleton plot of the 10 matching valley trees in the Ozernaya valley (Figure 7.6a), coincide on only two occasions, 1920 and 1928, with those of the event-plot trees (Figure 7.6b). These seven event trees, aged between 103 and 243 years (Table 7.1; Figure 7.3, trees A to G), showed visible signs of environmental alteration (e.g. tilting, exposed or buried roots, or other growth peculiarities), with these interpreted as reflecting geomorphological events (Table 7.3). Twelve events are characterized by anomalously low ring widths: 1827, 1878, 1885, 1887, 1904, 1911, at the end of 1920 or the beginning of 1921, 1928; 1935, 1950 and 1963. The presence in the tree-ring record of the 1921 event (which brought devastation to Almaty) is highlighted in Figure 7.7 showing averaged raw-ring widths of trees A, B and C. The plots for these three trees also show an earlier (unrecorded) event in 1878, and an event that impacted tree C in 1827.

Trees A and B, growing on the S8 delta-fan surface (Figure 7.5), show abrasion of bark near ground level and accumulated gravels around their stems. Excavation at the base of tree A revealed adventitious roots (roots growing out of the stem following sediment accumulation) 32 cm below ground level above a layer of close-packed cobbles. In addition, the poor crown condition and the suppressed growth of both trees from 1878 almost to the present indicate that sediment accumulation has been ongoing and is now probably close to the 1.6–1.9 m maximum for *Picea* survival proposed by Strunk (1997). Tree C is a 236-year-old specimen growing over a boulder flanking a tributary palaeochannel on the

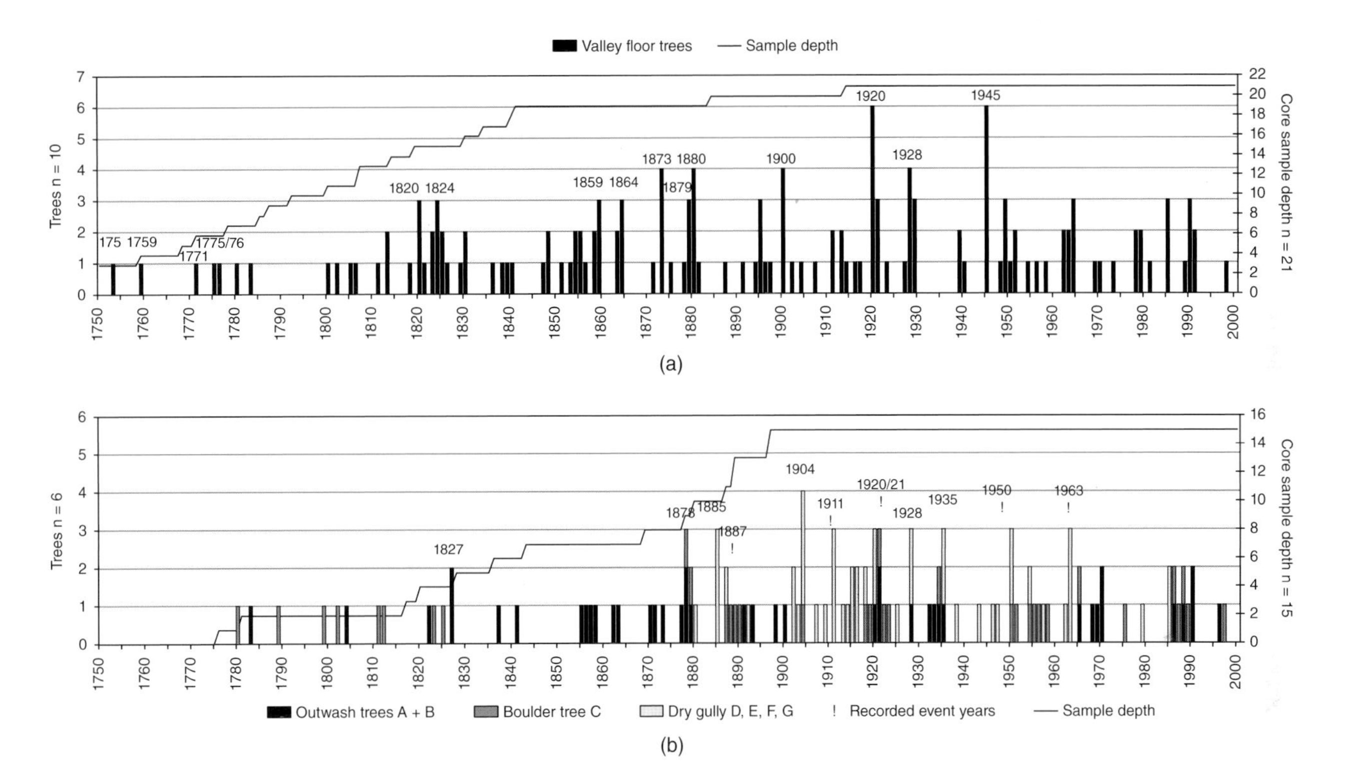

Figure 7.6 Skeleton plots of the Ozernaya valley showing pointer years 1750–2000 and sample depth. (a) Plots of cross-dating cores showing years of climate influence, with dated pointer years indicating years of low growth. (b) Event plots: pointer years suggest years of environmental disturbance

Figure 7.7 The boulder tree C showing root damage in 1878 and extensive soil loss on the channel side

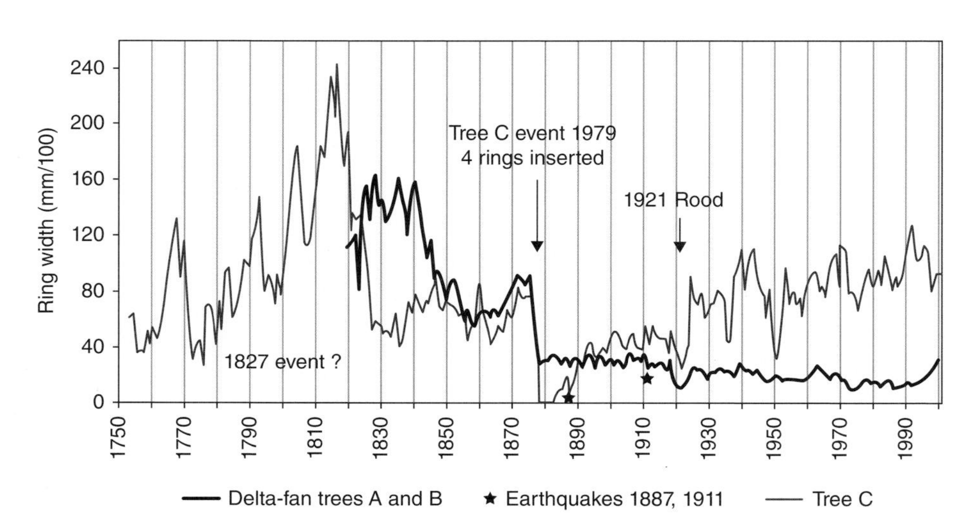

Figure 7.8 Mean of raw-ring widths of trees A, B and C showing major events impacting these trees in 1878 and 1921. Note that (i) the growths of trees A and B are highly correlated and have been averaged to produce a single plot, and (ii) the four rings inserted in the tree C record between 1878 and 1882 result in excellent cross-dating with the A and B tree records

northwestern margin of debris flow assemblage S3. The channel-side roots of this specimen are exposed and badly damaged, and a boulder (200 × 110 × 60 cm) lies up-ended against the southern side of its stem, 6 m from the gully's confluence with the Ozernaya River (Figures 7.3 and 7.7).

The likelihood that the boulder was emplaced and roots damaged during an event in 1878 is supported by cross-dating with trees A and B revealing that tree C has five missing rings between 1878 and 1882 and that all three trees reacted negatively to the 1921 event (Figure 7.8). East of tree C some 60 m, another sample tree D (Figure 7.3) growing in the same dry but vegetated channel has an ecesis date of c. 1860; this tree is unlikely to have survived later debris flows or flood events sourced from the main lateral gully leading up to the M. Almatinka divide, and hence the event of 1878 is associated with a large event from the Ozernaya River upstream of the study area.

Previous work in the northern ZailiyskiyAlatau has documented 23 major debris flow events between 1841 and 1999 (Gorbuhov & Severskiy, 2001; I.V. Severskiy, unpublished records). Figure 7.9 plots the dates of documented regional and local (specific to the study valleys) debris flows together with mean tree-ring widths and age estimates of debris flow events recorded in this study. This shows that with the exception of events in 1841 and 1854 in the Bolshaya and Malaya Almatinka valleys, the documentary record is confined to the post-1920 period and, since 1921, only two events (1950, 1963) are present in the Ozernaya skeleton plot record (Figure 7.6b). The documentary records (Table 7.4) show the main triggers for mud flows and lake outbursts are stormy summers coupled with high rates of glacier melting mostly during the months of July and August, with 70% following intense rainstorms and snow melt, 20% linked to intense melting and 10% due to the failure of moraine-dammed lakes, some of which may have been triggered by earthquakes.

Definition of seismic events is equivocal (Figures 7.6b, 7.7 and 7.9). Low growth occurred during the 1887 and 1911 earthquake years but, although the event skeleton plot shows pointer years for these dates, low growth is unexceptional and could equally suggest climatic and/or geomorphological influences. The distinction might be clearer if ring-width measurements showed a stronger negative anomaly and there was evidence from another source, such as datable mechanical tree damage due to surface rupture including stem cracking or tilting, broken branches or crown damage (Schweingruber, 1996).

7.5 Discussion

The topography of a valley plays an important role in the preservation of debris flow sediments and landforms. Narrowly confined valleys such as Kumbelsu Creek are particularly prone to erosion and stripping of unconsolidated sediments during large flood and debris flow events, and here it is perhaps not surprising that dating for the sedimentary record is limited to localized and discontinuous valley fills of 20th century origin. The wider valley floor of the Ozernaya study area, by contrast, has permitted preservation of several generations of 17th century and later debris flow assemblages at the confluence of steep tributary valleys. These landforms are laterally inset into older assemblages although their surfaces have been locally modified and aggraded by later events.

Caution is warranted in the analysis of debris flow records and long-term climate trends, not least because of the broad age spans established for 19th century and earlier events, and observational bias that is likely to have contributed to the relatively high incidence of debris flow events recorded in the mid–late 20th century (Figure 7.9 and Table 4). The latter reflects the advent of systematic meteorological and hazard monitoring since the 1920s paralleling

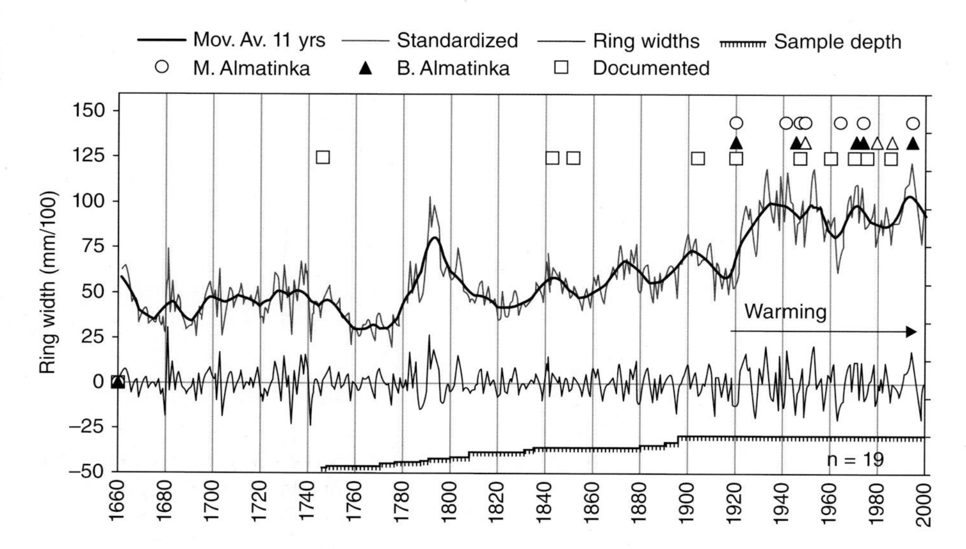

Figure 7.9 Graph showing (i) mean raw-ring widths of cross-dated tree cores, with (ii) an 11-year moving average showing growth trend in the study area; (iii) dates of documented debris flow events in the local and regional valleys, (iv) standardized curve of mean raw-ring widths showing change in frequency suggesting warming from 1920, and (v) sample depth: the reliability of the curve is directly related to sample depth, thus prior to 1750 the curve is for the record only

improved road access and the expansion of hydroelectric power generation, settlement, leisure facilities and scientific research in the high mountain valleys south of Almaty (Figure 7.1). Nevertheless, an enhanced frequency of debris flows is to be expected in high mountain environments experiencing glacier retreat during warming periods (e.g. Evans & Clague 1994; Chiarle *et al.*, 2007). The enhanced frequency reflects a combination of debuttressing of glacially over-steepened slopes, abundant unconsolidated and unvegetated

Table 7.4 Documented catastrophic debris flow events in the Zailiynski Alatau (I.V. Severskiy, unpublished records)

Origin	Event years (events mostly between May and August)	Trigger	Max. discharge (m^3/s)
Bolshaya Almatinka	1921, 1949, 1950, 1981	Rainstorm	(1950) 1000
Bolshaya Almatinka	1975, **1977**, **1994**, **1999**	Lake-break	(1977)[a] 10,000
Malaya Almatinka	1921, 1966, 1999	Rainstorm	
Malaya Almatinka	1944, 1951, 1956, 1973	Lake-break	(1973) 7000
Regionally documented	1921, 1947, **1958**, 1963, 1974, 1975, 1979, **1980**, 1993	Lake-break and rainstorms	(1921) 920 (1963) 7000
Undocumented	1750s, 1841, 1854	(Local tradition)	

[a]1977 Lake break at head of Kumbelsu Creek.

Note: documentation of debris flow events promoted following valley road construction after 1944. For highlighted dates see Table 7.1.

Figure 7.10 Tuyuksu glacier showing magnitude of retreat since the Little Ice Age maximum, as marked by the position of the moraines behind the hut compared with the positions of the present-day glacier tongues

sediments, and melting of buried ice in moraines, with these features promoting catastrophic drainage of moraine-dammed lakes. Records are particularly detailed for the Tuyuksu glacier in the M. Alamatinka valley, where some 750 m of ice retreat (Figure 7.10) has occurred since 1923 with this accompanied by four major debris flows, the largest occurring in 1951, 1956 and 1973 (Harrison *et al.*, 2004). The 1973 event is believed to have moved an estimated 3.8 million m^3 of material (I.V. Severskiy, unpublished data; Harrison *et al.*, 2004).

The classification of debris flow types in the documentary record permits an assessment of controlling mechanisms; Table 7.4 shows that 13 of the 23 large-scale documented debris flows in the region are associated with glacial lake outbursts, with these usually occurring during the summer months of July and August when they coincide with intense glacier ablation and meltwater production. Glacial lake outbursts triggered seven of the eight debris flows recorded between 1938 and 1999 in the Bolshaya and Malaya Almatinka valleys, with five of these coinciding with peak yearly May–August precipitation records. This pattern is consistent with the often weak correspondence noted between documented debris flows and records of total precipitation (VanDine & Bovis, 2002), although regional debris flow events associated with rainstorms tend to occur during the May–August periods of high summer rainfall and ice/snow melt conditions (Table 7.4). Events of this type in the Zailiyskiy Alatau are most probably the response to intense precipitation associated with the penetration of cold wet air masses from the northwest. In 1921 these conditions were of sufficient magnitude to trigger debris flows in both the Bolshaya and Malaya Almatinka valleys (Table 7.4), although these do not appear in our tree-ring record. Such events can lead to subsurface piping and a build up of pore water pressure on steep

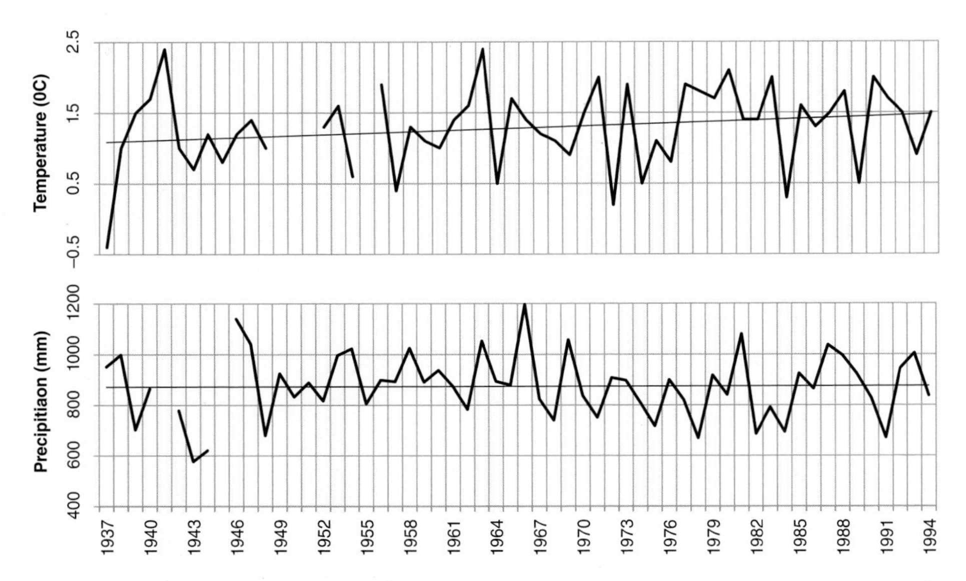

Figure 7.11 Plots of total precipitation and temperature over the months May–August for the period 1932–96 at Bolshaya Almatinka Lake meteorological station

unconsolidated slopes leading to sudden slope failure and debris flows (e.g. Rapp, 1963; Iverson *et al.*, 1997).

A detailed investigation of climate trends and the tree growth record is not pursued here since the number of tree samples currently available in the dendrochronological dataset is too low to permit a reliable analysis. It is interesting to note, however, that, beginning in 1920, there has been a marked increase in the growth trend together with increasing annual growth variability (Figure 7.10). The standardized graph (Figure 7.9) highlights the change in growth pattern: a pattern that may well be linked to climate warming in the region especially when seen in the context of an increase in temperature recorded (1932–1996) at the meteorological station in the Ozernaya valley (Figure 7.11) and a general increase in 20th century documentary temperature records in the northern Tien Shan, including the Zailiyskiy Alatau (Aizen *et al.*, 2007; Bolch, 2007).

While the documentary record of regional debris flows generally coincides with the 20th century period of climate warming, the geomorphological record of B. Almatinka debris flows spanning the 18th to 20th centuries indicates that high-magnitude events also occurred during the cooler climatic conditions towards the end of the Little Ice Age, possibly associated with glacial lake outbursts during short phases of glacier retreat. Available data on glacier fluctuations in the wider mountain regions of Central Asia suggest that this was a period of complex climatic change with short-lived (c. 20–30-year) periods of warming (Kotlyakov *et al.*, 1991).

7.6 Conclusions

Debris flows constitute one of the most serious landslide-type hazards to human safety, and records show that for the populations living in the valleys at the foot of the Zailiiskiy Alatau Mountain range debris flows are a regular occurrence. In the Ozernaya valley, based on the age of the oldest trees, we identified nine debris flow assemblages, with the earliest dating back to the early 17th century when there was a major period of slope activity affecting the whole valley. A second major period of activity occurred in the mid-1700s, with material largely fed through gullies descending from the eastern divide (extending and supporting a local tradition of slope movements at that time). Activity thereafter was mostly contained within and around the main Ozernaya river channel, the lateral gullies and their margins. In 1827 a debris flow, initiated near the eastern divide, probably descended the lateral gully between the S1 and S2 assemblages, with this also affecting the area covered by S5. Thereafter in 1878, a flow capable of moving boulders at least as large as 1.3 m^3 descended the Ozernaya valley. Valley sediments will have contributed to the regional debris flows in 1921, also clearing the upslope S5 surface of vegetation. That the valley has experienced multiple minor flows throughout the period is evidenced by large differences in age of the oldest trees and their younger cohorts on the assemblages, together with evidence of composite suites of debris flow and transitional-flow deposits accumulated around the oldest tree trunks.

Documentary evidence indicates that since the mid-19th century approximately two-thirds of large-scale debris flows are the result of glacial lake outbursts, with the remainder associated with high-intensity rainstorms. In the 20th century, evidence from Kumbelsu Creek and the lower Almatinka valley shows that, since 1921, flood levels, although increasing in frequency, have declined in magnitude implying that sediment supply has decreased over the last 90 years.

Although flow magnitudes appear to be diminishing, a major seismic movement could alter this trend. Thus, human vulnerability to debris flows and floods in the region persists, and building developments in the lower valleys are at risk as long as there are large quantities of unconsolidated sediments in the upper valleys and on the surrounding steep slopes ready to be destabilized by intense summer storms coupled with high rates of glacier and snow melting or outbursts from vulnerable moraine-dammed lakes. Climate warming could also promote an alternative threat to debris flows due to snow-pack melting on the summits of the Zailiyskiy Alatau and a consequent reduction in stream flow, with this affecting water and electricity supplies to Almaty and its surrounding region. We recommend careful monitoring: future geomorphological work could be directed towards identification of sediment supply areas and assessments of slope stability, sediment accumulation, erosion rates and meltwater supply.

Acknowledgements

This work forms part of an EU-funded INCO-COPERNICUS project, number IC15-CT98-0152 (DG12-MZEN). It was carried out with colleagues from the Ministry of Education and Sciences, Kazakhstan, and Germany. Personnel at the Institute of Geography, Academy of Sciences, Almaty, are thanked for their help, kindness and warm welcome.

References

Aizen, V.B., Aizen, E.M., Melack, J.M. (1995) Climate, snow cover, glaciers and runoff in the Tien Shan, Central Asia. *Water Resources Bulletin, American Water Resources Association* **31**: 1113–29.

Aizen, V.B., Kuzmichenok, V.A., Surazakov, A.B., Aizen, E.M. (2007) Glacier changes in the Tien Shan as determined from topographic and remotely sensed data. *Global and Planetary Change* **56**: 328–40.

Alestalo, J. (1971) Dendrochronological interpretation of geomorphological processes. *Fennia* **105**: 1–140.

Blair, T.C., McPherson, J.G. (1994) Alluvial fans and their natural distinction from rivers based on morphology, hydraulic processes, sedimentary processes, and facies assemblages. *Journal of Sedimentary Research* **A64**: 450–89.

Bolch, T. (2007) Climate change and glacier retreat in northern Tien Shan (Kazakhstan/Kyrgyzstan) using remote sensing data. *Global and Planetary Change* **56**: 1–12.

Bollschweiler, M., Stoffel, M., Ehmisch, M. Monbaron, M. (2007) Reconstructing spatio-temporal patterns of debris-flow activity using dendrogeomorphological methods. *Geomorphology* **87**: 337–51.

Chiarle, M., Iannotti, S., Mortara, G., Deline, P. (2007) Recent debris flow occurrences associated with glaciers in the Alps. *Global and Planetary Change* **56**: 123–36.

Clague, J.J., Evans, S.G. (2000) A review of catastrophic drainage of moraine-dammed lakes in British Columbia. *Quaternary Science Reviews* **19**: 1763–83.

Esper, J., Gärtner, H. (2001) Interpretation of tree-ring chronologies. *Erdkunde* **55**: 277–88.

Evans, S.G., Clague, J.C. (1994) Recent climatic change and catastrophic geomorphic processes in mountain environments. *Geomorphology* **10**: 107–28.

Fantucci, R., Sorriso Valvo, M. (1999). Dendrogeomorphological analysis of a slope near Lago, Calabria (Italy). *Geomorphology* **30**: 165–74.

Gasser, T., Müller, H.G. (1984) Estimating regression functions and their derivatives by the kernel method. *Scandinavian Journal of Statistics* **11**: 171–85.

Gorbuhov, A.P., Severskiy, I.V. (2001) *Mudflows Surrounding Almaty*. Alma-Ata: Nauka Publishers, 79 pp. [in Russian].

Gutsell, S.L., Johnson, A. (2002) Accurately ageing trees and examining their height-growth rates: implications for interpreting forest dynamics. *Journal of Ecology* **90**: 153–66.

Harrison, S., Passmore, D.G., Severskiy, I.V. (2004) Post-Little Ice Age fluctuations of the Tuyuksu Glacier, Zailiiskiy Alatau, Tien Shan mountains, Kazakhstan. In: Schröder, H., Severskiy, I.V. (eds), *Water Resources in the Basin of the Ili River (Republic of Kazakhstan)*. Berlin: Mensch & Buch Verlag, pp. 280–97.

Iverson, R.M. (1997) The physics of debris flows. *Reviews of Geophysics* **35**: 245–56.

Iverson, R.M., Reid, M.E., LaHusen, R.G. (1997) Debris flow mobilization from landslides. *Annual Review Earth Planetary Science* **25**: 85–138.

Jacoby, G.C. (1997) Application of tree-ring analysis to paleoseismology. *Reviews of Geophysics* **35**: 109–24.

Jones, P.D., New, M., Parker, D.E., Martin, S., Rigot, I.G. (1999) Surface air temperature and its changes over the past 150 years. *Reviews of Geophysics* **37**: 173–99.

Kotlyakov, V.M., Serebryanny, L.R., Solomina, O.N. (1991) Climate change and glacier fluctuation during the last 1000 years in the southern mountains of the USSR. *Mountain Research and Development* **10**: 1–12.

Marchenkoa, S.S., Gorbuhov, A.P., Romanovskya, V.E. (2007) Permafrost warming in the Tien Shan Mountains, Central Asia. *Global and Planetary Change* **56**: 311–27.

Passmore, D.G., Harrison, S., Winchester, V., Rae, A., Severskiy, I.V., Pimankina, N.V. (2008) Late Holocene debris flows and valley floor development in the northern Zailiiskiy Alatau, Tien Shan Mountains, Kazakhstan. *Arctic, Antarctic and Alpine Research* **40**: 548–60.

Rapp, A. (1963) The debris slides of Ulvadal, western Norway: an example of catastrophic slope processes in Scandinavia. *Nachrichten der Akademie der Wissenschaft in Gottingen II. Mathematisch-Physikalische Klasse* **13**: 197–210.

Rinn, F. (1988) Eine neue Methode zur Berechnung von Jahrringparametern. Auszug aus der Diplomarbeit Universität Heidelberg, 86 pp [thesis].

Rinn, F. (2003) TSAP-Win time series analysis and presentation for dendrochronology and related applications: version 0.53 for Microsoft Windows; user reference. Heidelberg: Rinn Tech, pp. 1–88.

Schweingruber, F.H. (1996) *Tree Rings and Environment: Dendroecology*. Berne: Paul Haupt Publishers, 602 pp.

Schweingruber, F.H., Echstein, D., Serre-Brachet, F., Bräker, O.U. (1990) Identification, presentation and interpretation of event years and pointer years in dendro series. *Dendrochronologia* **8**: 9–38.

Severskiy, I.V., Blagoveshchenskiy, V.P. (1983) *An Evaluation of Avalanche Hazard of Mountain Areas*. Alma-Ata: Nauka Publishers, 215 pp. [in Russian].

Severskiy, I.V., Severskiy, E.V. (1990) *Snow Cover and Seasonal Freezing of Grounds in the Northern Tien Shan*. 180 pp. [in Russian].

Severskiy, I.V., Blagoveshchenskiy, V.P., Severskiy, S.I., Xie Zichu (2000) *Snow Cover and Avalanches in Tien Shan Mountains*. Ministry of Education and Sciences of the Republic of Kazakhstan, Academy of Sciences of China. Almaty: VAC Publishing House, 180 pp.

Shroder, J.F. (1980) Dendrogeomorphology: review and new techniques of tree-ring dating. *Progress in Physical Geography* **4**: 161–88.

Solomina, O., Barry, R., Bodyna, N. (2004) The retreat of Tien Shan glaciers (Kyrgyzstan) since the Little Ice Age estimated from aerial photographs and historical data. *Geografiska Annaler* (A) **86**: 205–15.

Strunk, H. (1997) Dating of geomorphological processes using dendrogeomorphological methods. *Catena* **31**: 137–51.

Stefanini, M.C., Schweingruber, F.H. (2000) Annual and seasonal reconstruction of landslide activity from Turkey oak in the northern Apennines, Italy. *Dendrochronologia* **18**: 53–61.

VanDine, D.F., Bovis, M. (2002) History and goals of Canadian debris flow research, a review. *Natural Hazards* **26**: 69–82.

Weber, U., Schweingruber, F.H. (1995) A dendroecological reconstruction of western spruce budworm outbreaks (*Choristoneura occidentalis*) in the Front range, Colorado, from 1720–1986. *Trees* **9**: 204–13.

Winchester, V., Harrison S. (2000) Dendrochronology and lichenometry: an investigation into colonization, growth rates and dating on the east side of the North Patagonian Icefield, Chile. *Geomorphology* **34**: 181–94.

Winchester, V., Gärtner, H., Bezzi, M. (2007) Dendrogeomorphological applications. In: Kalvoda, J., Goudie, A.S. (eds), *Geomorphological Variations*. Prague: Nakladatelstvi P3K, pp. 183–203.

II
Land-use Change: Modelling and Impact Assessment

8 Regional Scenarios and Simulated Land-Cover Changes in Montane Mainland Southeast Asia

Jefferson Fox[1], John B. Vogler[2], Omer L. Sen[3], Alan L. Ziegler[4] and Thomas W. Giambelluca[5]

[1]*East-West Center, Honolulu, Hawaii, USA*
[2]*Center for Applied Geographic Information Science, University of North Carolina, Charlotte, USA*
[3]*Istanbul Technical University, Eurasia Institute of Earth Sciences, Istanbul, Turkey*
[4]*Department of Geography, National University of Singapore, Singapore*
[5]*Department of Geography, University of Hawaii, Manoa, Hawaii, USA*

8.1 Introduction

Montane mainland Southeast Asia (MMSEA), 300 m and above, is a large, ecologically vital region comprising about half of the land area of Cambodia, Laos, Myanmar, Thailand, Vietnam, and China's Yunnan Province (Figure 8.1). The region harbors a wealth of natural resources including globally important stocks of forest and biological diversity and the headwaters for several major river systems including the Mekong, Chao Praya, Irrawaddy, and Yuan-Hong. Much of MMSEA has only been reopened to outside influences within the last two decades, bringing profound and widespread changes to both its physical environment and to its local societies. Swidden agriculture, or swiddening (also called shifting cultivation), the dominant farming system in MMSEA where it has been practiced for at least a millennium, has greatly influenced land cover and land use throughout the region. Guo *et al.* (2002), however, asserted that 'from the mountains of Yunnan Province in southwestern China to the interior of Kalimantan' the ways of life of upland farmers throughout

Vulnerability of Land Systems in Asia, First Edition. Edited by Ademola K. Braimoh and He Qing Huang.

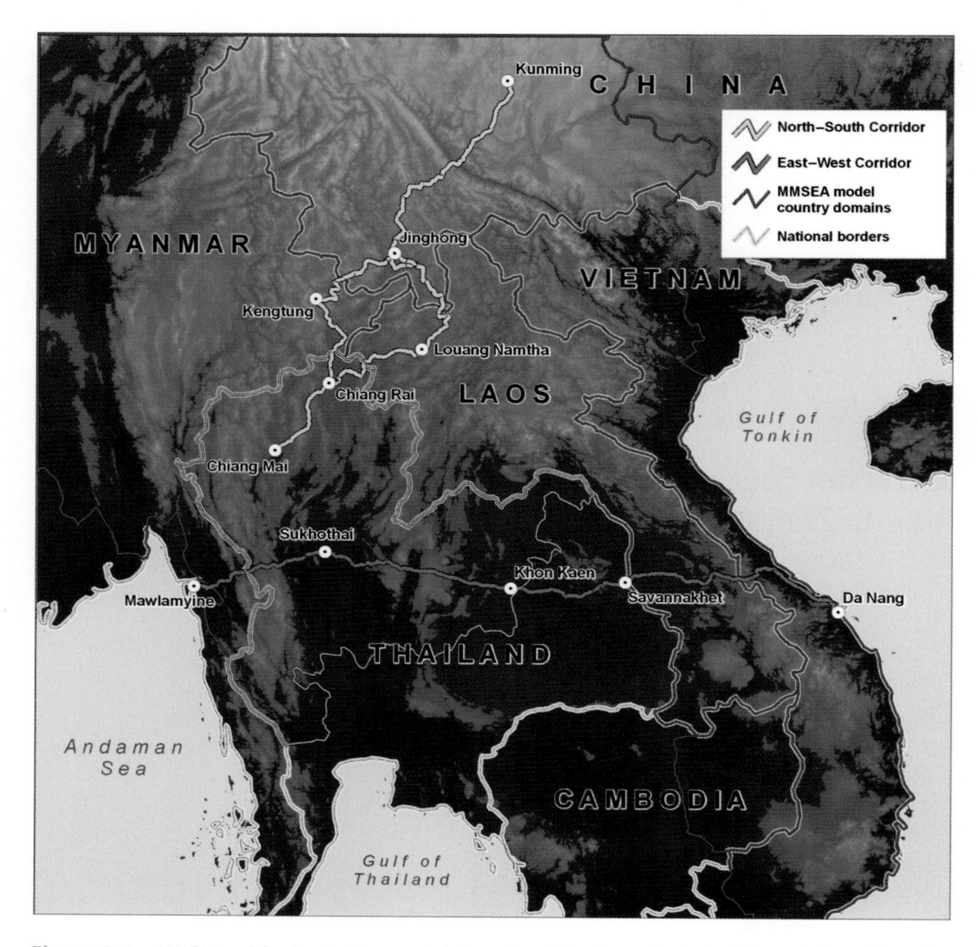

Figure 8.1 MMSEA with simulation model domains (blue boundaries), Chiang Mai to Kunming Highway Corridor (yellow lines) and East–West Corridor (red line). Areas shaded in green are 300 m and above

Southeast Asia are changing at rates and scales that 'are unprecedented.' They suggest that 'hill farmers with their hundreds of rice landraces, and their cyclic secondary forest fallows are disappearing throughout the region.'

In addition to the land-cover and land-use change (LCLUC) driven by political and economic forces, the Asian Development Bank is funding the building and improvement of roads running throughout the region. The Chiang Mai-Kunming Highway (Figure 8.1) provides the first reliable land-transport route connecting eastern China with central Thailand and the Malay Peninsula. This has created a 'North-South Economic Corridor' through the heavily forested region (Maekawa, 2002). Another new road, forming an 'East-West Economic Corridor', will begin at Mawlamyine, Myanmar, on the Andaman Sea, cross Thailand, Laos, and Vietnam, and end at Da Nang on the South China Sea in central Vietnam (Thant & Nair, 2002). In addition to the economic and social changes they produce, highways such as these, traversing forested regions, promote rapid LCLUC by providing access to formerly isolated areas and facilitating transport of timber and other products to market (Schneider, 1995). Construction of major new highways through the region sets the stage for wholesale forest conversion in MMSEA.

While it has become all too apparent that MMSEA is on the cusp of a major new upsurge in LCLUC, there is much uncertainty about the direction of change and the impacts it will have both on people's livelihoods and environmental variables such as biodiversity, carbon sequestration, watershed hydrology, and climate. A great deal of recent literature suggests that swidden farming is rapidly giving way to commercial agriculture driven by domestic demand and by regional trade agreements (Fox & Vogler, 2005; Thongmanivong *et al.*, 2005; Xu *et al.*, 2005; Padoch *et al.*, 2007; Fox *et al.*, 2008; Fujita & Phanvilay, 2008). In Xishuangbanna (the most southern prefecture in Yunnan Province), China, both semi-privatized state farms and minority farmers are planting rubber at rates that threaten to transform the landscape between 300 and 1000 m into an unbroken carpet of rubber (Xu *et al.*, 2005; Xu 2006; Li *et al.*, 2007; Sturgeon, 2010). In northern Thailand, rural people are becoming increasingly divorced from farming, with education and consumerism creating a context where rural people are disintensifying, even abandoning their land, in favor of non-farm pursuits (Rigg & Nattapoolwat, 2001; Rigg, 2006). In Laos and Cambodia, entrepreneurs have contracted farmers to grow corn, bananas and sugar cane for the Chinese and Vietnamese markets (Thongmanivong *et al.*, 2005; Fox *et al.*, 2008; Fujita & Phanvilay, 2008). In response to soaring prices for natural rubber, highlanders, usually ethnic minorities, are planting rubber trees on family plots. In Laos they are turning to relatives in China for advice; and in both countries, to merchants for seeds, grafts, and tapping tools (Thongmanivong *et al.*, 2005; Fujita & Phanvilay, 2008). In Vietnam researchers have reported the expansion of tree crops such as rubber, tea, and coffee in the Central Highlands (Thomas *et al.*, 2008) and fast-growing species for pulp and timber in the northern part of the country (Sunderlin & Huynh, 2005). An assessment of LCLUC in MMSEA, however, over the next few decades cannot merely be based on local case studies since case study location selection is often biased and results cannot be easily extrapolated to the broader region.

Conceptual models of land-cover change scenarios provide an alternative method of examining possible futures. Lebel (2006) proposed four possible scenarios of LCLUC in upper tributary watersheds in mainland Southeast Asia. These scenarios are arranged against different combinations of regional uncertainties relating to trade networks (varying from local to global) and agricultural development (varying from unified monocultures to diversified systems). Herein, we have adapted Lebel's framework to form the basis of change in land cover in MMSEA between 2001 and 2050 (Figure 8.2). Specifically, we sought to determine from regional experts what changes in land cover and land use (LCLU) in MMSEA are likely to occur over the next 50 years.

More specifically, this project sought to determine what LCLUC has occurred in recent decades and what significant LCLUC is likely to occur in MMSEA in the coming decades. To answer this question, the project developed a comprehensive, high-resolution database of land cover in MMSEA; developed scenarios of future LCLUC in the region; and simulated LCLUC in MMSEA to 2025 and 2050 based on these scenarios and expert knowledge.

8.2 Methods

The conversion of land use and its effects (CLUE-s) model (Verburg *et al.*, 1999a, 2002; Verburg, 2006) was used to simulate land-cover change in this work. Both CLUE-s and its

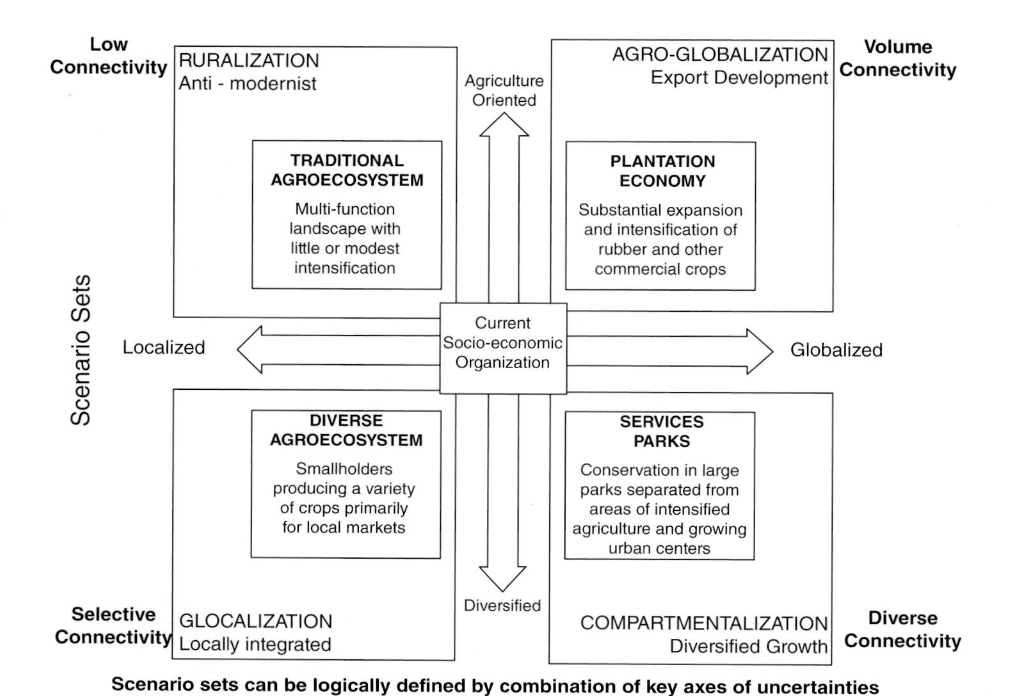

Figure 8.2 Regional scenarios of LCLUC in MMSEA (adapted from Lebel, 2006)

predecessor, CLUE, have proven application and validation over a wide range of scales of analysis in several regions worldwide, including Asia (Verburg *et al.*, 1999b), Central America (Wassenaar *et al.*, 2007), and Europe (Verburg *et al.*, 2006). In Southeast Asia, CLUE-s has been applied at the sub-national level in northern Vietnam (Castella & Verburg, 2007; Willemen, 2002) and Malaysia (Engelsman, 2002; Verburg *et al.*, 2002), and at the national and sub-national levels in the Philippines (Soepboer, 2001; Verburg & Veldkamp, 2004).

8.2.1 Baseline land-cover classification

A 2001 land-cover map was developed to serve as the model baseline. To do so, we first acquired a 1-km resolution global land-cover map from the USGS Land Processes Distributed Active Archive Center that was generated from 2000 to 2001 MODIS/Terra observations (Friedl *et al.*, 2002) and was made available in the 17-class International Geosphere-Biosphere Programme (IGBP) global vegetation classification scheme. The land-cover dataset was then clipped to the simulation model domain, which removed areas of the 'snow and ice' class. The IGBP classification was then reclassified to the 20-class Biosphere-Atmosphere Transfer Scheme (BATS) (Dickinson *et al.*, 1993) to support regional climate modeling of the study area in a related work (Sen *et al.*, 2012). Table 8.1 shows the translation table used during the reclassification. The resulting BATS class map for MMSEA excluded four BATS categories (tundra, ice caps and glaciers, water and land mixtures, and ocean categories) because they were not present in the study region. The BATS classification scheme does not recognize urban/built-up or sparsely vegetated areas,

Table 8.1 IGBP (International Geosphere-Biosphere Programme) to BATS (Biosphere-Atmosphere Transfer Scheme) reclassification translation table

IGBP class	BATS class
Croplands	Crops, mixed farming
Cropland/natural vegetation mosaic	
Grasslands	Short grass
Evergreen needleleaf trees	Evergreen needleleaf trees
Deciduous needleleaf trees	Deciduous needleleaf trees
Deciduous broadleaf trees	Deciduous broadleaf trees
Evergreen broadleaf trees	Evergreen broadleaf trees
Savannah	Tall grass
Urban/built-up	Urban/built-up
Croplands	Irrigated crops
Cropland/natural vegetation mosaic	
Barren or sparsely vegetated	Sparse vegetation
Permanent wetlands	Bogs and marshes
Water bodies	Inland water
Closed shrubland	Evergreen shrubs
Open shrubland	
Closed shrubland	Deciduous shrubs
Open shrubland	
Mixed forest	Mixed forest
Woody savannah	Forest/field mosaic

therefore we retained those areas from the IGBP class designation in the reclassified map. The final, 2001 BATS classification for the land-cover simulation region included 16 land-cover types (Figure 8.3).

8.2.2 CLUE-s model

CLUE-s employs an iterative spatial allocation procedure that involves parameterization of the following inputs to generate land-use/cover simulation maps at regular time steps: land-cover demands, location suitability, conversion characteristics, and location restrictions. MMSEA encompasses a region that extends into six countries and includes areas that lie above 300 m (Figure 8.1). These countries have unique social, political, and economic histories, and therefore potentially unique land-cover change trajectories. Hence, we developed a separate model for the portion of each country that lies within MMSEA (henceforth called country domain), allowing independent simulation of land-cover changes within each domain. This modeling approach provided the flexibility to parameterize situations that affect some, but not all, countries. Model inputs and the spatial allocation procedure are described in the following sections.

8.2.2.1 Land-cover demands Land-cover demands refer to the aggregate area occupied by each land-cover type at each simulation time step, in this case annual time steps. To determine these demands, we first interviewed experts with knowledge of LCLUC in each country domain. Experts were shown the 2001 land-cover classification of their country and asked to estimate the percentage of change in each land-cover category in 2025 and

Figure 8.3 Baseline (2001), 16-class land-cover map for the MMSEA simulation region

2050. Estimates were derived using a non-spatial, scenario-driven approach that involved consideration of four scenarios of change (Lebel, 2006) as outlined in Figure 8.2. These scenarios include:

1. **Plantation Economy Scenario.** Economic growth is led by agricultural businesses. Upland farmers are interested in planting a number of market crops, but rubber (*Hevea brasiliensis*) is the major commercial crop replacing traditional agriculture and secondary forests in the region, a direct result of strong market demands from China, the world's largest consumer. Newspaper reports in Laos suggest that over 50,000 ha of rubber had been planted there in 2000. In Cambodia, the Ministry of Agriculture plans to expand the area under rubber cultivation from 50,000 ha to as much as 800,000 ha by 2015. In Myanmar, rubber is expanding into border areas in Kachin and Shan States. In Thailand, rubber has expanded to include over 48,000 ha in the north and 64,000 ha in the northeast. The Thai Rubber Board predicts the total area of rubber in Thailand will increase from 1.9 to 2.4 million ha by 2020. Vietnam currently has c. 500,000 ha of

rubber, but little is known about where additional rubber trees are being planted or at what rate.

2. **Parks and Conservation Scenario.** Economic growth will unfold primarily through tourism that places a high value on forests, wetlands, rivers, and perhaps even 'ethnic' diversity. Large areas of northern Thailand have been set aside as national parks and wildlife sanctuaries; and a national program in the 1980s classified land according to watershed characteristics, placing strict restrictions on the ability of people to use sloping land (Thomas *et al.*, 2008). Cambodia has already established Virachey National Park in Ratanakiri and Stung Treng Provinces. Virachey is the largest park in the country covering 337,723 ha.
3. **Traditional Agroecosystem Scenario.** Farmers choose to reject outside change because of progressive lowering of private and public investments in regional infrastructure, either because these funds are targeted elsewhere or because of a prolonged global recession. This scenario could reflect anti-globalization movements, dwindling agricultural trade, and an expansion of local exchange systems.
4. **Diverse Agroecosystem Scenario.** Significant but diversified economic growth occurs, drawing on local comparative advantages in agriculture, tourism, and perhaps mining rather than on the adoption of more uniform technologies and production systems.

With these scenarios in mind, experts from each of the country domains characterized the drivers of change and estimated percent change in each land-cover class out to the years 2025 and 2050. Table 8.2 summarizes the number of experts we interviewed and their areas of expertise. By coupling these predictions with economic country profiles (Economic Intelligence Unit, 2004a, 2004b, 2005a, 2005b, 2005c, 2005d), we determined the final domain-specific land-cover demand values for each intermediate year of the simulation for a total of 50 years (2001–2050) of demand values. Table 8.3 summarizes for each country domain the observed land cover in 2001 and expected demands for change (expressed as percent of total country domain area) for simulation years 2025 and 2050. The percent change in land-cover demands from 2001 to 2050 is also indicated in Table 8.3.

8.2.2.2 Location suitability Location suitability refers to how well a particular location, in this case a grid cell, is suited to a particular land-cover type. Land-cover conversions typically occur at locations that possess the highest suitability for a particular land-cover type at a particular point in time. Determining the suitability of each land-cover type at each cell location in the landscape requires (i) consideration of the biophysical, geographic, and socio-economic factors and processes hypothesized to be driving different land-cover conversions, (ii) identifying a parsimonious set of factors influencing the locations of observed land-cover types and quantifying their relative influences by statistical means, and (iii) combining actual factor location values (e.g., elevation, distance to road) and their relative influences in a function that empirically quantifies the probability of each land-cover type at each location on the landscape. The land-cover type with the highest probability, or suitability, for a particular location at a particular time, is then placed on the landscape at that location and for all grid cell locations, and a resulting land-cover pattern emerges in an output map. This section describes the spatial datasets (potential driving factors and land cover), their preparation for input into CLUE-s and statistical software, and the statistical modeling used to determine the total suitability of landscape locations for each of the modeled land-cover types.

Based upon the drivers of change highlighted by experts from each country domain, a set of 1-km resolution raster datasets, including the 2001 land-cover map described above,

Table 8.2 Land cover (expressed as % of country in MMSEA domain) in 2001 and change demands as estimated from expert knowledge for 2025 and 2050

	Cambodia				Laos				Myanmar			
	59,579 km^2				283,363 km^2				462,495 km^2			
Area in MMSEA domain	2001	2025	2050	%ch	2001	2025	2050	%ch	2001	2025	2050	%ch
Land cover type												
Crops, mixed farming	1.4	4.5	7.5	6.1	1.3	2.0	3.0	1.7	2.7	4.0	5.0	2.3
Short grass	0.9	1.0	1.0	0.1	1.7	3.0	3.0	1.3	2.1	3.0	3.0	0.9
Evergreen needleleaf trees	0.1	0.1	0.1	0.0	0.2	0.0	0.0	−0.2	0.2	0.2	0.2	0.0
Deciduous needleleaf trees	na[a]	na[a]	na[a]		na[a]	na[a]	na[a]		0.0	0.0	0.0	0.0
Deciduous broadleaf trees	9.1	8.5	6.0	−3.1	13.5	16.0	18.0	4.5	15.5	18.0	20.0	4.5
Evergreen broadleaf trees	44.5	41.0	37.0	−7.5	63.3	55.5	51.0	−12.3	34.4	31.5	29.5	−4.9
Tall grass	7.0	6.0	5.0	−2.0	5.6	6.0	6.0	0.4	3.8	3.8	3.8	0.0
Urban/built-up	0.1	1.0	2.0	1.9	0.1	1.0	1.5	1.4	0.4	1.0	2.0	1.6
Irrigated crops	3.9	6.0	11.0	7.1	2.2	2.3	2.5	0.3	12.6	13.0	13.0	0.4
Sparse vegetation	0.3	0.3	0.3	0.0	0.2	0.2	0.2	0.0	1.1	1.1	1.1	0.0
Bogs and marshes	0.0	0.0	0.0	0.0	0.1	0.1	0.1	0.0	0.1	0.1	0.1	0.0
Inland water	1.5	1.5	1.5	0.0	0.6	0.6	0.6	0.0	0.8	0.8	0.8	0.0
Evergreen shrubs	0.1	0.1	0.1	0.0	0.2	0.5	1.0	0.8	0.3	0.3	0.3	0.0
Deciduous shrubs	0.1	0.1	0.1	0.0	0.1	0.1	0.1	0.0	0.3	0.3	0.3	0.0
Mixed forest	0.8	3.5	6.0	5.2	0.9	0.9	0.9	0.0	2.4	3.0	3.0	0.6
Forest/field mosaic	30.3	26.0	22.0	−8.3	10.1	12.0	12.0	1.9	23.3	20.0	18.0	−5.3

	Thailand				Vietnam				Yunnan (China)			
	303,093 km^2				285,271 km^2				337,532 km^2			
Area in MMSEA domain	2001	2025	2050	%ch	2001	2025	2050	%ch	2001	2025	2050	%ch
Crops, mixed farming	5.1	7.0	5.0	−0.1	6.7	10.0	11.0	4.3	4.5	8.0	7.0	2.5
Short grass	3.4	4.0	5.0	1.6	1.5	2.5	2.0	0.5	5.2	2.5	4.5	−0.7
Evergreen needleleaf trees	0.1	0.1	0.1	0.0	0.7	1.0	1.5	0.8	0.5	1.0	0.0	−0.5
Deciduous needleleaf trees	1.0	1.0	1.0	0.0	0.0	0.0	0.0	0.0	0.0	0.0	0.0	0.0
Deciduous broadleaf trees	9.7	13.0	15.5	5.8	8.9	8.0	7.0	−1.9	10.6	15.0	15.0	4.4
Evergreen broadleaf trees	22.7	25.0	27.0	4.3	42.2	38.0	35.0	−7.2	19.7	21.0	25.0	5.3
Tall grass	7.3	7.3	7.3	0.0	5.2	7.0	8.0	2.8	7.2	4.0	2.5	−4.7
Urban/built-up	0.6	1.0	2.0	1.4	0.5	1.0	2.0	1.5	0.6	2.0	4.0	3.4
Irrigated crops	27.2	24.0	22.0	−5.2	8.7	10.0	11.0	2.3	2.0	1.0	1.0	−1.0
Sparse vegetation	0.5	0.5	0.5	0.0	0.8	0.8	0.8	0.0	0.9	0.9	0.9	0.0
Bogs and marshes	0.1	0.1	0.1	0.0	0.2	0.2	0.2	0.0	0.1	0.1	0.1	0.0
Inland water	0.9	0.9	0.9	0.0	0.4	0.4	0.4	0.0	0.3	0.3	0.3	0.0
Evergreen shrubs	0.7	1.0	1.5	0.8	0.8	2.0	3.5	2.7	9.1	13.0	14.5	5.4
Deciduous shrubs	0.2	0.2	0.2	0.0	0.6	1.0	1.0	0.4	2.0	2.0	2.0	0.0
Mixed forest	0.5	1.0	1.0	0.5	3.9	3.0	2.5	−1.4	17.9	14.0	11.0	−6.9
Forest/field mosaic	21.0	15.0	12.0	−9.0	18.9	15.0	14.0	−4.9	19.2	15.0	12.0	−7.2

[a]Indicates land-cover type not present in model domain.

Table 8.3 Number of experts interviewed and areas of expertise

	Cambodia	Laos	Myanmar	Thailand	Vietnam	Yunnan (China)
No. of experts	1	3	1	11	3	5
Areas of expertise	International academic	Local academic (2), international academic (1)	International academic	Local academic (3), international academic (4), politician (2), land-use manager (1), farmer (1)	Local academic (2), international academic (1)	Local academic (3), international academic (2)

was developed in a geographic information system (GIS) and georeferenced to a common coordinate system. Table 8.4 lists the geospatial datasets, sources, years, and scales, and the type of variable (i.e., static, dynamic). It should be noted that the availability and scale of some datasets varied across the country domains, and some data layers were originally acquired in vector format (e.g., shapefile) and converted to raster grids. The raster grids were clipped to the country domains using administrative polygon boundaries.

A majority of the assembled datasets were stable in nature, that is, the grid cell values at each location represented a single snapshot in time and those values did not change over time during the simulations (e.g., elevation). To avoid a temporal mismatch of corresponding grid cell values across the assembled datasets, we attempted to obtain datasets that corresponded as closely as possible to the year of the baseline land-cover data, 2001. Two of the spatial variables, including the 'distance to road' accessibility layer and population density, were treated in the simulations as dynamic input variables that changed annually in the models. An updated distance to road layer was created for each model year and input at each annual time step, simulating annual step decreases in distance from each landscape grid cell to the nearest roads. In order to incorporate annual changing population density, we applied projected annual (2001–50) growth rates, obtained from the US Census Bureau's International Data Base (IDB), to the population density grid cell values of the baseline 2000 population density layer for each country domain and at each time step in the simulations.

CLUE-s uses a logit model to define a function that, for each modeled land-cover type, calculates a total probability (i.e., suitability) that a given land cover will occur in a grid cell based on a combination of factor (driver) values and their relative influences. For each country domain, we converted all raster grids to tables containing land-cover type presence/absence and corresponding location factors values. We used statistical software (SPSS) to run a series of binary logistic regressions in which each land-cover type was regressed against all potential driving/location factor values. A stepwise regression was performed on each land-cover type to identify a parsimonious set of the most significant, explanatory location factors for each type within each country domain.

For some land-cover types the probability of occurrence of that land-cover type at a particular location can be partially explained by the land-cover types found in the immediate surrounding area, or neighborhood, of the focal grid cell. Urban/built-up areas are a prime example where neighborhood interactions are important (Veldkamp & Fresco, 1996, 1997a, 1997b; Verburg *et al.*, 2003), with new urban growth typically developing at the edges of existing urban areas. Using the CLUE-s model we calculated an 'enrichment factor' to describe the neighborhoods of urban cells and to enhance probability of new urban cells growing around existing urban cells by applying weights to neighboring cells during the land-cover simulations.

Table 8.4 Geospatial datasets used to derive location factors

Variable	Data source	Baseline year	Scale or resolution	Static or dynamic
Land cover – land use	LPDAAC at USGS[a] (MODIS/Terra Land Cover) (http://lpdaac.usgs.gov/)	2001	1 km	Dynamic
Population density	SEDAC – CIESIN[b] (http://sedac.ciesin.org/gpw/)	2000	1 km	Dynamic
Distance to populated place	Derived from populated places UNEP-RRCAP[c] (http://www.rrcap.ait.asia/)	2001	1:100–1:250k	Static
Distance to domestic city (nearest domestic market)	Derived from cities GIS layer UNEP-RRCAP	2001	1:100–1:250k	Static
Distance to foreign city (nearest foreign market)	Derived from cities GIS layer UNEP-RRCAP	2001	1:100–1:250k	Static
Distance to major road	Derived from roads GIS layer UNEP-RRCAP	2001	1:100–1:250k	Dynamic
Distance to road	Derived from roads GIS layer UNEP-RRCAP	2001	1:100–1:250k	Dynamic

(continued)

Table 8.4 (*Continued*)

Variable	Data source	Baseline year	Scale or resolution	Static or dynamic
Distance to major river	Derived from major river GIS layer UNEP-RRCAP	2001	1:100–1:250k	Static
Distance to river	Derived from rivers GIS layer UNEP-RRCAP	2001	1:100–1:250k	Static
Parks and protected areas	WDPA[d] (http://www.unep-wcmc.org/world-database-on-protected-areas-wdpa_76.html/)	2005	National to global	Static
Elevation	SRTM at USGS[e] (http://srtm.usgs.gov/)	2000	1 km	Static
Slope	Derived from SRTM	2000	1 km	Static
Major landform	ISRIC[f] (http://www.isric.org/)	1997	1:5 m	Static
Human-induced soil degradation (major type and intensity)	ISRIC (http://www.isric.org/)	1997	1:5 m	Static

Sources:
[a]Land Processes Distributed Active Archive, US Geological Survey.
[b]Socioeconomic Data and Applications Center, Center for International Earth Science Information Network.
[c]United Nations Environment Programme, Regional Resource Center for Asia and the Pacific Region.
[d]World Database on Protected Areas.
[e]Shuttle Radar Topography Mission, US Geological Survey.
[f]International Soil Reference and Information Centre.

8.2.2.3 Conversion characteristics and restrictions Land-cover conversion characteristics refer to the temporal transition behaviors for individual land-cover types. Transition behaviors are defined in part by a transition matrix of possible 'from-to' land-cover conversions. Transition matrices were created for each of the modeled country domains, specifying which land-cover conversions were permitted or restricted (Table 8.5). An additional elasticity parameter quantifies how resistant or amenable each land-cover type is to change (Verburg *et al.*, 1999b). Table 8.6 shows the simulation elasticity values used for each of the modeled land covers in each country domain. Elasticity values ranged from 0 (easy conversion to other allowable types) to 1 (completely resistant, irreversible change). Values of 0 allow conversion without consideration of the current land cover or that in any adjacent cell. Values of 1 were typically assigned to land-cover types that are difficult or too costly to convert or revert to anything else (e.g., urban areas and forests).

Location restrictions define areas where land-cover conversions are either stimulated or restricted. We restricted land-cover change in our country domains in all areas designated as national parks and protected areas based on the World Database on Protected Areas (WDPA) 2005 spatial dataset (UNEP-WCMC, 2005). The vector format dataset was acquired and polygon boundaries of all protected areas in the MMSEA region were converted to raster grids. Protected area grid cells were coded such that change was completely restricted during simulations. In order to satisfy the annual land-cover demands, changes were then limited to non-restricted cells.

8.2.3 Land-cover allocation

Land-cover spatial allocation is an automated, iterative process in CLUE-s that attempts to generate a spatial pattern of land cover that best satisfies the annual land-cover demands (scenario-based) for a given simulation year. Up to 20,000 iterations are allowed for any given simulation year to reach a satisfactory solution. When a solution is reached, the land-cover pattern is saved and used as the input land cover for the next simulation year. Land-cover demand values by themselves are non-spatial, and thus the spatial patterns are driven and shaped by the combination of land-cover location suitability at the grid cell level, allowable land-cover transitions and cover type elasticities, and geographic restrictions as described above. We performed multiple 50-year land-cover simulation runs per country domain, tested sensitivity of input parameters, and evaluated resulting land-cover patterns against country domain scenario demands. In the end, we selected the simulation having overall land-cover patterns that most accurately reflected the plausible scenarios envisioned by the experts.

8.3 Results

Simulated land-cover change in the years 2025 and 2050 generally reflects the role of small-scale diversified farming, monocropping (by both large operators and smallholders), and establishment and strict maintenance of protected areas (national parks, state forests, and protected watersheds) in each country domain (Table 8.7). Across MMSEA, the following land covers are estimated to increase by 2050:

- diversified farming of crops and other mixed farming activities (2.45%, approximately 42,500 ha);
- rubber and other deciduous broadleaf trees (2.46%, approximately 42,500 ha);

Table 8.5 Land-cover type conversion matrix of permitted (1) and restricted (0) transitions

Land-cover type	Domain	Crops, mixed farming	Short grass	Evergreen needleleaf trees	Deciduous needleleaf trees	Deciduous broadleaf trees	Evergreen broadleaf trees	Tall grass	Urban/built-up	Irrigated crops	Sparse vegetation	Bogs and marshes	Inland water	Evergreen shrubs	Deciduous shrubs	Mixed forest	Forest/field mosaic
Crops, mixed farming	Cambodia	1	1	0	na[a]	1	0	0	1	1	0	0	0	0	0	0	1
	Laos	1	1	0	na[a]	1	0	0	1	1	0	0	0	1	0	0	1
	Myanmar	1	1	0	0	1	0	0	1	1	0	0	0	0	0	0	1
	Thailand	1	1	0	0	1	0	0	1	1	0	0	0	0	0	0	1
	Vietnam	1	1	0	0	1	0	0	1	1	0	0	0	1	0	0	1
	Yunnan	1	1	0	0	1	0	0	1	1	0	0	0	1	0	0	1
Short grass	Cambodia	1	1	0	na[a]	1	1	1	1	0	0	0	0	0	0	1	1
	Laos	1	1	1	na[a]	1	1	1	1	0	0	0	0	1	0	0	1
	Myanmar	1	1	0	0	1	1	0	1	0	0	0	0	0	0	1	1
	Thailand	1	1	0	0	1	1	0	1	0	0	0	0	1	0	1	1
	Vietnam	1	1	1	0	1	1	1	1	0	0	0	0	1	1	1	1
	Yunnan	1	1	1	0	1	1	1	1	0	0	0	0	1	0	1	1

Evergreen needleleaf trees	Cambodia	0	0	1	na[a]	0	0	0	0	0	0	0	0	0	0	0	0
	Laos	1	0	1	na[a]	1	0	0	1	0	0	0	0	1	0	0	0
	Myanmar	0	0	1	0	0	0	0	0	0	0	0	0	0	0	0	0
	Thailand	0	0	1	0	0	0	0	0	0	0	0	0	0	0	0	0
	Vietnam	1	0	1	0	1	0	0	1	0	0	0	0	1	0	1	0
	Yunnan	1	0	1	0	1	0	0	1	0	0	0	0	1	0	1	0
Deciduous needleleaf trees[a]	Myanmar	0	0	0	1	0	0	0	0	0	0	0	0	0	0	0	0
	Thailand	0	0	0	1	0	0	0	0	0	0	0	0	0	0	0	0
	Vietnam	0	0	0	1	0	0	0	0	0	0	0	0	0	0	0	0
	Yunnan	0	0	0	1	0	0	0	0	0	0	0	0	0	0	0	0
Deciduous broadleaf trees	Cambodia	1	1	0	na[a]	1	1	0	1	0	0	0	0	0	0	1	1
	Laos	1	1	0	na[a]	1	1	0	1	0	0	0	0	1	0	0	1
	Myanmar	1	1	0	0	1	1	0	0	0	0	0	0	0	0	1	1
	Thailand	1	1	0	0	1	1	0	1	0	0	0	0	1	0	1	1
	Vietnam	1	1	0	0	1	1	0	1	0	0	0	0	1	0	1	1
	Yunnan	1	1	0	0	1	1	0	1	0	0	0	0	1	0	1	1
Evergreen broadleaf trees	Cambodia	1	1	0	na[a]	1	1	0	1	0	0	0	0	0	0	1	1
	Laos	1	1	0	na[a]	1	1	0	1	0	0	0	0	1	0	0	1
	Myanmar	1	1	0	0	1	1	0	0	0	0	0	0	0	0	1	1
	Thailand	1	1	0	0	1	1	0	1	0	0	0	0	1	0	1	1
	Vietnam	1	1	0	0	1	1	0	1	0	0	0	0	1	0	1	1
	Yunnan	1	1	0	0	1	1	0	1	0	0	0	0	1	0	1	1

(*continued*)

Table 8.5 (*Continued*)

Land-cover type	Domain	Crops, mixed farming	Short grass	Evergreen needleleaf trees	Deciduous needleleaf trees	Deciduous broadleaf trees	Evergreen broadleaf trees	Tall grass	Urban/built-up	Irrigated crops	Sparse vegetation	Bogs and marshes	Inland water	Evergreen shrubs	Deciduous shrubs	Mixed forest	Forest/field mosaic
Tall grass	Cambodia	1	0	0	na[a]	1	1	1	1	0	0	0	0	0	0	1	1
	Laos	0	0	0	na[a]	0	0	1	0	0	0	0	0	0	0	0	0
	Myanmar	0	0	0	0	0	0	1	0	0	0	0	0	0	0	0	0
	Thailand	0	0	0	0	0	0	1	0	0	0	0	0	0	0	0	0
	Vietnam	1	0	1	0	1	1	1	1	0	0	0	0	1	1	1	1
	Yunnan	1	0	1	0	1	1	1	1	0	0	0	0	1	0	1	1
Urban/built-up	MMSEA[b]	0	0	0	0	0	0	0	1	0	0	0	0	0	0	0	0
Irrigated crops	MMSEA[b]	1	1	0	0	0	0	0	1	1	0	0	0	0	0	0	0
Sparse vegetation	MMSEA[b]	0	0	0	0	0	0	0	0	0	1	0	0	0	0	0	0
Bogs and marshes	MMSEA[b]	0	0	0	0	0	0	0	0	0	0	1	0	0	0	0	0
Inland water	MMSEA[b]	0	0	0	0	0	0	0	0	0	0	0	1	0	0	0	0
Evergreen shrubs	Cambodia	0	0	0	na[a]	0	0	0	0	0	0	0	0	1	0	0	0
	Laos	1	1	1	na[a]	1	1	0	1	0	0	0	0	1	0	0	1
	Myanmar	0	0	0	0	0	0	0	0	0	0	0	0	1	0	0	0
	Thailand	1	1	0	0	1	1	0	1	0	0	0	0	1	0	1	1
	Vietnam	1	1	1	0	1	1	0	1	0	0	0	0	1	0	1	1
	Yunnan	1	1	1	0	1	1	0	1	0	0	0	0	1	0	1	1

Deciduous shrubs	Cambodia	0	0	0	na[a]	0	0	0	0	0	0	0	0	0	1	0	0
	Laos	0	0	0	na[a]	0	0	0	0	0	0	0	0	0	1	0	0
	Myanmar	0	0	0	0	0	0	0	0	0	0	0	0	0	1	0	0
	Thailand	0	0	0	0	0	0	0	0	0	0	0	0	0	1	0	0
	Vietnam	1	0	0	0	1	0	0	1	0	0	0	0	0	1	1	0
	Yunnan	0	0	0	0	0	0	0	0	0	0	0	0	0	1	0	0
Mixed forest	Cambodia	1	1	0	na[a]	1	1	0	1	0	0	0	0	0	0	1	1
	Laos	0	0	0	na[a]	0	0	0	0	0	0	0	0	0	0	1	0
	Myanmar	1	1	0	0	1	1	0	1	0	0	0	0	0	0	1	1
	Thailand	1	1	0	0	1	1	0	1	0	0	0	0	1	0	1	1
	Vietnam	1	1	0	0	1	1	0	1	0	0	0	0	1	1	1	1
	Yunnan	1	1	0	0	1	1	0	1	0	0	0	0	1	0	1	1
Forest/field mosaic	Cambodia	1	1	0	na[a]	1	1	0	1	0	0	0	0	0	0	1	1
	Laos	1	1	1	na[a]	1	1	0	1	0	0	0	0	1	0	0	1
	Myanmar	1	1	0	0	1	1	0	1	0	0	0	0	0	0	1	1
	Thailand	1	1	0	0	1	1	0	1	0	0	0	0	1	0	1	1
	Vietnam	1	1	1	0	1	1	0	1	0	0	0	0	1	1	1	1
	Yunnan	1	1	1	0	1	1	0	1	0	0	0	0	1	0	1	1

[a]Indicates land-cover type not present in Cambodia and Laos model domains.
[b]Indicates land-cover transitions for each type coded the same in all six model domains.

Table 8.6 Land-cover conversion elasticity values by country domain

Land-cover type	Cambodia	Laos	Myanmar	Thailand	Vietnam	Yunnan (China)
Crops, mixed farming	0.8	0.5	1	0.75	0.75	1
Short grass	0.8	0.5	1	0.75	0.75	1
Evergreen needleleaf trees	1	1	1	1	1	1
Deciduous needleleaf trees	na[a]	na[a]	1	1	1	1
Deciduous broadleaf trees	1	1	1	1	1	1
Evergreen broadleaf trees	1	1	1	1	1	1
Tall grass	0.8	1	1	1	0.75	1
Urban/built-up	1	1	1	1	1	1
Irrigated crops	1	0.75	1	0.75	1	1
Sparse vegetation	1	1	1	1	1	1
Bogs and marshes	1	1	1	1	1	1
Inland water	1	1	1	1	1	1
Evergreen shrubs	1	0.5	1	0.75	1	1
Deciduous shrubs	1	1	1	1	1	1
Mixed forest	1	1	1	0.75	0.75	1
Forest/field mosaic	0.8	0.5	1	0.75	0.75	1

[a]Indicates land-cover type not present in model domain.

Table 8.7 Land cover (expressed as percent of region) in 2001 and simulation results aggregated to MMSEA region showing overall percent change for 2001–2025 and 2001–2050 time periods

	MMSEA		
	1,731,333 km^2		
Land-cover type	2001	2025	2050
Crops, mixed farming	3.69	+2.29	+2.45
Short grass	2.99	−0.06	+0.42
Evergreen needleleaf trees	0.31	+0.12	+0.01
Deciduous needleleaf trees	0.00	0.00	0.00
Deciduous broadleaf trees	12.82	+1.41	+2.46
Evergreen broadleaf trees	36.58	−2.96	−3.73
Tall grass	5.70	−0.27	−0.44
Urban/built-up	0.69	+0.51	+1.62
Irrigated crops	9.53	+0.56	+0.58
Sparse vegetation	1.00	0.00	0.00
Bogs and marshes	0.11	0.00	0.00
Inland water	0.64	0.00	0.00
Evergreen shrubs	2.30	+0.92	+1.63
Deciduous shrubs	0.6	+0.07	+0.07
Mixed forest	5.13	−0.66	−1.24
Forest/field mosaic	17.9	−1.67	−3.56

- tea and other evergreen shrubs (1.63%, or approximately 28,300 ha);
- urban and other built-up areas, such as those associated with peri-urbanization (1.62%, or approximately 28,300 ha).

These simulated increases take place largely at the expense of the following land-cover groups: (i) evergreen broadleaf trees, which are the most suitable habitat for rubber; and (ii) mixed forests, forest/field mosaics and tall grass – land covers that are historically associated with swidden cultivation. The overall decline in these four land-cover categories is about 9% (approximately 155,300 ha). The predicted decline in native tree cover is somewhat offset by a 4% increase in deciduous broadleaf trees (i.e., rubber), tree crops, tea, and other evergreen shrubs. Figure 8.4 highlights the areas in the MMSEA region where simulated changes occur during the 2001–2025 and 2001–2050 time periods. Again, these results only include conversions between the modeled land-cover classes used in the study; other conversions that do not change the classification of land cover, e.g., among crop types or between residential and industrial areas, are not counted.

Among the six countries in MMSEA, Cambodia has the least amount of land within the region (59,579 km^2) and Myanmar the most (462,495 km^2). Laos, Thailand, and Vietnam have roughly the same amount of land in the region (283,363 to 303,093 km^2) (Table 8.8). Model results suggest that Myanmar will undergo the least amount of change throughout the 50-year period (less than 10%). This is evident in the map of change/no change for the 2001–2050 time period in Figure 8.4. This result, as well as that from Cambodia, however, should be interpreted cautiously as we were only able to interview one expert on the future of land-cover change in those countries. In Laos, Thailand, and Vietnam, the model suggests that land-cover changes will range from 9.38 to 11.9% during the first 25-year period, and 14.46 to 19.17% of the landscape during the 50-year period. These robust results suggest land-cover experts in these countries see relatively similar futures. In Cambodia and Yunnan, the model suggests that land-cover change will affect approximately one-quarter of the upland landscape over the 50-year period. Overall, the model simulated change across the entire region of approximately 10% of the landscape over the first 25-year period and 16% over the 50-year period.

8.4 Discussion and conclusions

This project used the CLUE-s model to simulate LCLUC patterns that might emerge in MMSEA in 2025 and 2050 based upon an amalgam of (i) expert knowledge of historical LCLUC; (ii) expert knowledge and observations of current regional trends and narratives of land-cover change that included consideration of agricultural intensification, road development, and market growth in the region; and (iii) economic forecasts. Land-change simulations are at best future explorations, and true validation of future projections is not possible (Wassenaar *et al.*, 2007). We were not able to validate these models in a quantitative manner because of a lack of historical land-cover data at multiple time points using similar classification techniques and consistent land-cover types for MMSEA. Qualitative assessments by experts from each country indicated the simulated land-cover patterns were reasonable representations of the scenario within each country domain. Visual inspection of the output maps revealed some localized anomalies in the location of individual grid cells of particular land-cover types. However, when the six country domain simulation maps were appended, the resulting land-cover patterns along and across borders between domains were consistent, with no artificial or unexpected abrupt changes in land-cover patterns at national borders.

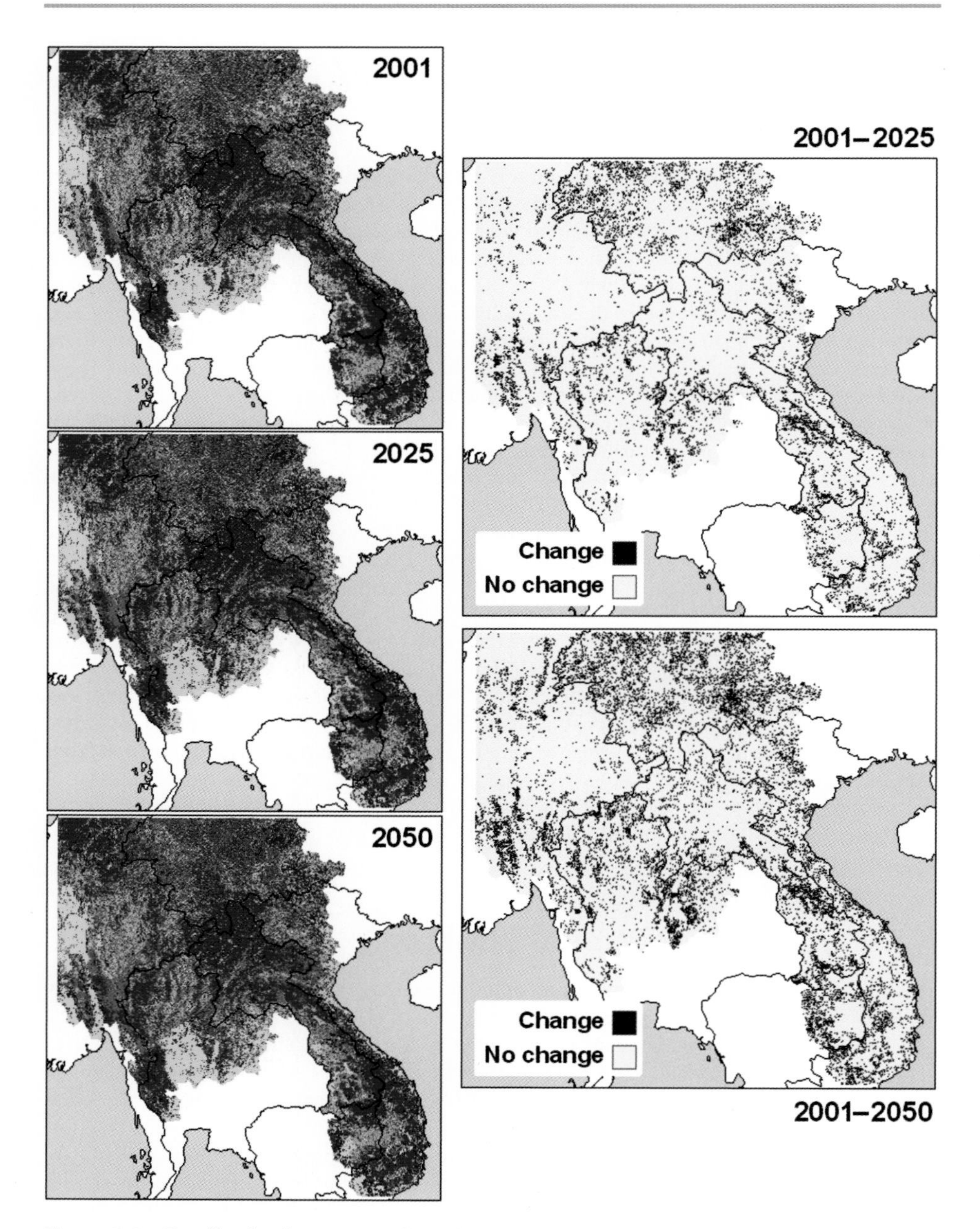

Figure 8.4 Baseline land-cover map (2001), simulation output maps for years 2025 and 2050, and maps showing areas of change/no change for 2001–2025 and 2001–2050 time periods

Table 8.8 Cumulative land-cover change for the six countries in MMSEA and the region in terms of area (km²) and percent calculated in 5-year increments over the 25- and 50-year periods

		2001–2025		2001–2050	
Country	Total upland area (km²)	Cumulative area change (km²)	Cumulative % change	Cumulative area change (km²)	Cumulative % change
Cambodia	59,579	6142	10.31	14,963	25.11
Laos	283,363	26,589	9.38	40,982	14.46
Myanmar	462,495	27,096	5.86	44,819	9.69
Thailand	303,093	28,583	9.43	50,032	16.51
Vietnam	285,271	33,954	11.90	54,694	19.17
Yunnan	337,532	49,189	14.57	78,560	23.27
MMSEA	1,731,333	171,553	9.91	284,050	16.41

We attribute the positive results we obtained to quality input datasets and model testing; plausible, scenario-based land-cover demands that reflect regional expert knowledge of historical, current, and future trends in policies (and how those policies influence land-cover change); expert consensus regarding the future of MMSEA; and the ability of the CLUE-s model to produce realistic, complex, non-linear land-cover patterns. While policies are not explicit inputs to the models, the exception being the change in restrictions placed on parks and protected areas, policy is a primary consideration in determining the future land-cover demands.

Our estimates of total cumulative LCLUC in the six countries of the MMSEA simulation region ranged from 5.86 to 14.57% during the first 25-year period, and from 9.69 to 25.11% during the 50-year period from 2001 to 2050 (Table 8.8 and Figure 8.4). These changes are two to three times greater than the 5 to 8% simulated change predicted to take place in Europe between 2000 and 2030 (Verburg *et al.*, 2006). The models suggest that MMSEA may lose as much as 9% of native cover (secondary trees, shrub, and grass) by 2050, but this loss will be somewhat offset by the increase in rubber and other tree crops as well as tea and other evergreen shrubs (about 4% in total). These predicted changes in the composition of the tree cover have implications for watershed hydrology and regional precipitation patterns.

At the watershed level, Guardiola-Claramonte *et al.* (2008) showed that an increase in rubber cropping caused a decrease in soil water availability in Xishuangbanna Prefecture, China, during the dry season. While the root-water uptake of the other land covers (secondary forest, shrub, tea) declined throughout the dry season, water extraction in 1- and 2-m soil layers under rubber increased sharply in mid-February and remained high through March. This response to decreasing soil water availability coincided with the annual leaf shedding and new leaf-flushing period. Guardiola-Claramonte *et al.* (2008) suggest that conventional land-atmosphere models, which generally modulate evapotranspiration (ET) as a function of soil moisture, are not appropriate for predicting the hydrological impacts of expanding rubber cultivation.

At the regional scale, climate change models using these scenarios predict little change in regional precipitation patterns (Sen *et al.*, 2012). Sen's work suggests that global warming-related effects will have a greater impact on precipitation than our predicted 16.41% changes in land cover. Much of the uncertainty regarding the hydrological impacts of tropical deforestation stems from a failure to develop and use realistic LCLU projections in climate simulations (Giambelluca *et al.*, 1996). This study clearly demonstrates the need to integrate realistic LCLUC simulations based on expert opinion with climate change simulations.

In terms of the socioeconomic implications of the expansion of rubber throughout this region, Fox and Castella (2010) found that the largest rubber-producing countries in the world (Thailand, Vietnam, Indonesia, Malaysia, and India) have all made conscious institutional decisions to support smallholder rubber production. The reasons for this vary but are linked to ongoing land reform policies in the different countries, interest from smallholders in establishing rubber, and the inability of governments to control large estates. Drawing upon the experiences of Xishuangbanna and northeast Thailand, Fox and Castella conclude that smallholder rubber production is a viable and effective proposition in moving households and communities out of poverty. In Laos, Cambodia, Myanmar, and Vietnam, however, despite the drafting of laws and ordinances that could assist smallholders to maintain control over their land and to invest in commercial crops such as rubber, a lack of financial and human resources and competing policy and political agendas have prevented these laws from being effectively implemented. Consequently in many communities villagers are selling their land and migrants are moving into the areas; and in other communities farmers are struggling to maintain community lands and forests in the face of growing pressures from investors and government institutions to impose concession arrangements. In order to promote the establishment of a vibrant smallholder rubber sector the state needs to establish and effectively implement national policies and institutional structures to support smallholder rubber cultivators.

Acknowledgements

This research was supported through NASA Grant # NNG04GH59G (Project ID: IDS/03-0365-0079) and Asia Pacific Network grants ARCP2006-06NMY & ARCP2007-01CMY.

References

Castella, J., Verburg, P.H. (2007) Combination of process-oriented and pattern-oriented models of land-use change in a mountain area of Vietnam. *Ecological Modelling* **202**: 410–20.

Dickinson, R.E., Henderson-Sellers, A., Kennedy, J.P. (1993) Biosphere-atmosphere transfer scheme (bats) version 1e as coupled to the NCAR community climate model. Technical report. National Center for Atmospheric Research.

Economic Intelligence Unit (2004a) *Country Profile 2004: Myanmar (Burma)*. London: The Economic Intelligence Unit Ltd.

Economic Intelligence Unit (2004b) *Country Profile 2004: Vietnam*. London: The Economic Intelligence Unit Ltd.

Economic Intelligence Unit (2005a) *Country Profile 2005: Cambodia*. London: The Economic Intelligence Unit Ltd.

Economic Intelligence Unit (2005b) *Country Profile 2005: China*. London: The Economic Intelligence Unit Ltd.

Economic Intelligence Unit (2005c) *Country Profile 2005: Laos*. London: The Economic Intelligence Unit Ltd.

Economic Intelligence Unit (2005d) *Country Profile 2005: Thailand*. London: The Economic Intelligence Unit Ltd.

Engelsman, W. (2002) Simulating land-use changes in an urbanizing area in Malaysia: An application of the CLUE-S model in the Selangor river basin. Research thesis paper, Department of Environmental Sciences, Wageningen University, The Netherlands.

Fox, J., Castella, J.C. (2010) Expansion of rubber (*Hevea brasiliensis*) in mainland Southeast Asia: What are the prospects for small holders? Paper presented at the Conference on Revisiting Agrarian Transformations, 13–15 May 2010, Chiang Mai, Thailand.

Fox, J., Vogler, J. (2005) Land-use and land-cover change in montane mainland Southeast Asia. *Environmental Management* **36**: 394–403.

Fox, J., McMahon, D., Poffenberger, M., Vogler, J. (2008) *Land Use and Tenure Change in Ratanakiri: 1989–2007*. Phnom Penh, Cambodia: Community Forestry International.

Friedl, M.A., McIver, D.K., Hodges, J.C.F., *et al.* (2002) Global land cover mapping from MODIS: algorithms and early results. *Remote Sensing of Environment* **83**: 287–302.

Fujita, Y., Phanvilay, K. (2008) Land and forest allocation in Lao People's Democratic Republic: Comparison of case studies from community-based natural resource management research. *Society and Natural Resources* **21**: 120–33.

Giambelluca, T.W., Tran, L.T., Ziegler, A.L., Menard, T.P., Nullet, M.A. (1996) Soil-vegetation-atmosphere processes: Simulation and field measurement for deforested sites in northern Thailand. *Journal of Geophysical Research (Atmospheres)* **101**: 25867–25885.

Guardiola-Claramonte, M., Troch, P.A., Ziegler, A.D., Giambelluca, T.W., Vogler, J.B., Nullet, M.A. (2008) Local hydrologic effects of introducing non-native vegetation in a tropical catchment. *Ecohydrology* **1**: 13–22; doi:10.1002/eco.3.

Guo, H., Padoch, C., Coffey, K., Aiguo, C., Yongneng, F. (2002) Economic development, land use and biodiversity change in the tropical mountains of Xishuangbanna, Yunnan, Southwest China. *Environmental Science and Policy* **5**: 471–9.

Lebel, L. (2006) Multi-level scenarios for exploring alternative futures for upper tributary watersheds in mainland Southeast Asia. *Mountain Research and Development* **26**: 263–73.

Li, H.M., Aide, T.M., Ma, Y.X., Liu, X.J., Cao, M. (2007) Demand for rubber is causing the loss of high diversity rainforest in SW China. *Biodiversity and Conservation* **16**: 1731–45.

Maekawa, T. (2002) Road link makes dream possible. *ADB Review* Nov–Dec: 9–10.

Padoch, C., Coffey, K., Mertz, O., Leisz, S., Fox, J., Wadley, R. (2007) The demise of swidden in Southeast Asia? Local realities and regional ambiguities. *Danish Journal of Geography* **107**: 29–42.

Rigg, J. (2006) Land, farming, livelihoods and poverty: Rethinking the links in the Rural South. *World Development* **34**: 180–202.

Rigg, J., Nattapoolwat, S. (2001) Embracing the global in Thailand: Activism and pragmatism in an era of de-agrarianisation. *World Development* **29**: 945–60.

Schneider, R.R. (1995) Government and the economy on the Amazon frontier. In: *Latin America and the Caribbean Technical Dept*. Washington, DC: World Bank.

Sen, O.L., Bozkurt, D., Vogler, J.B., Fox, J., Giambelluca, T.W., Ziegler, A.D. (2012) Hydroclimatic effects of future land-cover/land-use change in montane mainland Southeast Asia. *Climate Change* doi:10.1007/s10584-012-0632-0.

Soepboer, W. (2001) CLUE-s: an application for Sibuyan Island, the Philippines. Research thesis paper, Department of Environmental Sciences, Wageningen University, The Netherlands.

Sturgeon, J. (2010) Rubber transformations: Post-socialist livelihoods and identities for Akha and Dai farmers in Xishuangbanna, China. In: Michaud, J., Forsyth, T. (eds), *Moving Mountains: Ethnicity and Livelihoods in Highland China, Vietnam, and Laos*. Vancouver: University of British Columbia Press.

Sunderlin, W., Huynh, T.B. (2005) *Poverty Alleviation and Forests in Vietnam*. Jakarta: CIFOR, 73 pp.

Thant, M., Nair, O. (2002) Linking nations coast to coast. *ADB Review* Nov–Dec: 14–16.

Thomas, D., Ekasingh, B., Ekasingh, M., *et al.* (2008) *Comparative Assessment of Resource and Market Access of the Poor in Upland Zones of the Greater Mekong Region*. Chiang Mai, Thailand: World Agroforestry Center, 350 pp.

Thongmanivong, S., Fujita, Y., Fox, J. (2005) Resource use dynamics and land-cover change in Ang Nhai Village and Phou Phanang National Reserve Forest, Lao PDR. *Environmental Management* **36**: 382–93.

UNEP-WCMC (2005) World Database on Protected Areas. URL: http://www.wdpa.org/ [accessed 6 May 2014].

Veldkamp, A., Fresco, L. (1996) CLUE-CR: An integrated multi-scale model to simulate land use change scenarios in Costa Rica. *Ecological Modeling* **91**: 231–48.

Veldkamp, A., Fresco, L. (1997a) Exploring land use scenarios: An alternative approach based on actual land use. *Agricultural Systems* **55**: 1–17.

Veldkamp, A., Fresco, L. (1997b) Reconstructing land use drivers and their spatial scale dependence for Costa Rica (1973 and 1984). *Agricultural Systems* **55**: 19–43.

Verburg, P.H. (2006) Conversion of land use and its effects (CLUE) model. In: Geist, H. (ed.), *Our Earth's Changing Land: An Encyclopedia of Land-Use and Land-Cover Change*, Vol. **1** (A–K). Westport, CT: Greenwood Press, pp. 144–6.

Verburg, P.H., Veldkamp, A. (2004) Projecting land use transitions at forest fringes in the Philippines at two spatial scales. *Landscape Ecology* **19**: 77–98.

Verburg, P.H., de Koning, G.H.J., Kok, K., Veldkamp, T.A., Bouma, J. (1999a) A spatial allocation procedure for the modeling the pattern of land use change based upon actual land use. *Ecological Modelling* **116**: 45–61.

Verburg, P.H., Veldkamp, A., Fresco, L.O. (1999b) Simulation of changes in the spatial pattern of land use in China. *Applied Geography* **19**: 211–33.

Verburg, P.H., Soepboer, W., Veldkamp, A., Limpiada, R., Espaldon, V., Mastura, S.S.A. (2002) Modeling the spatial dynamics of regional land use: the CLUE-S model. *Environmental Management* **30**: 391–405.

Verburg, P.H, de Groot, W.T., Veldkamp, A. (2003) Methodology for multi-scale land-use change modelling: concepts and challenges. In: *Global Environmental Change and Land Use* Dolman, A.J., Verhagen, A., Rovers, C.A. (eds), Dordrecht/Boston/London: Kluwer Academic Publishers, pp. 17–51.

Verburg, P.H., Schulp, C.J.E., Witte, N., Veldkamp, A. (2006) Downscaling of land use change scenarios to assess the dynamics of European landscapes. *Agriculture, Ecosystems and Environment* **114**: 39–56.

Wassenaar, T., Gerber, P., Verburg, P.H., Rosales, M., Ibrahim, M., Steinfeld, H. (2007) Projecting land use changes in the Neotropics: The geography of pasture forest expansion. *Global Environmental Change* **17**: 86–104.

Willemen, L. (2002) Modelling of land-cover changes with CLUE-s in Bac Kan province, Vietnam. Research thesis paper, Department of Environmental Sciences, Wageningen University, The Netherlands.

Xu, J.C. (2006) The political, social and ecological transformation of a landscape: The case of rubber in Xishuangbanna, China. *Mountain Research and Development* **26**: 254–62.

Xu, J., Fox, J., Zhang, P., *et al.* (2005) Land-use/land-cover change and farmer vulnerability in Xishuangbanna. *Environmental Management* **36**: 404–13.

9 Land-use Change and its Impacts on Agricultural Productivity in China

Huimin Yan, He Qing Huang, Xiangzheng Deng and Jiyuan Liu

Institute of Geographic Sciences and Natural Resources Research, Chinese Academy of Sciences, Beijing, China

9.1 Introduction

Land use activities, primarily for agricultural expansion and economic growth, have transformed one-third to one-half of our planet's land surface in the form of forest clearance, agricultural practices and urban expansion, exerting profound impacts on ecosystem services, food production and the eco-environment (Vitousek *et al.*, 1997; Foley *et al.*, 2005). In China, food security has always been a concern because of lack of cropland, increasing population and water shortage (Tao *et al.*, 2009). Over the past 50 years, remarkable achievements in agricultural production have occurred, although China is facing a great challenge in feeding the world's largest population with a cropland per capita far below the world average (Chen, 2007). Cereal production has increased steadily with an annual growth rate of 3.7%, which is substantially higher than the world mean growth rate of 2% during the same period (Fan *et al.*, 2012). Although the increase in cereal production mainly resulted from yield increases, some has been attributed to cropland expansion, especially in northeast and northwest China (Liu *et al.*, 2005a, 2010; Wang *et al.*, 2009).

The land system in China has been experiencing dramatic modifications due to the unprecedented combination of economic and population growth since the early 1980s (Chen, 1999; Liu *et al.*, 2003). The loss of cultivated land due to widespread construction is one of the most serious problems affecting China's food security (FAO, 2002). As

Vulnerability of Land Systems in Asia, First Edition. Edited by Ademola K. Braimoh and He Qing Huang.

a result, considerable concerns have arisen in recent years about the impacts of land use change and whether or not China can maintain food self-sufficiency (Brown, 1995; Rozelle and Rosegrant, 1997; Heilig, 1999; Feng *et al.*, 2005). Owing to the spatial heterogeneity of land use change caused by highly diverse natural and socio-economic conditions across China, accurate information on cropland area and spatially explicit quantification of agricultural productivity are both of critical importance for policy makers to assess the impacts of land use modification on national food security (Fischer *et al.*, 1998; Verburg *et al.*, 2000).

Satellite remote sensing technology has played a significant role in monitoring and assessing changes in land use and land cover at large scales over the past two decades. Derived from Landsat TM images taken during the period 1990–2000, China's National Land-Use/Land-Cover Dataset (NLCD hereafter) with 1:100,000 mapping scale developed by the Chinese Academy of Sciences (CAS) provides an unambiguous and spatially explicit data source for understanding the degree to which the whole nation's cropland has been converted. Using the dataset, the area and spatial pattern of cropland transformation during the 1990s was analyzed (Liu *et al.*, 2005a,b), focusing on five widely occurring land transformation types: conversions of cropland to urban area; cropland to forest; cropland to grassland; grassland to cropland; and forest to cropland. In addition to the determination of the quantity of cropland transformation through analyzing land use change data from NLCD, the influences of cropland transformation on the agricultural productivity of China during the 1990s have been evaluated in terms of the net primary productivity (Yan *et al.*, 2009) and the potential outputs of cropland (Deng *et al.*, 2006).

9.2 Land-use data

Satellite sensors, such as Landsat TM, and the French SPOT system, have been used successfully for measuring deforestation, biomass burning and other land cover changes, including the expansion and contraction of deserts (Skole & Tucker, 1993). Remote sensing techniques have also been used widely to monitor the conversion of agricultural land to infrastructure (Palmera & Lankhorst, 1998; Woodcock *et al.*, 2001; Milesi *et al.*, 2003; Ogud *et al.*, 2003).

To characterize the spatial and temporal patterns of land use/land cover changes across China during the 1990s, the Chinese Academy of Sciences initiated a nationwide land cover/land use project in the late 1990s. Land use datasets for the time around the end of the 1980s (NLCD80) and the end of the 1990s (NLCD90) with a mapping scale of 1:100,000 were developed. NLCD80 was derived from Landsat TM images at the end of the 1980s and/or the beginning of the 1990s, and NLCD90 was produced from remotely sensed images taken during 1999 to 2000. According to the land cover classification system used for NLCD, land cover is categorized into six types: cropland, forest, grassland, water bodies, unused land, and built-up land including urban areas (Liu *et al.*, 2002, 2003, 2005a).

The accuracy of NLCD dataset was evaluated with about 8000 pictures taken by cameras equipped with a global positioning system (GPS) from an extensive field survey along transects extending for more than 75,000 km across China. The accuracy of TM images derived from cropland is 94.9%, while the identification of built-up areas has the highest accuracy of 96.3%. For forest and grassland, accuracies of 90.1% and 88.1% are attained, respectively. A more detailed description of the data sources and methods used for developing the NLCD dataset and for evaluating the dataset's accuracy are given in the work of Liu *et al.* (2002, 2003, 2005a).

9.3 Methods for estimating changes in agricultural productivity

In addition to estimating the quantity of cultivated land conversions, there are two ways to estimate changes in the productivity of cultivated land, which can be measured with the summed net primary productivity (NPP), defined as the production over a crop area in each geographic unit, with the units of g C for pixels or Mt C (10^{12} g C) at a regional level, and the total potential outputs of cropland at a provincial level. One way is to estimate changes in NPP, and another way is to evaluate changes in the potential outputs of cultivated land. For the former, the GLO-PEM model is deployed, whereas for the latter the Agro-ecological Zones (AEZ) model is applied.

9.3.1 NPP estimation with the GLO-PEM model

Satellite remote sensing directly detects actual vegetation dynamics top-down over a large area, and the observations provide a significant data source for quantifying vegetation productivity. Measures of photosynthetic production from remote sensing observations, typically the NPP, provides a common unit of production among different crop types, and therefore has been treated as a useful 'common currency' of land productivity for quantifying the impacts of land transformation (Bounoua *et al.*, 2003).

For establishing the link between satellite data and NPP, a production efficiency model (GLO-PEM) has been developed to estimate plant photosynthetically active radiation (PAR) absorption, light use efficiency and autotrophic respiration (Prince, 1991; Prince & Goward, 1995; Cao *et al.*, 2004). GLO-PEM was designed to run with both biological and environmental variables derived from satellite images (Prince & Goward, 1995; Goetz & Prince, 2000). Hence, it is a suitable tool to incorporate the effects of both climate change and crop management measures on crop growth activity into NPP estimation at the same resolution as satellite data. GLO-PEM consists of linked components that describe the processes of canopy radiation absorption, utilization, autotrophic respiration, and the regulation of these processes by environmental factors such as temperature, water vapour pressure deficit and soil moisture (Prince & Goward, 1995; Goetz & Prince, 2000; Cao *et al.*, 2004). In the model, the following relationship is used to determine NPP:

$$\mathrm{NPP} = \sum_{t} \left[(S_t \cdot N_t)\, \varepsilon_g - R \right] \tag{9.1}$$

where S_t is the incident PAR in time t, N_t is the fraction of incident PAR absorbed by the vegetation canopy (Fapar) at time t (as demonstrated by Prince & Goward, 1995; N_t can be calculated as a linear function of the Normalized Difference Vegetation Index (NDVI)), ε_g is the light utilization efficiency of PAR absorbed by vegetation within the context of gross primary production, and R is the autotrophic respiration calculated as a function of standing above-ground biomass, air temperature and photosynthetic rate (Prince & Goward, 1995; Goetz & Prince, 2000).

Plant photosynthesis depends on both the capacity of photosynthetic enzymes to assimilate CO_2 (Farquhar *et al.*, 1980; Collatz *et al.*, 1991) and the stomatal conductance for CO_2 from the atmosphere into the intercellular spaces (Gollan *et al.*, 1992; Harley *et al.*, 1992). The two factors are affected by environmental factors, such as air temperature, water vapour pressure deficit (VPD), soil moisture and atmospheric CO_2 concentration. For this, the following relationship can be used to determine ε_g:

$$\varepsilon_g = \varepsilon_g^* \cdot \sigma \tag{9.2}$$

where ε_g^* is the maximum possible light utilization efficiency of PAR absorbed by vegetation determined by photosynthetic enzyme kinetics, which is a function of photosynthetic pathway, temperature and CO_2/O_2 ratio, and σ is the reduction of ε_g^* caused by environmental factors that control stomatal conductance and can be determined from the following relationship:

$$\sigma = f(T) f(\delta q) f(\delta \theta) \tag{9.3}$$

where $f(T), f(\delta q)$ and $f(\delta\theta)$ represent the stresses of air temperature, VPD and soil moisture to stomatal conductance, respectively.

GLO-PEM is driven with the Pathfinder AVHRR (Advanced Very High Resolution Radiometer) data at resolutions of 8 km and 10 days derived from AVHRR sensors aboard the National Ocean and Atmosphere Administration (NOAA) satellites; algorithms to calculate NDVI, Fapar (fraction of absorbed photosynthetically active radiation), biomass, air temperature and VPD from the AVHRR data are included in the model (Prince & Goward, 1995; Goetz & Prince, 2000). Detailed descriptions of the algorithms for determining the interrelationship among model variables and satellite data preprocessing are given in the studies of Prince and Goward (1995), Goetz and Prince (2000), and Cao *et al.* (2004). In this study, we used GLO-PEM and the NOAA/NASA Pathfinder AVHRR Land (PAL) data at resolutions of 8 km and 10 days to estimate NPP for the periods at the end of the 1990s (1998 and 1999) and the end of the 1980s (1988 and 1989) for the purpose of matching with land use change data in temporal dimensions.

9.3.2 Agro-ecological zones (AEZ) model

The Food and Agriculture Organization of the United Nations (FAO), in collaboration with the International Institute for Applied Systems Analysis (IIASA), has developed a commonly used method for calculating the potential output of cropland, the agro-ecological zones (AEZ) methodology. The AEZ methodology can serve as an evaluative framework for biophysical limitations and production potential of major food and fibre crops under various levels of inputs and management scenarios at global and regional scales (Keyzer, 1998; Fischer *et al.*, 2000, 2005; Heilig *et al.*, 2000; Fischer and Sun, 2001; Albersen *et al.*, 2002). In its simplest form, the AEZ framework contains three elements: selected agricultural production systems with defined input/output relationships; geo-referenced land resources data (including climate, soil and terrain data); and procedures for calculating potential yields, matching environmental requirements for crop (by land units and grid cells) with the corresponding environmental characteristics available in the land resources database. The LUC (Land Use Change) group of IIASA has applied the AEZ methodology to assess the potential output of China's cultivated land.

9.3.3 Calculating agricultural productivity change caused by land use change

Changes in agricultural production originate from changes either in cropland area or in crop production per unit land area. In this study, total agricultural production is defined as the multiple of crop area and NPP in the form of:

$$P = A \cdot NPP = c \cdot A_g \cdot NPP \tag{9.4}$$

where P is the total agricultural production, NPP is the production per unit crop area (g C/m^2/year), and A is the cropland area equal to the multiple of the proportion of cropland in each pixel (c) and pixel area (A_g= 64 km^2). In calculating changes in P, this study takes the value of NPP in 1988 and 1989 as the NPP level at the end of the 1990s (NPP_1) so as to maintain consistency with the land cover change data. Similarly, the average value of NPP in 1998 and 1999 is regarded as the NPP level at the end of the 1980s (NPP_2).

To evaluate the effect of cropland area changes on agricultural production, the following method proposed by Hicke *et al.* (2004) is applied to determine the fractional contributions of cropland area (A) and/or NPP to P:

$$\Delta A = A_2 - A_1 \tag{9.5}$$

$$\Delta NPP = NPP_2 - NPP_1 \tag{9.6}$$

$$\begin{aligned} \Delta P = P_2 - P_1 &= A_2 \cdot NPP_2 - A_1 \cdot NPP_1 \\ &= (A_1 + \Delta A) \cdot (NPP_1 + \Delta NPP) - A_1 \cdot NPP_1 \\ &= \Delta NPP \cdot A_1 + \Delta A \cdot NPP_1 + \Delta A \cdot \Delta NPP \end{aligned} \tag{9.7}$$

where A_1 and A_2 represent the cropland areas at the ends of the 1980s and 1990s, respectively. Rearranging Equation 9.7 and dividing ΔP by P_1 results in:

$$\Delta P/P_1 = \Delta A/A_1 + \Delta NPP/NPP_1 + \Delta A \cdot \Delta NPP/P_1 \tag{9.8}$$

Thus, the fractional increase in P is the sum of the fractional increases in area A and NPP plus an interaction term.

9.4 Agricultural productivity change caused by cropland transformation

9.4.1 Cropland transformation

Using the methods and data described above, our analysis shows that China's cropland had a net increase of 2.79 Mha during the 1990s, with 5.97 Mha of newly cultivated cropland and 3.18 Mha of existing cropland modified for other uses, typically for urban expansion, grassland and forest.

Figure 9.1 shows the spatial distribution of cropland transformation during the 1990s. Of the land that was converted from cultivated area, the most – about 38% – was converted to built-up areas. In addition, 17% of the cultivated area was converted to forestry, 30% into grasslands and 16% to others. Among the areas newly converted to cultivated land, 55% came from grasslands, 28% from forested areas, and around 20% from a combination of wetlands, the reclamation of unused land and other uses. China's newly cultivated cropland is mainly distributed in the Northeast Region (Zone I), Inner Mongolia Region (Zone II) and Gan-Xin Region (Zone VIII). In contrast, the cropland that was modified for other uses is largely in southern China.

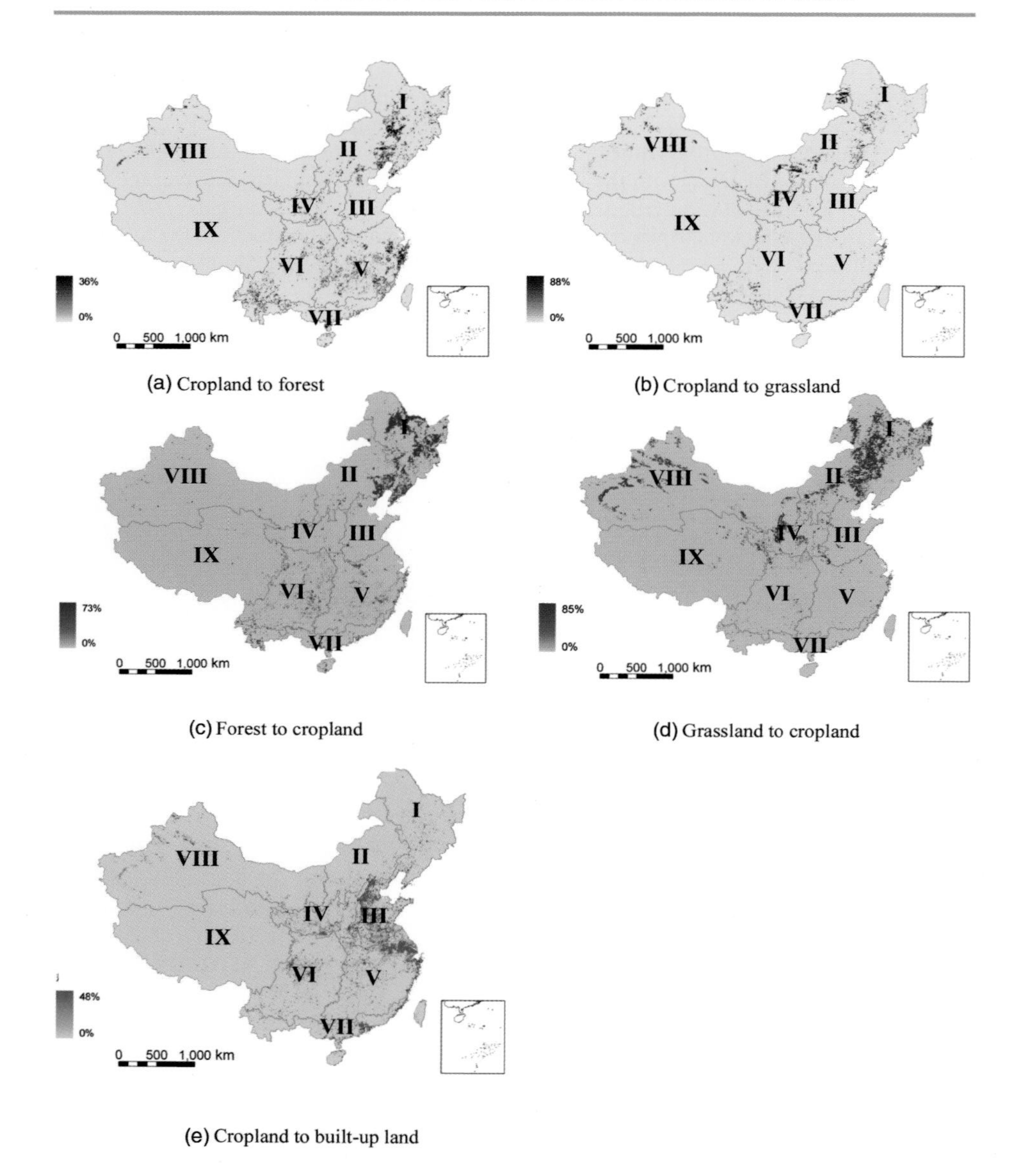

Figure 9.1 Spatial distribution of cropland transformation in China during the 1990s

9.4.2 Contributions of cropland area change to agricultural productivity

Given that China's total agricultural productivity increased by 150.16 Mt/year during the 1990s, it can be calculated from Equation 9.8 that 6.96 Mt/year of the productivity can be attributed to cropland transformation due to the increased cropland area, while 142.99 Mt/year results from the increased NPP, and a negligible fraction is due to the area–NPP

Table 9.1 Effects of changes in cropland area and net primary productivity (NPP) on total production (unit: Mt C)

Zone[a]	Agricultural region	Cropland transformation	NPP	NPP–area interaction
I	Northeast Region	11.78	9.78	0.91
II	Inner Mongolia and the Great Wall Region	2.05	7.20	0.28
III	Huang-Huai-Hai Region	−2.31	32.09	−0.38
IV	Loess Plateau Region	0.12	−0.45	−0.003
V	Middle and lower reaches of the Yangtze River drainage basin	−3.23	57.35	−0.54
VI	Southwest Region	−0.25	21.33	−0.03
VII	South China Region	−1.33	1.59	−0.09
VIII	Gan-Xin Region	0.08	−0.22	0.08
IX	Tibet Region	0.08	−0.002	−0.003
	All China	6.96	142.99	0.21

[a]Refer to Figure 9.2 for map of zones.

interaction (Table 9.1). The spatial variation patterns of total agricultural productivity and its two fractions (area-induced and NPP-induced) across China in the 1990s are presented in Figure 9.2. Although land conversion accounted for the crop production increase in northern China and the decrease in southern China, its effect on the total agricultural production is spatially scattered due to the NPP-induced total agricultural production change. Most of the cropland with reduced agricultural production is located in northwestern China, besides some areas in the deltas of the Yangtze River and the Pearl River (parts of Zones V and VII) (Figure 9.2a). By comparing the maps of the variation in agricultural productivity induced respectively by changes in cropland area (Figure 9.2b) and NPP (Figure 9.2c), the relative contributions of cropland area and NPP to the change in agricultural productivity can be determined. In the deltas of the Yangtze and Pearl rivers, land conversion exerted strong negative effects, making the total production decrease significantly, while positive effects of land conversion occurred on the western side of Song-Nen Plain (part of Zone I). The negative effects of land conversion in most southern regions were offset by the increased NPP. In the Loess Plateau Region (Zone IV), however, the newly cultivated cropland was not sufficient to compensate for the loss of productivity caused by reduced NPP.

It can also be seen in Figure 9.2 that the area-induced decrease in agricultural production occurred largely in the northern and eastern regions of China where NPP showed an increasing trend, typically in Huang-Huai-Hai Plain Region (Zone III) and the Yangtze River drainage basin (Zone V). In the western part of the Song-Nen Plain (part of Zone 1), however, NPP decreased although a large amount of land was newly cultivated.

Reduction in cropland area occurred widely in Huang-Huai-Hai Plain Region (Zone III), the lower and middle areas of the Yangtze River (Zone V), South China Region (Zone VII) and the Southwest Region (Zone VI). The cropland area in the four regions had decreased by 4684, 5174, 587 and 1963 km^2, respectively, which reduced the total production by 2.32, 3.24, 0.26 and 1.33 Mt C, correspondingly. Nevertheless, the increased NPP in the four regions not only offset the negative effects of cropland transformation but also enhanced crop production substantially, reflected in total production improvements of

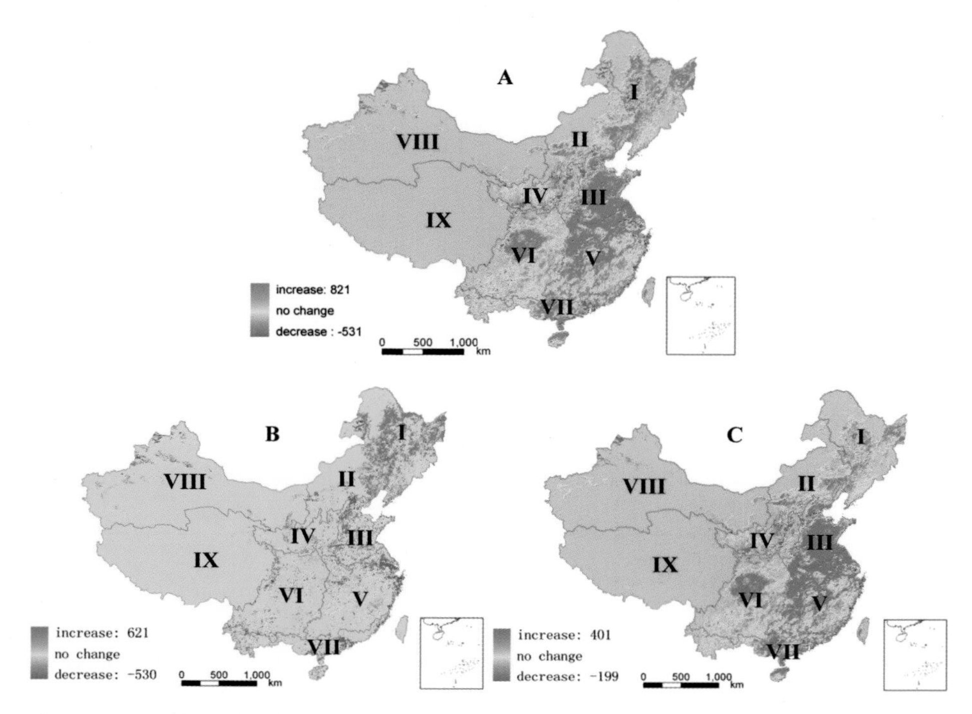

Figure 9.2 Effects of changes in land use and resultant net primary productivity (NPP) on total productivity in China during the 1990s: (a) total productivity change; (b) land use transformation-induced change; and (c) NPP-induced change. Adapted from Lebel, 2006

18.6%, 23.7%, 12.7% and 15.2%, respectively. In the Loess Plateau Region (Zone IV) and Gan-Xin Region (Zone VIII), the land resources are generally of poor quality with significant soil erosion, lower NPP and susceptibility to degradation. Although there were 1376 and 3389 km^2 of newly cultivated cropland in the two regions, respectively, the reduced NPP still made total production fall slightly. In the Northeast Region (Zone I) and Inner Mongolia Region (Zone II), both cropland area and NPP exerted positive effects, but the effects of newly cultivated cropland were more significant than the increase in NPP in the Northeast Region (Zone I).

9.4.3 Agricultural productivity change caused by major land use change types

Among the various forms of land transformation, there are five types that played a dominant role: the conversions of cropland to urban area, forest and grassland, of grassland to cropland, and of forest to cropland (Liu *et al.*, 2005b). During 1990–2000, urbanization caused crop production to decrease by 7.84 Mt, while transformation of cropland into forest and grassland reduced crop production by 2.64 and 1.69 Mt C, respectively. Conversely, the increase in crop production from newly cultivated cropland transformed from forest and grassland reached 8.34 and 10.13 Mt C, respectively.

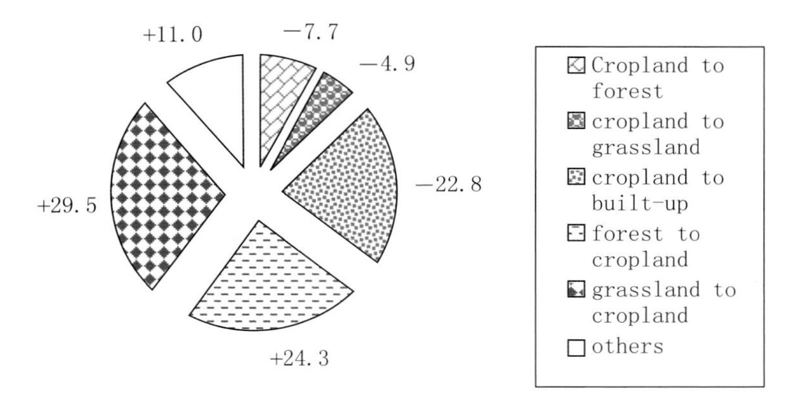

+ indicates land use changes make agricultural productivity increase

− indicates land use changes make agricultural productivity decrease

Figure 9.3 Impacts of different forms of land use change on total agricultural productivity in China during the 1990s

Among the above mentioned five types of land use change, the increase in grassland area exerted the most significant positive effect on food production, followed by the increase in cropland transformed from forest. Meanwhile, urban sprawl appears to be the most influential of human activities, having a negative impact on crop production almost double that of the loss from the transformation of cropland to forest and to grassland (Figure 9.3). Of the resulting productivity changes due to cropland transformation, 22.8% was lost to urbanization, while 29.5% was gained from the newly cultivated land transformed from grassland. However, all the cropland lost to urbanization accounts for only 16.6% of the total transformed area, while the total cropland transformed from newly cultivated grassland constitutes 38.8% of the total transformed cropland area. Generally, the occupation of 1 ha of cropland by urban development requires the cultivation of 1.8 ha grassland to offset the resulting production loss.

9.4.4 Changes in potential agricultural productivity due to cropland conversions

Using the results from the AEZ (agro-ecological zone) model in conjunction with land use data, the total potential agricultural productivity change due to cropland conversion was found to fall by 5855 billion kcal, or by only 0.3%. In total the conversions of cultivated land to other uses led to a net loss of 34,829 billio kcal, or 1.8% of total potential productivity, at the end of the 1990s. Of this total amount, a decrease of 20,489 billion kcal, or about 59% of the total decreased production potential (i.e., 20,489/34,829), is due to the conversion of cultivated land into built-up areas. The high percentage is due in large part to the fact that the land being thus converted is of higher quality than the other types of land. This higher quality is due to the fact that the converted land is in the south and east of China, so it can be cultivated during two or more seasons. Land in the south and east is also less steep and receives more precipitation. In addition, of the total reduction in cultivated area production potential due to conversion, 16% (or 5623 billion kcal) is due to conversion to forestry. The total newly converted land increased production potential by more than 28,971 billion kcal, meaning that newly converted land raised production potential by 1.5%. Of the total, conversions from grasslands (48%, or 13,879 billion kcal) and forests (36%, or 10,335 million kcal) account for most of the increased production potential. Hence, although the quality of land that was converted into cultivated areas was lower than the land converted from cultivated areas, especially that converted into built-up areas, the increase in land that could be cultivated significantly offset the fall in production potential due to the conversion to built-up areas.

9.5 Summary

The nationwide impacts of China's cropland transformation on agricultural productivity are evaluated by integrating TM-derived land use data and the NPP model driven with satellite remote-sensing data at a spatial resolution of 8 km. It is demonstrated that in the 1990s, cropland transformation in China led to an increase in agricultural productivity of 6.96×10^6 tonnes, which made the nation's total agricultural productivity increase by 5%. Impacts of land use change on agricultural productivity estimated from both the GLO-PEM model and AEZ (agro-ecological zones) model demonstrate that land use conversions in China during the 1990s did not negatively affect total agricultural productivity due to the enlarged cropland area. However, the change in China's arable land is characterized by a significant loss to urban sprawl in southern China and a significant gain from the newly cultivated land in northern China, and the productivity of arable land occupied by urban expansion is 80% higher than that of the newly cultivated land. This implies that the increase in the nation's agricultural production induced by land transformation during the 1990s resulted mainly from the transformation of poor quality land into cropland. This is not a healthy, sustainable way for development and, if no appropriate measures are implemented, the food production potential and sustainability of China's agro-ecosystems will suffer to a significant degree in the near future.

Acknowledgements

This study is supported by an international collaborative project of the Ministry of Science and Technology of China (Grant No. 2013DF91700).

References

Albersen, P., Fischer, G., Keyzer, M., Sun, L. (2002) *Estimation of Agricultural Production Relation in the LUC Model for China*. Laxenburg, Austria: International Institute for Applied Systems Analysis.

Bounoua, L., DeFries, R.S., Imhoff, M.L., Steininger, M.K. (2003) Land use and local climate: A case study near Santa Cruz, Bolivia. *Meteorology and Atmospheric Physics* doi:10.1007/s00703-00300616-8.

Brown, L.R. (1995) *Who Will Feed China? Wake-Up Call for a Small Planet*. New York: Norton.

Cao, M.K., Prince, S.D., Small, J. (2004) Satellite remotely sensed interannual variability in terrestrial net primary productivity from 1980 to 2000. *Ecosystems* **7**: 233–42.

Chen, B. (1999) The existing state, future change trends in land-use and food production capacities in China. *Ambio* **28**: 682–6.

Chen, J. (2007) Rapid urbanization in China: A real challenge to soil protection and food security. *Catena* **69**:1–15.

Collatz, G.J., Ball, J.T., Grivet, C., Berry, J.A. (1991) Physiological and environmental regulation of stomatal conductance, photosynthesis and transpiration: a model that includes a laminar boundary layer. *Agricultural and Forest Meteorology* **54**: 107–36.

Deng, X., Huang, J., Rozelle, S., Uchid, E. (2006) Cultivated land conversion and potential agricultural productivity in China. *Land Use Policy* **23**: 372–84.

Fan, M., Shen, J., Li, X., *et al.* (2012) Improving crop productivity and resource use efficiency to ensure food security and environmental quality in China. *Journal of Experimental Botany* **63**: 13–24.

FAO (2002) *Food Insecurity: When People Must Live with Hunger and Fear of Starvation. The State of 2002.* Rome: Food and Agriculture Organization of the United Nations.

Farquhar, G.D., Von Caemmerer, S., Berry, J.A. (1980) A biochemical model of photosynthetic CO_2 assimilation in levels of C3 species. *Planta* **149**: 78–90.

Feng, Z., Yang, Y., Zhang, Y., Zhang, P., Li, Y. (2005) Grain-for-green policy and its impacts on grain supply in west China. *Land Use Policy* **22**: 301–12.

Fischer, G., Sun, L. (2001) Model based analysis of future land-use development in China. *Agriculture, Ecosystems and Environment* **85**: 163–76.

Fischer, G., Chen, Y., Sun, L. (1998) The balance of cultivated land in China during 1988-1995. IR-98-047. Laxenburg: International Institute for Applied Systems Analysis.

Fischer, G., van Velthuizen, H.T., Nachtergaele, F.O. (2000) *Global Agroecological Zones Assessment: Methodology and Results*. Laxenburg: International Institute for Applied Systems Analysis.

Fischer, G., Prieler, S., van Velthuizen, H. (2005) Biomass potentials of miscanthus, willow and poplar: results and policy implications for Eastern Europe, Northern and Central Asia. *Biomass and Bioenergy* **28**: 119.

Foley, J.A., DeFries, R., Asner G.P., *et al.* (2005) Global consequences of land use. *Science* **309**: 570–4.

Goetz, S.J., Prince, S.D. (2000) Interannual variability of global terrestrial primary production: results of a model driven with satellite observations. *Journal of Geophysical Research* **105**: 20077–91.

Gollan, T., Schurr, U., Shulze, E.C. (1992) Stomatal response to drying soil in relation to changes in the xylem sap composition of *Helianthus annuus*: I. The concentration of cations, anions, amino acids in and pH of, the xylem sap. *Plant, Cell and Environment* **15**: 551–9.

Harley, P.C., Thomas, R.B., Reynolds, J.F., Strain, B.R. (1992) Modeling photosynthesis of cotton grown in elevated CO_2. *Plant, Cell and Environment* **15**: 272–82.

Heilig, G.K. (1999) Can China feed itself? A system for evaluation of policy options. Laxenburg: International Institute for Applied Systems Analysis.

Heilig, G.K., Fischer, G., van Velthuizen, H.T. (2000) Can China feed itself? An analysis of China's food prospects with special reference to water resources. *International Journal of Sustainable Development World Ecology* **7**: 153–72.

Hicke, J.A., Lobell, D.B., Asner, G.P. (2004) Cropland area and net primary production computed from 30 years of USDA agricultural harvest data. *Earth Interactions* **8**: 1–20.

Keyzer, M. (1998) *Formulation and Spatial Aggregation of Agricultural Production Relationships Within the Land-use Change (LUC) Model*. Laxenburg, Austria: International Institute for Applied Systems Analysis.

Liu, J., Liu, M., Deng, X., Zhuang, D., Zhang, Z., Luo, D. (2002) The land use and land cover change database and its relative studies in China. *Journal of Geographical Sciences* **12**: 275–82 [in Chinese].

Liu, J., Liu, M., Zhuang, D., Zhang, Z., Deng, X. (2003) Study on spatial pattern of land-use change in China during 1995–2000. *Science in China Series D* **46**: 373–84.

Liu, J., Liu, M., Tian, H., *et al.* (2005a) Spatial and temporal patterns of China's cropland during 1990–2000: An analysis based on Landsat TM data. *Remote Sensing of Environment* **98**: 442–56.

Liu, J., Tian, H., Liu, M., Zhuang, D., Melillo, J.M., Zhang, Z. (2005b) China's changing landscape during the 1990s: Large-scale land transformations estimated with satellite data. *Geophysical Research Letters* **32**: L02405; doi:10.1029/2004GL021649.

Liu, J., Zhang, Z., Xu, X., *et al.* (2010) Spatial patterns and driving forces of land use change in China during the early 21st century. *Journal of Geographical Sciences* **20**: 483–94.

Milesi, C., Elvidge, C.D., Nemani, R.R., Running, S.W. (2003) Assessing the impact of urban land development on net primary productivity in the southeastern United States. *Remote Sensing of Environment* **86**: 401–10.

Ogud, Y.Y.S., Takeuchi, S., Gawa, K., Tsuchiya, K. (2003) Monitoring of a rice field using Landsat-TM and Landsat ETM+ Data. *Advances in Space Research* **32**: 2223–8.

Palmera, J.F., Lankhorst, J.R.-K. (1998) Evaluating visible spatial diversity in the landscape. *Landscape and Urban Planning* **43**: 65–78.

Prince, S.D. (1991) A model of regional primary production for use with coarse resolution satellite data. *International Journal of Remote Sensing* **12**: 1313–30.

Prince, S.D., Goward, S.N. (1995) Global primary production: a remote sensing approach. *Journal of Biogeography* **22**: 815–35.

Rozelle, S., Rosegrant, M.W. (1997) China's past, present, and future food economy: Can China continue to meet the challenges? *Food Policy* **22**: 191–200.

Skole, D., Tucker, C. (1993) Tropical deforestation and habitat fragmentation in the Amazon: satellite data from 1978 to 1988. *Science in China* (Series D) **260**: 1905–9.

Tao, F., Yokozawa, M., Liu, J., *et al.* (2009) Climate change, land use change, and China's food security in the twenty-first century: an integrated perspective. *Climatic Cange* **93**: 433–45.

Verburg, P.H., Chen, Y.Q., Veldkamp, T.A. (2000) Spatial explorations of land use change and grain production in China. *Agriculture, Ecosystem and Environment* **82**: 333–54.

Vitousek, P.M., Mooney, H.A., Lubchenco, J., Melillo, J.M. (1997) Human domination of earth's ecosystems. *Science* **277**: 494–9.

Wang, Z., Song, K., Zhang, B., *et al.* (2009) Shrinkage and fragmentation of grasslands in the West Songnen Plain, China. *Agriculture Ecosystems & Environment* **129**: 315–24.

Woodcock, C.E., Macomber, S.A., Pax-Lenney, M., Cohen, W.B. (2001) Monitoring large areas for forest change using Landsat: Generalization across space, time and Landsat sensors. *Remote Sensing of Environment* **78**: 194–203.

Yan, H.M., Liu, J.Y., Huang, H.Q., Tao, B., Cao, M.K. (2009) Assessing of land use change consequences on agricultural productivity in China. *Global and Planetary Change* **67**: 13–19.

10 Long-Term Land-Cover Change in the Amur River Basin

Shigeko Haruyama[1], Yoshitaka Masuda[2] and Akihiko Kondoh[3]

[1]*Mie University, Graduate School and Faculty of Bioresource, Mie, Japan*
[2]*Nippon Telegraph and Telephone East Corporation, Tokyo, Japan*
[3]*Center for Environmental Remote Sensing, Chiba University, Chiba, Japan*

10.1 Introduction

The Amur River is an international river with a total length of 4350 km and a huge drainage area of 2,051,500 km^2 (Figure 10.1). Its upstream region is in Mongolia, which has a steppe climate (Köppen's Bs world climate zone). Its midstream and downstream regions flow through China and the vicinity of the Russian border, which lies in the cool-temperate zone with low winter rainfall (Dw zone). Recently, environmental changes originating in the land-cover change have been occurring in the Amur River basin. For instance, the wetland habitat of the crane has been shrinking because of land development, increased sedimentation of rivers through the development of arable land, increased chemical loading of rivers through the expansion of agricultural activity, and air pollution due to forest fires.

In particular, large-scale agricultural reclamation has been done by national corporations over the last 20 years in Heilongjiang Province (Ganzey, 2005) and the area has changed into an important food production region (Motoki, 2001). In contrast, the Russian territory through which the river passes has been in a state of economic depression. Therefore, the regional differences in land-cover change in the Amur River basin are great. An objective evaluation of the factors influencing anthropogenic land-cover change in the basin is needed (Shiraiwa, 2005). The only research data that we have on land-use and land-cover change are the results of statistical material analyses of individual regions

Vulnerability of Land Systems in Asia, First Edition. Edited by Ademola K. Braimoh and He Qing Huang.

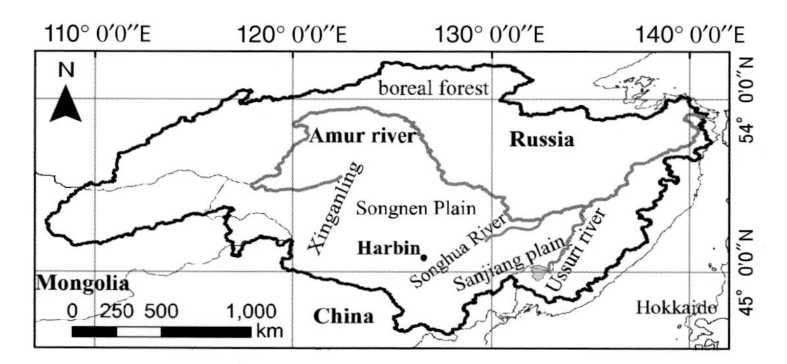

Figure 10.1 The Amur River basin

(Himiyama, 2001). There has been no research covering the entire Amur River basin; nor has there been any research on temporal variations in land cover.

We therefore performed satellite remote-sensing research on the temporal variations in land cover over the entire Amur River basin. By using the NDVI (Normalized Difference Vegetation Index), we aimed to pinpoint regions that are experiencing marked land-cover changes and to explain the trends in these changes.

10.2 Outline of study area

The Amur River basin is located at 41.42°–55.56°N and 107.32°–141.70°E. The gradient of the Amur River changes as the river courses eastwards from midstream to downstream. The river is joined by the Ussuri River, which flows from south to north around Khabarovsk in Russia, and by the large branch Songhua River, which joins the main Amur 250 km east of Khabarovsk (Figure 10.1). Structural geomorphologic changes are seen in the Amur River basin as the river flows from west to east. The mountainous district more than 1000 m above sea level (a.s.l.) extends to the west, while plains where the altitude is below 100 m extend in the lesser Khingan Mountain range to the east (Figure 10.2).

Peat wetlands are distributed in the floodplains and on the plains of the valley bottoms in the hilly country south of the Amur River. There is much agricultural activity in the eastern

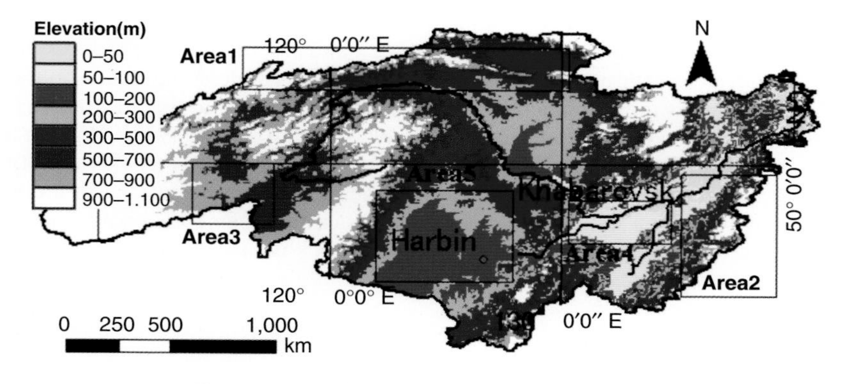

Figure 10.2 Elevation map of Amur river basin

Amur River basin. However, the climate is very cold: in the city of Harbin the annual mean temperature is 4.6°C, and the temperature in January is –20°C (China Map Publishers, 2001). The mean temperature in July is 25°C, and rainfall is usually concentrated in the summer. The farming calendar in the lower Amur River basin involves the planting of one crop a year (Jyo & Hasegawa, 2005).

10.3 The dataset

10.3.1 NOAA/AVHRR PAL dataset

Pathfinder AVHRR (Advanced Very High Resolution Radiometer) Land NDVI (PAL) data offered by the DAAC (Data Active Archive Center) of the National Aeronautics and Space Administration/Goddard Space Flight Center was used to analyze land-cover changes in the Amur River basin. The PAL dataset was collected by the AVHRR sensor installed on the National Oceanic Atmospheric Administration weather satellite; Channel 1 (visible light), Channel 2 (near-infrared rays), Channel 4, Channel 5 (heat infrared rays), and Normalized Difference Vegetation Index (NDVI) data were included. The PAL dataset from July 1981 to 2000 was available to be analyzed over a 19-year timescale for vegetation of the world. Yearly datasets from 1982 to 2000 were used in this analysis of land-cover change research of the Amur basin. Moreover, because the PAL dataset is collected globally, the entire Amur River basin could be analyzed, making this dataset the most suitable for large-scale and long-trend analysis of land-cover change.

The PAL dataset contains composite data on every tenth day, and divides the year into 36 seasons. The spatial resolution is 0.1° (about 10 km) converted into equal latitude and longitude. The NDVI relates to the density of green coverage, and active growth or revitalization of green plants, and it can be calculated from differences in the spectral reflection of chlorophyll in the visible light and near-infrared ranges (Yamamoto, 2001) NDVI is widely used for the observation and evaluation of vegetation, and it is assumed to be related to vegetation parameters such as land cover (Sannier *et al.*, 1998), leaf area index (LAI) (Spanner *et al.*, 1990), and biomass (Box *et al.*, 1989). In PAL data, NDVI is calculated by the following formula:

$$NDVI = (Ch.2 - Ch.1)/(Ch.2 + Ch.1)$$

where Ch.1 is channel 1 and Ch.2 is channel 2 of the Pathfinder AVHRR Land sensor.

10.3.2 Statistical materials used in the agricultural and field investigation

The amount of land in Heilongjiang Province transformed for irrigation between 1980 and 2000 was determined from the Heilongjiang Province statistical yearbook (China Statistics Publishers, 2001). Moreover, the area sown to the main commercial crops between 1978 and 2000 and the patterns of transition of production were determined from a statistical data book (China Statistics Publishers, 2001). In September 2005 we performed a field investigation in which we used GPS and digital cameras to observe and record the land cover around the middle reaches of the Amur basin.

10.4 Method of study

10.4.1 Analysis of secular variation from 1982 to 2000

We used the technique proposed in 2004 by Kondoh to analyze the global-scale vegetation and land-cover changes from 1982 to 2000 from the PAL data (Kondoh, 2004) and thus to clarify the land-cover changes over the entire Amur River basin. The following four parameters were used for the analysis: multiplication value of NDVI (ΣNDVI) during the year; maximum NDVI ($NDVI_{max}$), standard deviation ($NDVI_{std}$) of ΣNDVI, and the trajectory on the *T*s (surface temperature)–NDVI scatter chart (TRJ) (Nemani & Running, 1997).

The flow chart in Figure 10.3 was used for PAL data analysis. $NDVI_{max}$ is used as an index related to the production of commercial crops, because it shows the growth situation of the crop in every year. ΣNDVI is an index that corresponds to the biomass each year. $NDVI_{std}$ is used as an index for the disturbance of vegetation, because it shows the level of biomass increase and decrease every year. For instance, the differences in both ΣNDVI and $NDVI_{std}$ change throughout the year in regions where floods and forest fires occur frequently and in regions where there is a large amount of vegetation growth by the time the snow has melted. The TRJ is an index that tracks each pixel drawn in a year to reveal differences in land cover in two-dimensional space. NDVI is shown on the horizontal axis and *T*s on the vertical axis. Land-cover change can be calculated by applying a straight line to the tracks for every year, and analyzing the change in inclination of the straight line over 19 years. The inclination of the TRJ increases when the land cover changes from meadow to bare ground, and the inclination of the TRJ declines when the land cover changes into forest. As a threshold between vegetated and non-vegetated, the commonly used value of NDVI = 0.1 was applied (Kondoh, 2004). Pixels for which NDVI was lower than 0.1 were judged as non-vegetated regions for each particular season and excluded from the calculation. Moreover, the correspondence with NDVI derived from PAL data analysis results was verified by statistical material analysis and the regional field investigation; thus

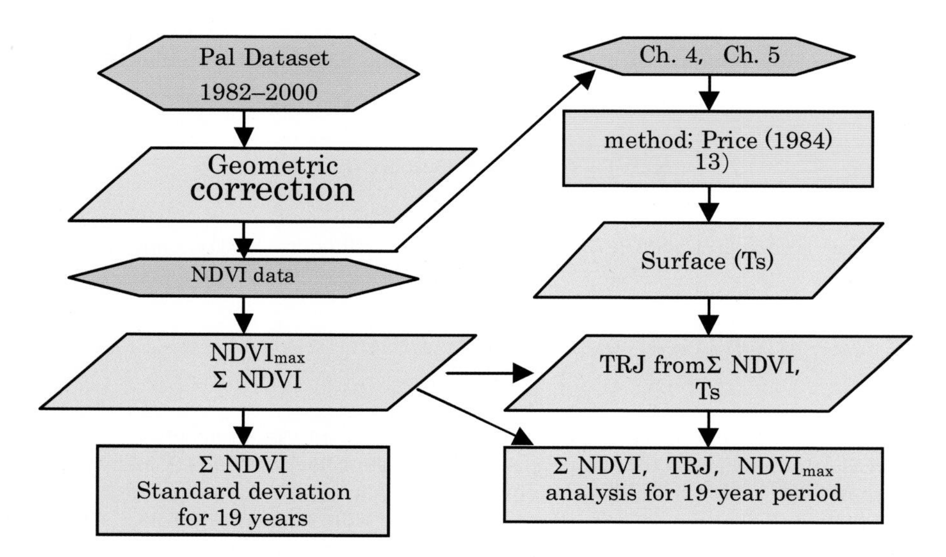

Figure 10.3 Flow chart of PAL data analysis. See text for definitions of symbols and abbreviations

the areas in which the influence of artificial land alteration was greatest were determined (Figure 10.3).

10.5 Results and consideration

10.5.1 Analysis of secular variation in land cover from 1982 to 2000

The NDVI value falls below the threshold in winter because the Amur River basin is covered with snow. Therefore, the secular variation in each parameter represents the NDVI changes from early spring to autumn. Figures 10.4 to 10.7 show the changes in each NDVI index in the Amur River basin over the 19-year period (1982–2000). From among these results we selected and interpreted five regions in which the changes in each parameter were clear.

10.5.1.1 Area 1 Area 1 is located between 52.75–54.85°N and 116.75–130.35°E and is covered by extensive coniferous forest. Figure 10.5 shows that the value of ΣNDVI (Figure 10.4) increased from east to west over 19 years. The area of ΣNDVI ≥ 0.05 extended both east and west and was distributed in a low mountainous district 500 to 800 m a.s.l. $NDVI_{std}$ was also high ($\geq$1.5) (Figure 10.5) in the area where ΣNDVI is high. No remarkable change in $NDVI_{max}$ (Figure 10.6) or TRJ (Figure 10.7) was seen across the region. In 1997, Myneni found active vegetation activity with relationship of the activity under global warming in northern Eurasia, Alaska, and the Canadian northwest (latitude 45°00–70°00N) during the period 1982–1990 (Myneni *et al.*, 1997). Moreover, the mean temperature rose by 4°C in winter in these three regions during 1961–1990 (Chapman & Walsh, 1993). Therefore, this temperature change appears to have been reflected in an increase in ΣNDVI in that the vegetation growth period was extended: the snow melted earlier in spring because the temperature rose earlier in winter. Moreover, the value of $NDVI_{std}$ increased because the annual difference in ΣNDVI was large. Area 1 can thus be interpreted as a region influenced readily by both annual changes in meteorological conditions and global climate change.

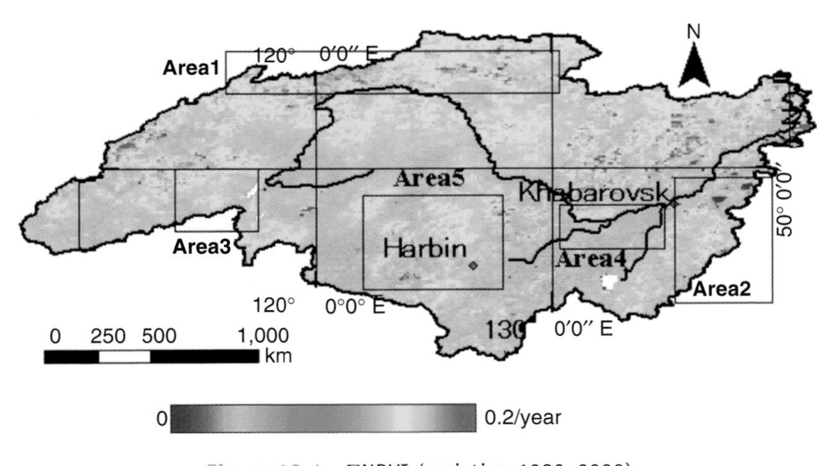

Figure 10.4 ΣNDVI (variation 1982–2000)

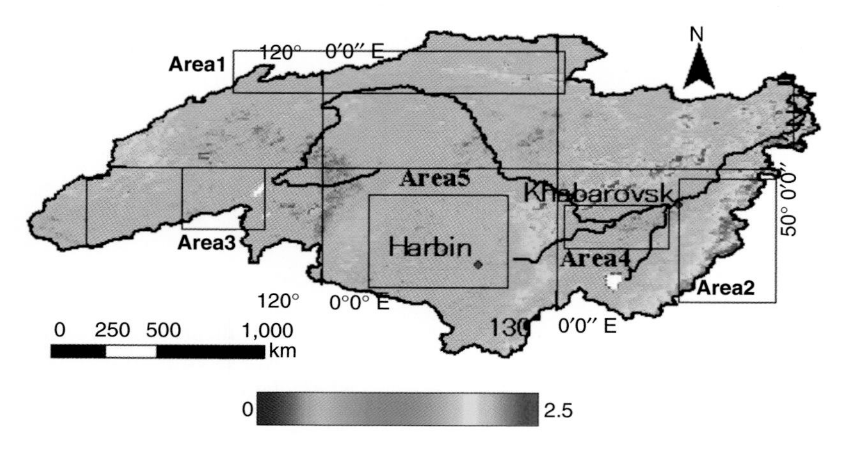

Figure 10.5 $NDVI_{std}$ (variation 1982–2000)

10.5.1.2 Area 2 Area 2 is located at 43.65–50.15°N and 134.65–140.75°E. In this region coniferous forests extend from 500 to 1300 m a.s.l. An area where $NDVI_{std} > 1.5$ and $\Sigma NDVI = 0.05$ was distributed from the southwest to the northeast along the basin's boundary. No marked change was seen in $NDVI_{max}$ or TRJ. However, an area of $\Sigma NDVI = -0.21$ (decreased biomass) extended from Khabarovsk to about 250 km eastward. Large-scale forest fires occurred in Khabarovsk Territory in 1976, 1998–1999, and 2001 (Makhinova, 2003). We consider that this decrease in ΣNDVI in Area 2, and thus the decrease in biomass, resulted from the frequent forest fires.

10.5.1.3 Area 3 Area 3 is located at 47.00–50.00°N, 114.50–118.00°E. In this region the plateaus and hills extend to 550–900 m a.s.l. The area has a steppe climate, and the main region is covered by meadow vegetation. The whole area had a TRJ = 3, but ΣNDVI, $NDVI_{max}$, and $NDVI_{std}$ showed no marked changes. Therefore, the increase in TRJ must

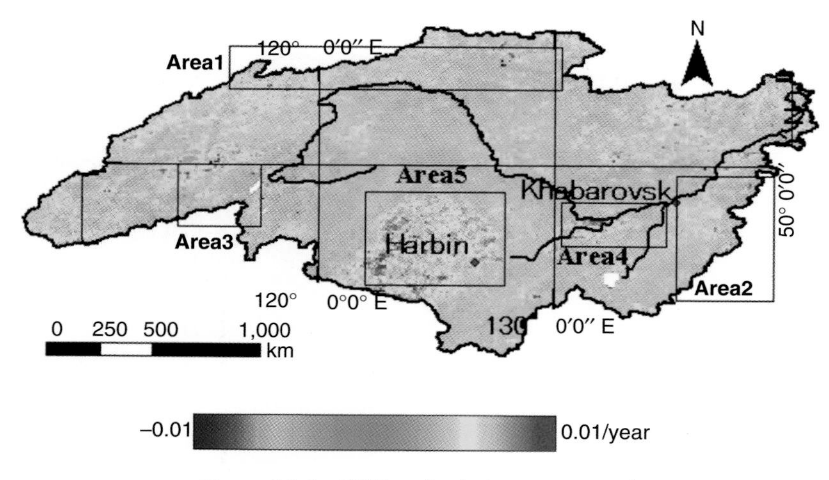

Figure 10.6 $NDVI_{max}$ (variation 1982–2000)

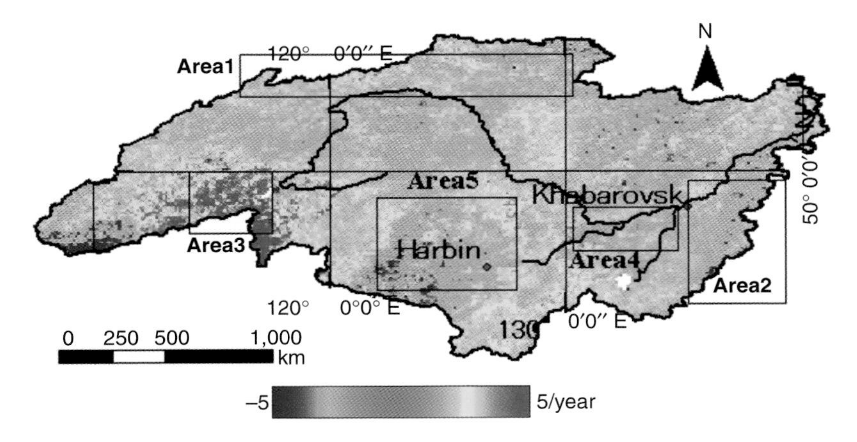

Figure 10.7 TRJ (variation 1982–2000)

be derived from an increase in *T*s. Generally, with meadow vegetation such as that in Area 3, there is a period of bare ground between snow melting and foliation. Consideration of ΣNDVI, $NDVI_{max}$, $NDVI_{std}$, and the vegetation of Area 3 leads to the interpretation that the period of bare ground increased over the 19 years, and the *T*s rose because snow melt occurred increasingly early; as a result, the TRJ increased.

10.5.1.4 Area 4 Area 4 is located at 46.55–47.85°N and 129.55–134.05°E. This region is known as the Sanjiang Plain, and here the floodplain is about 50–70 m a.s.l. In the Soughuag River basin, the $NDVI_{max}$ = 0.008 and the TRJ was about −3.0. Changes in these parameters on the floodplain between the Songhunag River and the Amur River are obvious. When biomass increases and the conditions for vegetation growth improve, transpiration becomes active and *T*s decreases. Moreover, the heat budget at ground level is changed by water transmission via irrigation (Kondoh, 2004). Therefore, the secular variation in each parameter in Area 4 can be interpreted as showing land-cover change through the development of agricultural activity (Yamamoto, 2001).

10.5.1.5 Area 5 Area 5 is located at 44.75–48.75°N and 121.85–128.05°E and is called the Songnen Plain. An increase in $NDVI_{max}$ was clear on the floodplain and on the hills, which are at an altitude of about 130–280 m. The $NDVI_{max}$ was 0.005 in the hill zone to the north of the city of Harbin. The hill zone to the west or southwest of the city of Qiqihar had a larger $NDVI_{max}$ (0.006). ΣNDVI and TRJ also showed obvious changes in the floodplain and hill zones: TRJ was below −1 and ΣNDVI = 0.1. In Heilongjiang Province, an area that extends for 200 km north of the city of Harbin and has seen a marked expansion in the area under rice cultivation since the 1980s (Himiyama, 2001), the ΣNDVI was 0.08 and the $NDVI_{max}$ was 0.005. The trends in each parameter were similar to those in Area 4 and can be interpreted as indicative of land-cover change in response to the development of agricultural activity. ΣNDVI increased in the low mountainous districts (especially those higher than 200 m a.s.l), suggesting that increases in afforestation, as well as in agricultural activity, increased the ΣNDVI (Nemani & Running, 1997).

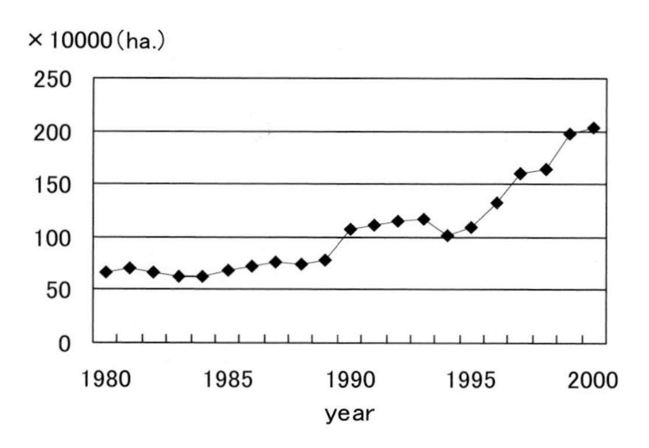

Figure 10.8 Transition of area under irrigation in Heilongjiang Province

10.5.2 Verification of validity of PAL data analysis

The validity of the PAL data analysis was verified by analyzing agricultural statistics from Heilongjiang province. Statistical data for Areas 4 and 5 were used, because we considered that the influence of artificial land alteration was the greatest in these two areas out of the whole Amur River basin. These regions are important food production regions of Heilongjian Province. The area of Heilongjiang Province under irrigation expanded rapidly, by 300% or more (from 670,500 to 2,032,000 ha) in the 21 years from 1980 through 2000 (Figure 10.8). The area sown to rice and upland crops and the production of rice increased greatly in the 23 years from 1978 through 2000 (750%, or 214,100 to 1,606,000 ha of sown area; and 1460%, or 715,000 to 10,422,000 t in the case of rice) (Figures 10.9 and 10.10). In our September 2005 field investigation, we confirmed that floodplains were being used for rice fields, hillside terraces for soybean fields, and hills for cornfields. Only parts of the wetlands remained. Recently, the Sanjiang Plain and the Songnen Plain, located in Areas 4 and 5, respectively, have become important food production regions. We confirmed that the secular variation in each parameter was associated with the development of arable land for irrigation and with land-cover change resulting from rice farming development, as mentioned in the results of the PAL data analyses.

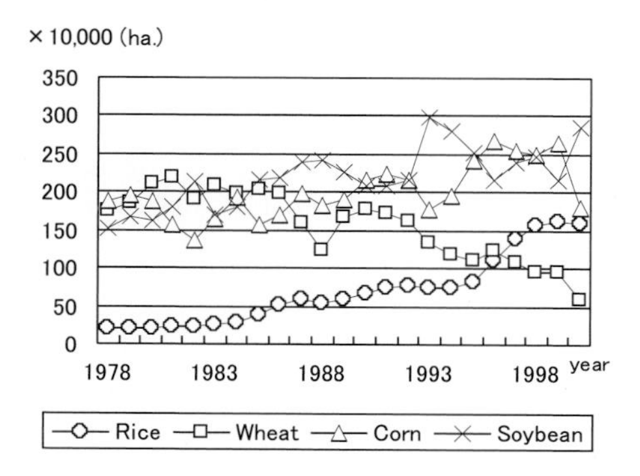

Figure 10.9 Transition of area sown to rice, wheat, corn, and soybean in Heilongjiang Province

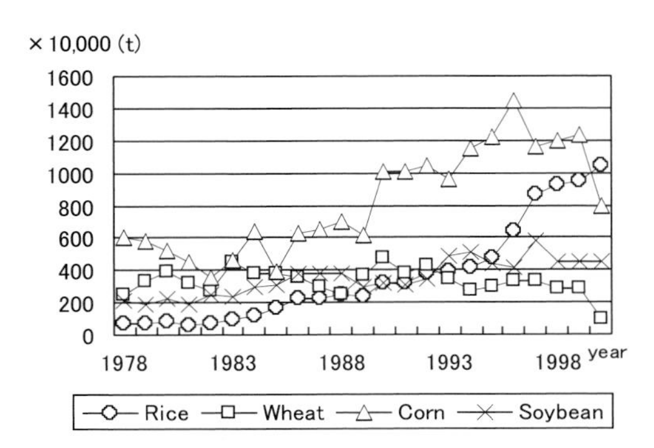

Figure 10.10 Transition in production of rice and upland crops in Heilongjiang Province

10.6 Summary

We demonstrated the trends in land-cover change in the Amur Basin from 1982 to 2000 spatially by using four indices calculated from the NDVI. Moreover, we were able to explain the land-cover change in five selected areas by combining and interpreting these indices. Land-cover change in the Amur River basin is especially remarkable in the grain producing region of Heilongjiang Province in China. To validate the results of the PAL data analysis, we performed a statistical material analysis and field investigation for Areas 4 (Sanjiang Plain) and 5 (Songnen Plain). We confirmed that the secular variation in each parameter in these regions was associated with arable land development for irrigation and with land-cover changes due to the development of rice farming. Thus, the transformation of the region from past to present was clarified by this land-cover change study. This research should be helpful in planning the future development and administration of the basin. By using data on land-cover change and elevation it should also be possible to approximate the changes in volumes of materials transported into the river.

Acknowledgements

This research is part of a project on 'Evaluation of the influence of human activity in northeast Asia on the production of biota in the North Pacific' by the Research Institute for Humanity and Nature (RIHN).

References

Box, E.O., Holben, B.N., Kalb, V. (1989) Accuracy of the AVHRR vegetation index as a predictor of biomass, primary productivity and net CO2 flux. *Vegetatio* **80**: 71–89.

Chapman, W.L., Walsh, J.E. (1993) Recent variations of sea ice and air temperatures in high latitudes. *Bulletin of the American Meteorological Society* **74**: 33–47.

China Map Publishers (2004) *Atlas of China*. Beijing: China Map Publishers.

China Statistics Publishers (2001) *Statistics of Heilongjiang Province*. Beijing: China Statistics Publishers.

Ganzey, S.S. (2005) *Transboundary Geo-Systems in the South of the Russian Far East and in Northeast China*. Vladivostok: Dalnauka.

Himiyama, Y. (2001) Land use change in northeast China – GIS database and evaluation for sustainable development. Land Use Project Report VII, pp. 83–98.

Jyo, S., Hasegawa, K. (2005) Rice field development and its problems for agricultural management in Heilongjiang Province. *Bulletin of Department of Agriculture, Mie University* **32**: 61–78 [in Japanese].

Kondoh, A. (2004) Analysis of vegetation and land cover change using global remote sensing. *Journal of Hydrology and Water Research* **17**: 459–67.

Makhinova, A. (2003) *Smoke from Forest Fires and Innocent Hostages to Experiments with Nature, Siberia and East of Russia*. Russian Academy of Science, pp. 47–52.

Motoki, Y. (2001) Agricultural land use change of north east China – view of role of rice production. Land Use Project Report VII, pp. 83–98.

Myneni, R.B., Keeling, C.D., Tucker, C.J., Asrar, G., Nemani, R.R. (1997) Increased plant growth in the northern high latitudes from 1981 to 1991. *Nature* **386**: 698–702.

Nemani, R., Running, S. (1997) Land cover characterization using multi-temporal red, near-IR and thermal-IR data from NOAA/AVHRR. *Ecological Applications* **7**: 79–90.

Spanner, M.A., Pierce, L.L., Running, S.W., Peterson, D.L. (1990) Seasonal AVHRR data of temperate coniferous forests: relationship with leaf area index. *Remote Sensing of Environment* **33**: 97–112.

Sannier, C.A.D., Taylor, J.C., du Plessis, W., Campbell, K. (1998) Real-time vegetation monitoring with NOAA/AVHRR in southern Africa for wildlife management and food security assessment. *International Journal of Remote Sensing* **19**: 621–39.

Shiraiwa, T. (2005) The Amur Okhotsk Project. Report on Amur-Okhotsk Project, No. 3, December 2005, Research Institute of Humanity and Nature, pp. 1–2.

Yamamoto, H. (2001) A study on biomass estimation in Mongolian grassland using NOAA AVHRR LAC data and ground based data. *Journal of the Japan Society of Photogrammetry and Remote Sensing* **40**: 25–37.

11 Simulating Land-use Change in China from a Global Perspective

Xuefeng Cui[1,2,3], Mark Rounsevell[1], Yuan Jiang[3,4], Muyi Kang[4], Paul Palmer[1], Wen Chen[5] and Terence Dawson[6]

[1]*School of GeoSciences, University of Edinburgh, Edinburgh, UK*
[2]*College of Global Change and Earth System Research, Beijing Normal University, Beijing, China*
[3]*State Key Laboratory of Earth Surface Processes and Resource Ecology, Beijing Normal University, Beijing, China*
[4]*College of Resources Science and Technology, Beijing Normal University, Beijing, China*
[5]*Institute of Atmospheric Physics, Chinese Academy of Sciences, Beijing, China*
[6]*School of Environment, University of Dundee, Dundee, Southampton, UK*

11.1 Introduction

Land use activities, primarily driven by agricultural expansion and economic growth, have transformed between one-third and one-half of the Earth's land surface – in the form of forest clearance, agricultural practices and urban expansion. This has had profound impacts on ecosystem services, food production and the environment (GLP, 2005) and is closely associated with socio-economic change and human activities (Foley *et al.*, 2005). Significant changes in land use and land management have taken place in China over the last two decades as the country embarked on market reforms and active participation in the global economy (Lin & Ho, 2003). Sustained growth in agricultural productivity and stable relationships with global food suppliers are important determinants of food security in China (Khan *et al.*, 2009). Shortfalls in domestic food production can have impacts on

Vulnerability of Land Systems in Asia, First Edition. Edited by Ademola K. Braimoh and He Qing Huang.

international food markets, and turbulence in global energy markets can affect food prices and supply costs. Therefore, a systematic assessment of land use change in China needs to address both physical and socio-economic driving factors at the global scale. Thus, this chapter explores land use change in China from a global perspective using a simplified, integrated assessment model.

11.2 Land use in China

China has a total land area of 932,748,000 ha, of which 14% was arable land, 18% permanent meadows and 22% forest in the year 2000. Agricultural land use is central to any debate about sustainability in China, since the country needs to feed 22% of the global population with less than 9% of the world's cultivated land (Chen, 2007). Furthermore, heavy degradation of arable land has occurred in China over the past decades due to desertification, secondary salinisation (Zhang *et al.*, 2007) and the conversion of fertile land to non-agricultural uses (Lichtenberg & Ding, 2008).

Relying on its limited arable land resources, China has achieved considerable success in feeding its people by shifting from a long-term shortage to a balance of basic food staples against total demand; consequently, the number of undernourished people fell from 250 million in 1978 to 29 million in 2003 (Chen, 2007). This success has mostly derived from the rapid increase of domestic cereal production, which has outpaced the growth of consumption demand associated with population growth and economic development. Figure 11.1, from the World Development Chart 2003, illustrates the change in food

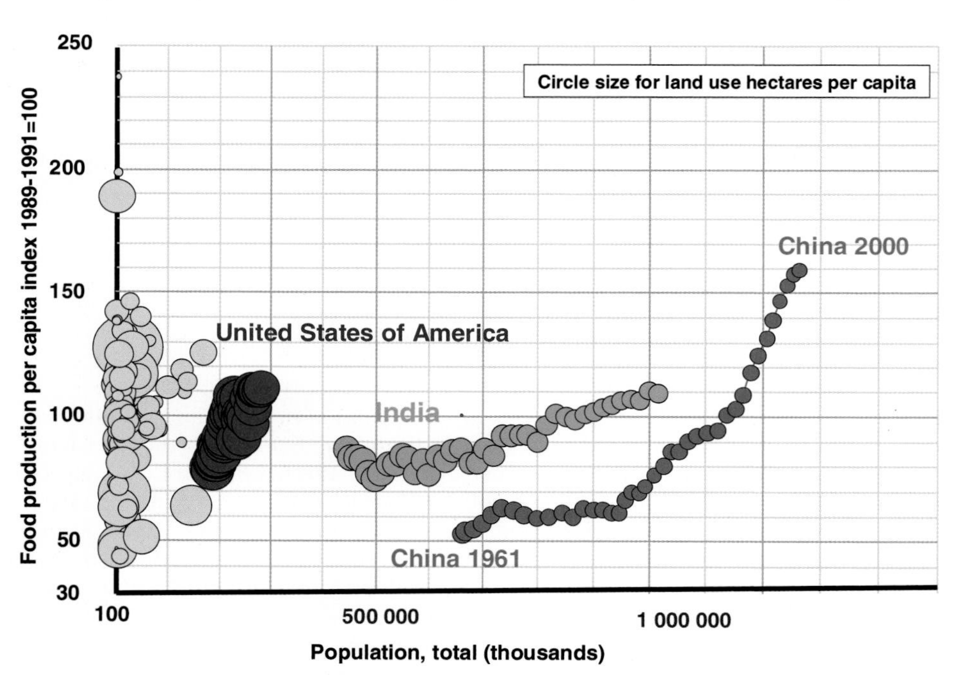

Figure 11.1 Change in food production per capita and population for the United States of America (blue), India (green) and China (red) from 1961 until 2000. The grey circles indicate the positions of other countries in the year 2000. Circle size represents arable land use per capita in each country. The chart was produced by the software tool World Development Chart 2003

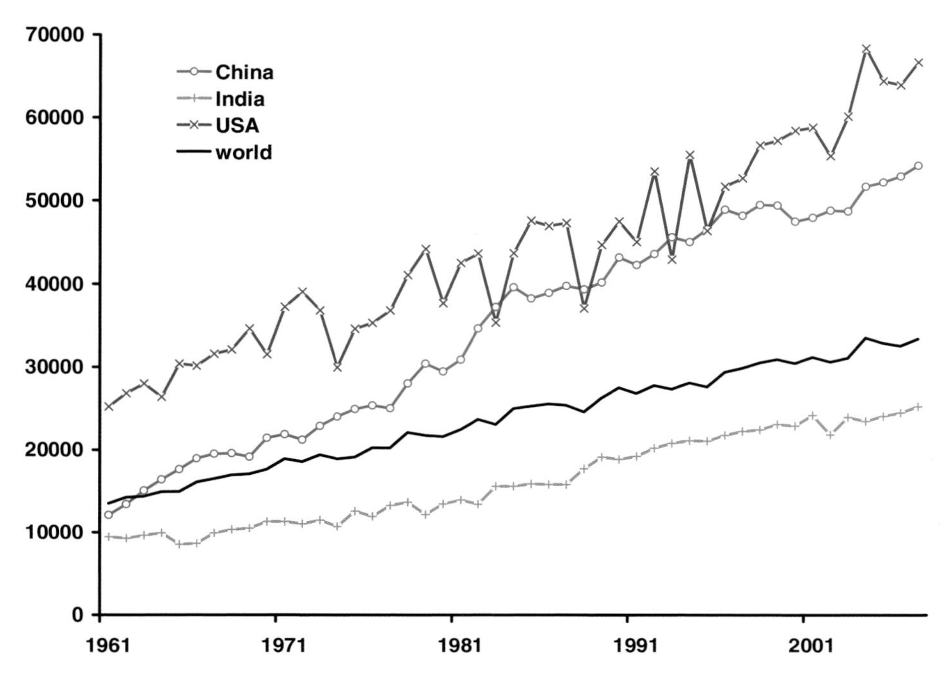

Figure 11.2 Cereal yields in USA, China, India and the world average from 1961 to 2006

production per capita against population for China from 1961 to 2000. Population growth is the primary driver of increasing food consumption and has exerted great pressure on land resources. As indicated by the circle size in Figure 11.1, people from China and India have lived on very small areas of arable land per person. China (and India) doubled its population during the past 40 years, increasing from 672 (454) million in 1961 to 1.27 (1.04) billion in 2000. As for many developed countries, the USA has, for example, a relatively high level of arable land per capita, but the population of the USA has only increased from 189 million to 284 million between 1961 and 2000. While food production per capita increased in the USA and India by about 30% from 1961 to 2000, China tripled its food supply per capita.

Given the limits to agricultural land expansion in China, the increase in production has derived from an increase in productivity, as shown in Figure 11.2. World cereal yields grew on average from 1353 to 3347 kg/ha between 1961 and 2006, but yields in China outperformed the world average by increasing four-fold, from 1211 to 5432 kg/ha. The most rapid increase in cereal yields in China occurred from 1978 when the Chinese government introduced the household responsibility system for agriculture, which encouraged greater production by rural households (Du, 2006) and had dramatic impacts on food security and poverty reduction (Fan *et al.*, 2009). It has been suggested that most of the yield increases in the world during the past decades were due to the so-called 'green revolution' (Evenson & Golin, 2003). For China, the growth in grain yields was determined primarily by the use of fertilizer and new technologies aided by strong institutional support (Wang *et al.*, 1996; Khan *et al.*, 2009). Fertilizer use per ha increased from 10 kg in 1960 to about 330 kg in 2002. Chemical fertilizer use increased rapidly with the rural economic reforms initiated in 1978, surpassing the use of organic fertilizer by 1982 (Liu & Chen, 2007). From 1997 to 2003, yields in China remained almost unchanged (even decreasing slightly), concurrent with unchanging cereal production. Productivity might have increased in China from 2004, but volatility in cereal yields has a big impact on food security in China and elsewhere.

Cereal productivity in the USA is much higher (generally double) than the world average and has generally increased over the past 40 years with quite large inter-annual variations. At the same time, cereal productivity in India is relatively lower than the world average and has a similar growth rate to the world average. Yield is determined by soil quality, water availability, climate, technology and several other factors. It is difficult to explain the large inter-annual variations seen in yields in the USA but not in China or India, although climate and price variability may play important roles. The different rates of yield change between China and India probably explain the different rates of growth in food production per capita for the two countries shown in Figure 11.1. Yield increase is possibly the only sustainable solution to improve production in order to meet the consumption demand. The Chinese government views agricultural biotechnology as a tool to help China improve the nation's food security, increase agricultural productivity and farmer incomes, foster sustainable development and improve its competitive position in international agricultural markets (State Science and Technology Commission, 1990). Future changes in the climate are likely to increase the need to find new technologies to maintain yield growth, such as genetically modified crops that are resistant to flooding (Ashikari, 2009) or droughts.

11.3 Global perspectives

Land use change occurs locally, but has direct or indirect consequences elsewhere and is influenced by drivers that operate at global scale (Held *et al.*, 1999). China, as with most countries of the world, tries to meet its food needs by improving domestic agricultural production (Brown & Funk, 2008). But, with such a large population, China has to make up any shortfall in domestic production through trade with other countries and in this way will influence global agricultural markets.

In principle, the quantity of a product that a nation needs to trade (imports minus exports) with other countries is determined by the difference between the domestic production and consumption (demand) with full access to the market. Figure 11.3a shows that China has imported more cereals than it has exported until very recently (year 2000). During the 1960s and the early part of the 1970s, China imported about 5 million tonnes of cereal more than it exported each year. Despite the rapid increase in production from 1978 when China started its economic reforms, population and economic growth exerted greater pressure on consumption demand leading to higher net trade quantities of cereals in the late 1970s and 1980s, with a peak of 17.5 million tonnes in 1983. The relatively low net trade years (1984–1986) are probably associated with bigger harvests (1982–1984). In the 1990s (except 1995), China imported less than in the 1980s and finally started exporting more than it imported from 2000. Generally speaking, China imports 5–8% of the world's cereal total, but exported only about 2% in the 1960s and 1970s and 3–4% in the 1980s and 1990s. Since 2000, China has exported 5% of the total world cereals, but imports only 4%. International trade is normally influenced by the difference between the world trade price and the domestic trade price. In China, the domestic trade prices of cereals have been controlled by the Chinese central government (Holton & Sicular, 1991). For example, the government raised cereal prices in 1994, which may have led to the extremely high import value and relatively low export value in 1995. Because of its massive trade volume, China has a great influence on the global cereal market especially after joining the World Trade Organization (WTO) in 2001.

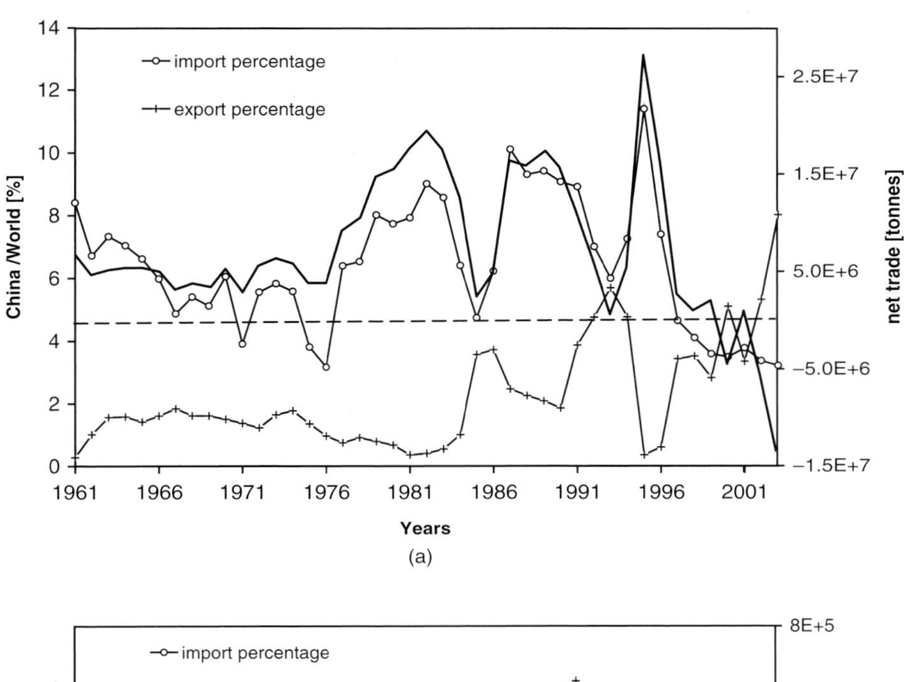

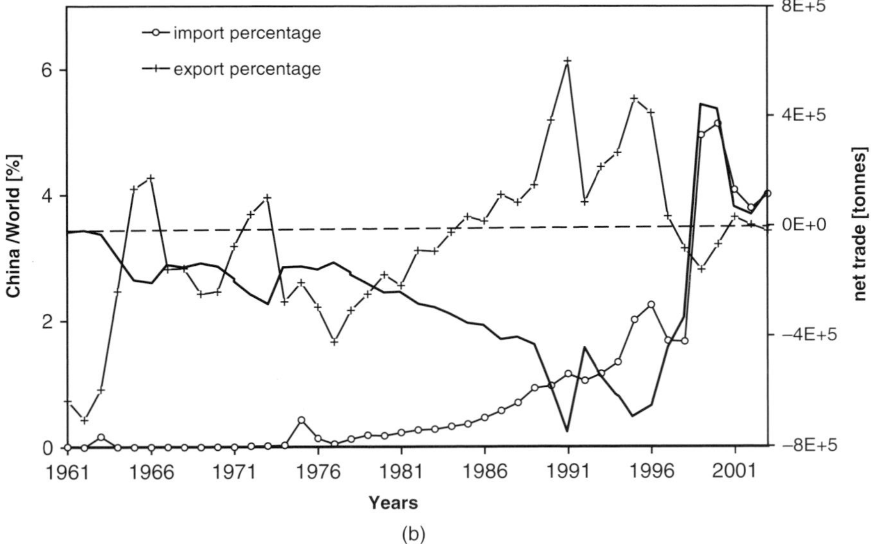

Figure 11.3 Change in the net trade of (a) cereals and (b) meat in China (black solid line), and the percentage of cereals imported to China and quantities of meat exported relative to the world total (dotted line with circle symbols). Dashed line indicates the zero line for net trade quantity; grey lines indicate the net trade quantities respectively

Chinese diets have also changed in recent years with more consumption of meat products associated with rising incomes over the last two decades. In contrast to cereals, Figure 11.3b shows that during the 1960s and 1970s China exported meat products (200,000 tonnes of meat), with almost no imports. Since the late 1970s, China started to import meat. However, the rate of exports also increased rapidly during the 1980s with a 4–6% share of the total world trade in meat in the late 1980s to early 1990s at a time when only 1–2% was imported. This situation changed in 1999. Chinese meat exports dropped to 3% of the world total while imports increased to 4–5% of the world total and China began to import more meat than it exported. The recent change in trade in cereals and meat clearly reflects changing food consumption habits of the Chinese population associated with rapid economic development and migration from rural to urban areas. If this trend continues, which seems quite possible, then China will probably need to import more meat from the global market in addition to increasing its domestic production. As 4–8 tonnes of cereal are needed on average to produce 1 tonne of meat then these changes will be profound, and we need to understand the impact of dietary change on food production and consequent land use change in China and its influence on the global food market. Furthermore, it is important to understand how other drivers such as climate change or a global oil crisis, will impact on China's food security and land use.

11.4 Model and data

The study of land use has to deal with the complexity and uncertainty of human-environment interactions. Integrated assessment models are able tackle this complexity within a global framework as they combine the effects of many drivers and land-use change processes across different parts of the world (Gibbs, 2000).

The work presented here is based on the application of a Global Food System Model (GLOBFOOD). GLOBFOOD was developed within the SIMILE declarative modelling framework (Muetzelfeldt & Massheder, 2003; http://www.simulistics.com). The model comprises 145 country units in which four modules (consumption, production, land use and biofuel products) are simulated respectively. Consumption demand is calculated from the consumption per capita, which is influenced by GDP per capita in each country. The land use module calculates the major transition of land use between agricultural arable land and other land sectors, i.e. forest, grassland and land use for biofuel production. The production module translates the consumption demand into agricultural land use in each country. A global trade module simulates the balance of agricultural commodities between production and consumption at the global level through trade flows between countries (e.g. Van Tongeren *et al.*, 2001). Crops used for biofuels divert land resources from food production. Developing viable alternatives to fossil fuels that reduce or negate the demand for biofuels would reduce the pressure on land and water resources in ways that also maintain food security. A more detailed description of the model structure and evaluation at the global level can be found in Jiang *et al.* (2014). This chapter is a case study to assess the model performance in a single country. It does this by evaluating the model against China's historical consumption, production and land use records. The scenario applied in this chapter is a simple attempt to test the model's ability to simulate future changes, but further development of the model will include probabilistic projections.

Data (except GDP) for model initialisation, drivers (population, GDP and yield) and evaluation were obtained from the UN Food and Agriculture Organisation (FAO) FAOSTAT database (http://faostat.fao.org/). As meat consumption per country is not available directly from the FAO, it was calculated – from meat primary production plus meat net trade in order to include all processed meat consumption, which is increasingly important (Van Tongeren *et al.*, 2001). GDP data were derived from the Global Social Change Research Project (GSCRP; available from http://gsociology.icaap.org). Land use and other socio-economic data generally contain large statistical biases, but it is difficult to evaluate data quality due to data limitation. The main objective of the work reported here was to simulate land use change in China within a global context and to explore the linkages between consumption and production. We will not evaluate the accuracy of the data used and apply the same dataset from the FAO to model evaluation to reduce the systematic errors of data uncertainties.

A simulation experiment was initialised in 1970 and run until 2000 driven by actual population, GDP and yield changes from the FAO data. Jiang *et al.* (2014) evaluated this experiment at the global scale and found that the model performed reasonably well in simulating world consumption, production and land use change over the 30-year simulation period. The focus here is on an evaluation of the model output for China. The experiment was stopped in 2000 since China's trade relationship in cereal and meat changed dramatically in 1999/2000 as discussed above.

11.5 Model results

11.5.1 Historical simulation

Figure 11.4 shows the simulated and observed change in cereal consumption and production, and milk and meat consumption in China from 1970 to 2000. The model reproduces the increase in cereal consumption with some underestimation particularly during

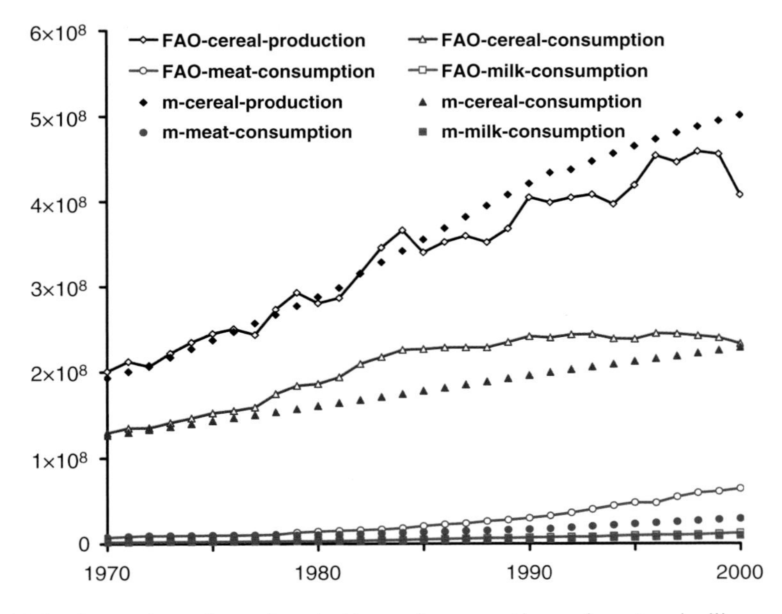

Figure 11.4 Comparison of cereal production and consumption and meat and milk consumption for China by simulation (solid symbols) and observed statistical data (line with hollow symbols) from 1970 to 2000. m, model simulation

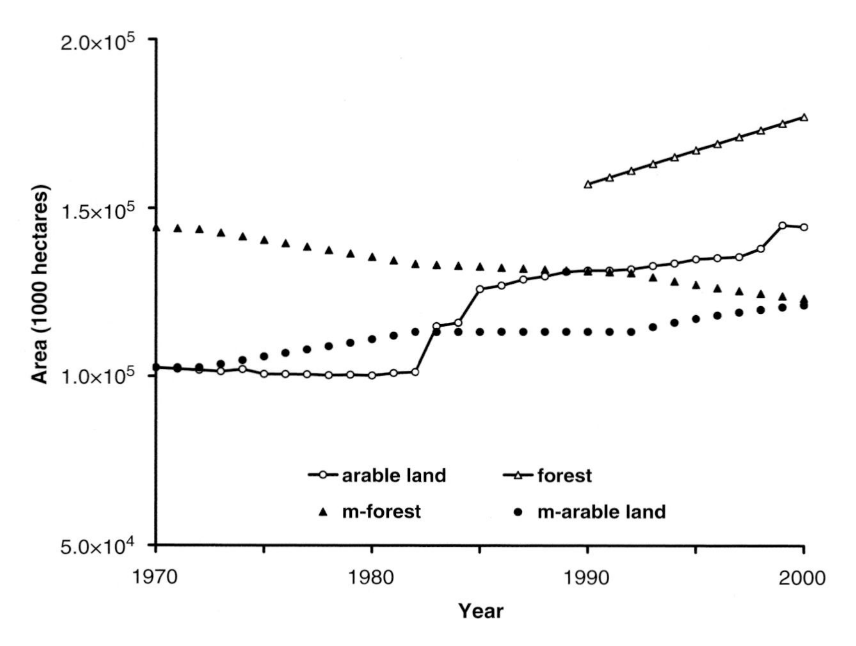

Figure 11.5 Comparison of arable land and forest for China by simulation (solid symbols) and observed statistical data (line with hollow symbols) from 1970 to 2000

the 1980s. In general it underestimates the consumption of meat, and this underestimation increases through the simulation run. This is probably because some of the model parameters, e.g. the ratio between meat/milk consumption per capita and GDP per capita, were based on the situation in 1970. Milk makes up only a small proportion of the Chinese diet and little change is observed in both the model simulations and the FAO data. The production of cereals is calculated in the model from the demand in human cereal consumption and animal feed plus trade between other countries. This shows that GLOBFOOD can simulate the increasing trend of cereal production in China relatively well. Relatively large gaps between the simulation and the actual FAO data occur in the late 1980s and 1990s, mainly influenced by the simulation of cereal and meat consumption. Another factor in the overestimation of cereal production is that the model does not consider grain storage. The data show that the world stock of cereals has declined recently mainly due to the change in stocks of grain in China (Headey & Fan 2008).

Figure 11.5 shows the simulated and observed land use change in China over the last 30 years. Unlike the good agreement for production and consumption in Figure 11.4, there are large differences between the modelled and observed land use changes. Some of these differences could be explained by uncertainties in the observed data. For example, the FAO records of forest land in China only started in 1990, since when the area increased in a steady linear way. The growth of new forest mainly derives from new planting resulting from the Chinese national forest policy (Zhang *et al.*, 2000). At the same time, the area of arable land also increased during the 1980s and 1990s after a marked rise in the first half of the 1980s. It is widely accepted that changes in the area of China's arable land were caused by significant losses to urban sprawl in southern China, but by significant gains from the cultivation of new land in the north (e.g. Deng *et al.*, 2006; Lichtenberg & Ding 2008;

Yan *et al.*, 2009). However, the change in the total area of arable land is highly uncertain. As discussed above, several studies have shown that arable land in China has degraded in recent decades (Chen, 2007). Conversely, some studies have recorded a net increase in cultivated land between 1986 and 2000 (Deng *et al.*, 2006). This demonstrates the difficulty in accurately estimating the arable land area in a country as large as China. As discussed above, the intention of this paper is not to evaluate data quality, but simply to use the FAO data as a reference against which to compare the model simulations. However, potential data inconsistencies need to be kept in mind when analysing the model results.

The model simulates transformations between arable land, forest, grassland and land for biofuel products in response to the demand from domestic consumption and international trade requirements. China is classified as a developing country in the model, with the assumption that the country will try to improve domestic production to meet consumption demand rather than importing from other countries (in contrast to developed countries). As new forest plantation is not included in surplus land within the model, the transformation between arable land and forest only reflects the requirements of domestic production, not forest change as a whole. As shown in Figure 11.5, the model simulates an increasing trend in arable land at the expense of forest. For this experiment, China's trade rate in the global market was assumed to be constant throughout the whole simulation period, i.e., 5% imported cereal, 2% exported cereal, 3% exported meat and no imported meat. Therefore, the increase in the arable area is mostly required because of the increase in domestic consumption as shown in Figure 11.4. As the model does not distinguish between land of different quality it cannot allocate spatially where these land use transitions would occur. In China, the cultivated land lost to urban expansion in the south is 80% higher in productivity than the new arable land brought into agricultural production in the north (Yan *et al.*, 2009). The model application presented here assumes equal productivity on all arable land, which means that larger areas of arable land are needed for the same crop production.

Overall, the model simulates the consumption and production of basic food staples in China reasonably well, but with a poorer representation of land use change. This could be because the model is a simplified representation of a highly complex system. Future model development will seek to improve the land use simulation module. This could include, for example, the simulation of land use change transitions between arable land, forest, grassland and other land use types; better spatial differentiation of land productivity; and inclusion of variable cereal stocks and trade rates.

11.5.2 Future 'business as usual' scenario

A simulation experiment was undertaken to demonstrate the utility of the model in projecting future trends. The historical simulation showed that the land use system could be modelled using only population, GDP and yield as drivers, and this gives some confidence in using the model to explore future socio-economic development pathways. A 'business-as-usual' scenario was assumed in which the parameter values remain unchanged from the current situation with the exception of changing projections of the three drivers: population, GDP and yield. Figure 11.6 shows the future scenarios of population and GDP per capita for China. In the past, GDP per capita increased rapidly due to the economic development of the 1990s, and we assume that this rate of development will continue until 2030. We also assume that cereal yields will continue to increase at the same rate as from 1990 to 2000. The population in China increased from 0.83 billion in 1970 to 1.26 billion in 2000, contributing to the world total population increase from 3.68 billion in 1970 to 6.11 billion in 2000 (UN, 2001). The FAO also projects that the population in China will reach

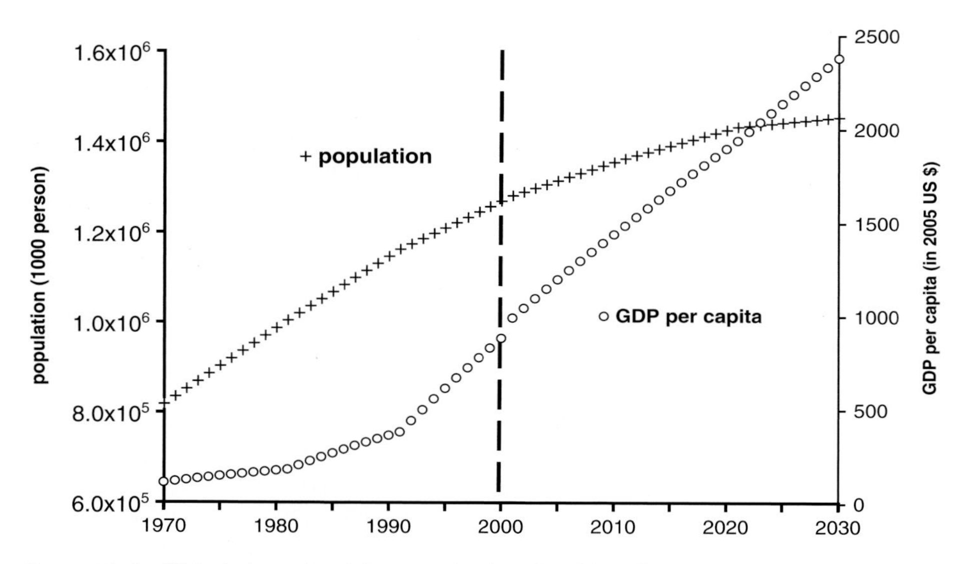

Figure 11.6 Historical trend and future projections for China of population (crosses) (FAO) and GDP per capita (circles) from 1970 to 2000 with a linear interpolation from 1990 to 2000 for 2000 to 2030. Dark dashed line indicates the end of the statistical record

1.43 billion by 2020 and 1.46 billion by 2030, which is considered by some to be the peak (Lutz *et al.*, 2003). The scenario assumptions presented here are not future forecasts, but explorations of one plausible development pathway.

Figure 11.7 illustrates the simulation of consumption and production from 2000 to 2030. The historical simulation is shown for reference purposes. The consumption of cereals and meat continues to increase whilst the consumption of milk remains relatively low. The increase is driven by population growth and higher meat consumption per capita resulting from larger and rising incomes. As China is classified as a developing country in the future experiment, it is assumed that it will meet its consumption demand by increasing domestic production through expansion of arable land from forest (not more than 20% of its total area each year) if forest land is available. Trade rates remain constant in the historical simulation. It is expected that the increase in cereal production and arable land will occur at the expense of forests (Figure 11.8). Although we should not read too much into these model results, the simulated trends explore plausible changes in production and consequent pressures on land resources. In addition to increases in production, the model simulates the need for more land for transport infrastructure, housing and energy generation to support the increasing urban population (Chen, 2007) thus exerting more pressure on already degrading land resources. The demand for meat and dairy products, which require substantially more water to produce than cereals, will continue to increase (Molden, 2007). These results suggest that the careful management of future land resources is essential if China is to continue to be able to feed its population. This simple business-as-usual simulation sheds some light on the relationship between population growth and economic development and land use change. Further development of the model to undertake probabilistic ensemble simulations will enable exploration of the relationships between domestic production and the global market.

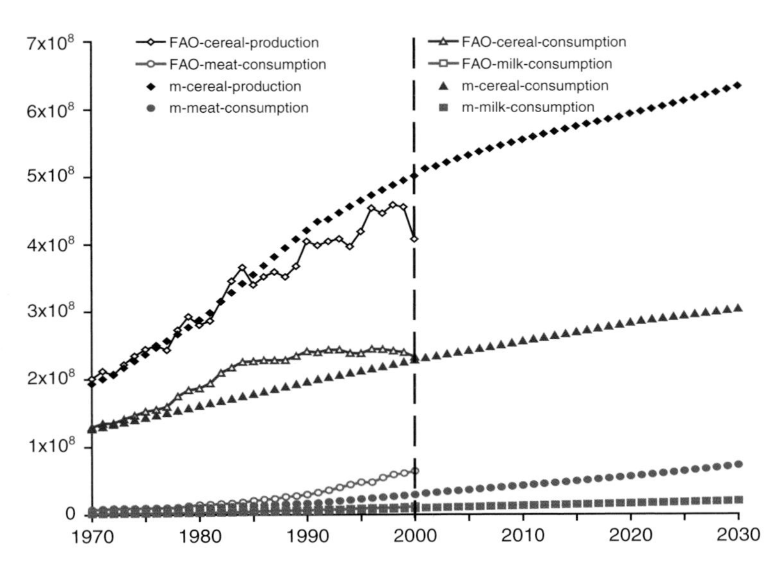

Figure 11.7 Simulated future projections of cereal production and consumption, and meat and milk consumption from 2000 to 2030. The FAO data are shown from 1970 to 2000 for reference. Note that the curve for 'FAO milk consumption' is covered by the line of 'm-milk-consumption' as they are similar

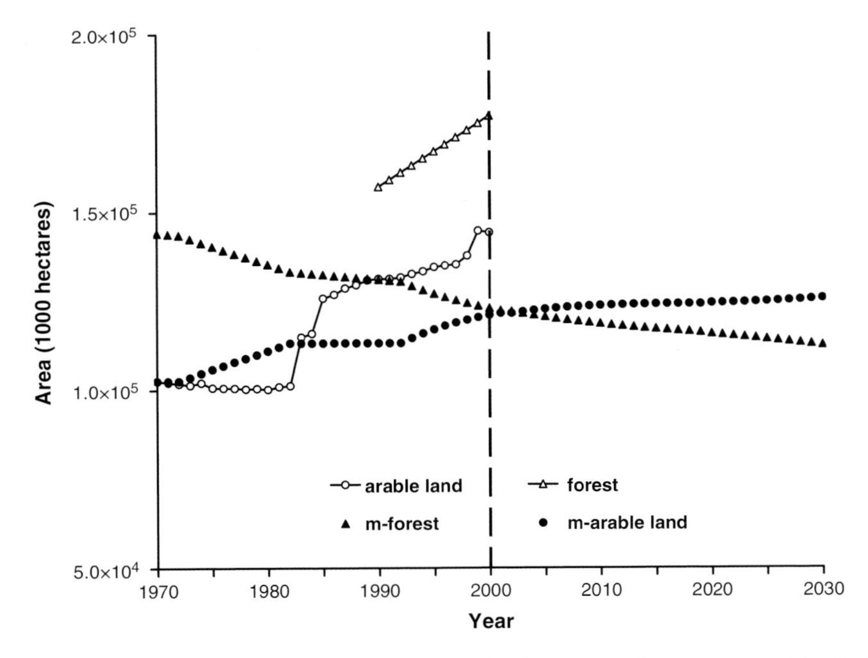

Figure 11.8 Simulated future projections of arable land and forest from 2000 to 2030. The FAO data are shown from 1970 to 2000 for reference

11.6 Discussion and conclusions

China has achieved great success in feeding 20% of the world's population with only 9% of the world's arable land. However, continued population growth, migration from rural to urban areas and rapid economic development suggest increasing pressure on China's already degrading land and water resources. Modelling studies can assist in exploring the consequences for China's land use of these types of development pathways. A global perspective is also important because of the trading relationships between countries as well as issues such as food security, climate change and environmental management.

The parsimonious land use model, GLOBFOOD, applied in this study was developed to simulate land use futures probabilistically by linking food consumption and production. Simulation of the past (1970–2000) shows that GLOBFOOD is able to simulate cereal and meat consumption in China quite well with some underestimation. Cereal production for both human consumption and animal feed is also reproduced well by the model. Animal feed is calculated for cereal equivalents needed to produce demanded quantities of meat and milk. The model is still being developed, but shows great potential in simulating land use from very few input parameters. The simplicity of the model structure also permits probabilistic simulations to explore uncertainties in future environmental changes, including climate change, and parameter assumptions.

Balancing food production and environmental quality has been a major issue for China, especially in recent years. The simple business-as-usual simulation experiment applied here suggests that the model could be useful in identifying the relationships between the population, the economy, food and land use. In addition to simulating land use change, the model could also be used to assess the role of agricultural policy, for example, the effects on food security of the Chinese 'land for ecology' programme, which limits arable land reclamation and maintains ecological conservation. The results presented here also demonstrate that a global, parsimonious model can be used to address land use change in a single country, especially by modelling the interactions between that country and the rest of the world.

Acknowledgements

The authors would like to acknowledge the financial support of the National Basic Research Development Program of China (grant no. 2011CB952001) and National Science Foundation of China (grant no. 41271542).

References

Ashikari, M. (2009) Flood-survival genes surface after years of fieldwork in rice paddies. *Nature* **460**: 932.

Brown, M., Funk, C. (2008) Food security under climate change. *Science* **319**: 580–1.

Chen, J. (2007) Rapid urbanization in China: A real challenge to soil protection and food security. *Catena* **69**: 1–15.

Deng, X., Huang, J., Rozelle, S., Uchida, E. (2006) Cultivated land conversion and potential agricultural productivity in China. *Land Use Policy* **23**: 372–84.

Du, R. (2006) The course of China's rural reform. International Food Policy Research Institute.

Evenson, R.E. & Gollin, D. (2003) Assessing the impact of the Green Revolution, 1960 to 2000. *Science* **300**: 758–62.

Fan, S., Kanbur, R., Zhang, X. (eds) (2009) *Regional Inequality in China: Trends, Explanations and Policy Responses*. Routledge.

Foley, J., DeFries, R., Asner, G. *et al.* (2005) Global consequences of land use. *Science* **309**: 570–4.

Gibbs, D. (2000) Globalisation: the bioscience industry and local environmental responses. *Global Environmental Change* **10**: 245–57.

GLP (2005) Global land project. Science plan and implementation strategy. IGBP Report no. 53/IHDP Report no. 19. Stockholm: IGBP Secretariat.

Headey, D., Fan, S. (2008) Anatomy of a crisis: the causes and consequences of surging food prices. *Agricultural Economics* **39**: 375–91.

Held, D., McGrew, A., Goldblatt, D., Perraton, J. (eds) (1999) *Global Transformations. Politics, Economics and Culture*. Oxford: Polity Press.

Holton, R., Sicular, T. (1991) Economic reform of the distribution sector in China. *American Economic Review* **81**: 212–18.

Jiang, L., Cui, X., Xu, Y., Jiang, M., Rounsevell, D., Murray-Rust (2014) A simple global food system model. *Agricultural Economics* **60**: 188–97.

Khan, S., Hanjra, M., Mu, J. (2009) Water management and crop production for food security in China: A review. *Agricultural Water Management* **96**: 349–60.

Lichtenberg, E., Ding, C. (2008) Assessing farmland protection policy in China. *Land Use Policy* **25**: 59–68.

Lin, G., Ho, S. (2003) China's land resources and land-use change: insights from the 1996 land survey. *Land Use Policy* **20**: 87–107.

Liu, X., Chen, B. (2007) Efficiency and sustainability analysis of grain production in Jiangsu and Shaanxi Provinces of China. *Journal of Cleaner Production* **15**: 313–22.

Lutz, W., Scherbov, S., Sanderson, W. (2003) The end of population growth in Asia. *Journal of Population Research* **20**: 125–41.

Molden, D. (2007) Water responses to urbanization. *Paddy and Water Environment* **5**: 207–9.

Muetzelfeldt, R., Massheder, J. (2003) The Simile visual modelling environment. *European Journal of Agronomy* **18**: 345–58.

State Science and Technology Commission [China] (1990) *Development Policy of Biotechnology*. Beijing: The Press of Science and Technology.

UN (2001) World population prospects, the 2000 revision – highlights. Doc. No. ESA/P/WP.165. New York: United Nations.

Van Tongeren, F., Meijl, H., Surry, Y. (2001) Global models applied to agricultural and trade policies: a review and assessment. *Agricultural Economics* **26**: 149–72.

Wang, Q., Halbrendt, C., Johnson, S. (1996) Grain production and environmental management in China's fertilizer economy. *Journal of Environmental Management* **47**: 283–96.

Yan, H., Liu, J., Huang, H., Tao, B., Cao, M. (2009) Assessing the consequence of land use change on agricultural productivity in China. *Global and Planetary Change* doi:10.1016/j.gloplacha.2008.1012.1012.

Zhang, K., Yu, Z., Li, X., Zhou, W., Zhang, D. (2007) Land use change and land degradation in China from 1991 to 2001. *Land Degradation and Development* **18**: 209–19.

Zhang, P., Shao, G., Zhao, G., *et al.* (2000) China's forest policy for the 21st century. *Science* **288**: 2135–6.

12 Sustainable Land Use Planning in West Asia Using MicroLEIS Decision Support Systems

Farzin Shahbazi[1], Maria Anaya-Romero[2], Ademola K. Braimoh[3] and Diego De la Rosa[4]

[1]*Department of Soil Science, Faculty of Agriculture, University of Tabriz, Tabriz, Iran*
[2]*Evenor-Tech, Spin-off from IRNAS-CSIC, Seville, Spain*
[3]*The World Bank, Washington, DC, USA*
[4]*Institute for Natural Resources and Agrobiology (IRNAS), Spanish Science Research Council (CSIC), Seville, Spain*

12.1 Introduction

In many parts of the world, land-use planning has been devolved to local government agencies that are expected to consult and involve a wide array of stakeholders from diverse sectors in identifying development options for their regions. Countries that are signatories to the Convention on Biological Diversity are compelled to adopt the principles embedded in Local Agenda 21, namely that local decision-making for integrated development planning (IDP) is democratic, and based on the goal of achieving social, economic and environmental sustainability (United Nations Conference on Environment and Development, 1992; Pierce *et al.*, 2005).

The role of soils in this kind of assessment is essential for evaluating optimal land use for each particular zone based on its own particular capability and susceptibility to degradation (FAO, 1976, 1978). Soil is a vital natural resource that performs key environmental, economic and social functions. It is non-renewable within human timescales. It develops slowly

Vulnerability of Land Systems in Asia, First Edition. Edited by Ademola K. Braimoh and He Qing Huang.

and changes gradually over time, showing great spatial variability. In the past, soil quality evaluation was biased towards agricultural production rather than for purposes related to the broad range of functions and services that it performs. The new concept of soil quality as the capacity of a specific soil to function with its surroundings, sustain plant and animal productivity, maintain or enhance soil, water and air quality, and support human health and habitation (Karlen *et al.*, 1997), based on data collected in standard soil surveys, appears to be the most appropriate framework. The quality of urban and agro-ecological soil should be evaluated to support public services for environmental quality management. Planners should also adjust their decisions towards more sustainable land use design (Vrscaj *et al.*, 2008).

Soil evaluation is of special importance in developing and emerging countries where land-use change occurs more rapidly than in developed countries. This is the case of most countries in Asia. Since 1990, Asia has witnessed impressive economic growth just as it has experienced diverse sustainable development challenges. Population growth and socioeconomic activities are important drivers of demand for ecosystem services, whilst global change has adversely affected food, soil and water security. Global change is predicted to profoundly affect agriculture, exacerbate water resource scarcity, and increase the threats to biodiversity as it compounds the pressure on ecosystem resources associated with urbanization and economic growth (Global Land Project, 2009).

In this context, decision support systems (DSSs) are indispensable for the planning and management process, at local and national levels for the development of specific plans of action. Decision support systems are computer programs that can be used to support complex decision-making and problem-solving (Shim *et al.*, 2002). Many agricultural and environmental DSS are already in place but need to be more widely applied, further developed and strengthened in Asia.

MicroLEIS has been widely applied in different parts of the world to support development policies. For example, application of the MicroLEIS DSS in the Pampean region of Argentina with special reference to a humid or semihumid subtropical climate showed that conversion of grassland into cropland had been the major land-cover process over the last 10 years, accounting for about a 28% increase of cultivated area (Moscatelli & Sobral, 2005). This DSS was also applied in Egypt, for 30 soil units of newly reclaimed areas (Darwish *et al.*, 2006). These results prompted the application of an agro-ecological land evaluation decision support system, MicroLEIS DSS (De la Rosa *et al.*, 2004), in a hitherto unstudied semi-arid region in West Asia.

The present chapter is intended to show the possibilities of using an agro-ecological land evaluation decision support system in this part of the world. The main aims are to identify the best agricultural lands; restore marginal areas; diversify crop rotation; and predict productivity of areas with certain selected soils. The study is underpinned by two principles: the conservation of soil and its services forms the basis of environmental, social and economic sustainability. Additionally, but still important, the goal of the present chapter is to contribute information that will help bridge the communication gap between soil scientists, urban planners and decision-makers.

Although biological quality indicators of soil are not considered in land evaluation, this agro-ecological approach can be useful for analyzing a soil's physical and chemical quality from the viewpoint of long-term changes (Ball & De la Rosa, 2006). In a more operational sense, suitability expresses how well the biophysical potentials and limitations of the land unit match the requirements of the land use type. Therefore, new investigations must be based on a sound understanding of past studies (De la Rosa & Sobral, 2007).

Figure 12.1 Location of the study area (East Azarbaijan, Iran)

12.2 Materials and methods

12.2.1 Study area

This study was conducted in Ahar province, East Azarbaijan, Iran (Figure 12.1), which has different kinds of land use associated with different parent materials such as limestone, old alluvium and volcano-sedimentary rocks. It is about 9000 ha in area and extends between 47°00′00″ and 47°07′30″E and between 38°24′00″ and 38°28′30″N. Its slopes range from <2 to 30%, with elevation ranging from 1300 to 1600 m a.s.l. Flat alluvial plain, hillsides and mountains are the main physiographical units in the study area.

12.2.2 Climate

Climate data such as mean average maximum and minimum temperatures for each month and total annual precipitation for the last 20 consecutive years (1986–2006) were collected from Ahar meteorological station. Data were integrated in the MicroLEIS CDBm program (De la Rosa *et al.*, 2004, 2009). Several climatic parameters were calculated for use in the study (Table 12.1).

A graphical representation of results for Ahar station using the CDBm (Monthly Climate Database) program of MicroLEIS is shown in Figure 12.2.

12.2.3 Benchmark soils

Soil data were extracted from 44 soil profiles representative of Ahar zone. These sample points were identified by applying an exhaustive grid survey method based on geology and slope (Figure 12.3).

Table 12.1 Results of the CDBm (Monthly Climate Database) program of MicroLEIS for the Ahar synoptic station (1986–2006)

Month	Tm (°C)	Tmax (°C)	Tmin (°C)	P (mm)	ETo (T, mm)	Hui	Ari	GS	Mfi	Aki
Jan	0.1	4.1	−4	18.6	0.1	–	–	–	–	–
Feb	0	3.3	−5.2	18	0	–	–	–	–	–
Mar	3	7.9	−1.9	25.4	9.3	–	–	–	–	–
Apr	8.3	14.2	2.6	38.2	35.5	–	–	–	–	–
May	12.9	18.9	6.9	58.9	68.2	–	–	–	–	–
Jun	17.8	24.7	10.9	27.1	102.2	–	–	–	–	–
Jul	21	27.3	14.6	11.5	126.4	–	–	–	–	–
Aug	22.3	28.6	15.9	8.9	127.7	–	–	–	–	–
Sep	19.3	25.9	12.8	9.3	95.2	–	–	–	–	–
Oct	14.5	20.9	8.2	23.7	61.7	–	–	–	–	–
Nov	8.3	13.3	3.2	34.6	26.8	–	–	–	–	–
Dec	2.9	7.2	−1.3	20.2	7.3	–	–	–	–	–
Annual	10.8	16.3	5.3	294.4	660.3	0.45	6	8	32	76.1

Tm, mean temperature; Tmax, maximum temperature; Tmin, minimum temperature; P, precipitation; ETo, evapotranspiration calculated by Thornthwaite method; Hui, humidity index; Ari, aridity index; GS, growing season; Mfi, modified Fournier index; Aki, Arkley index.

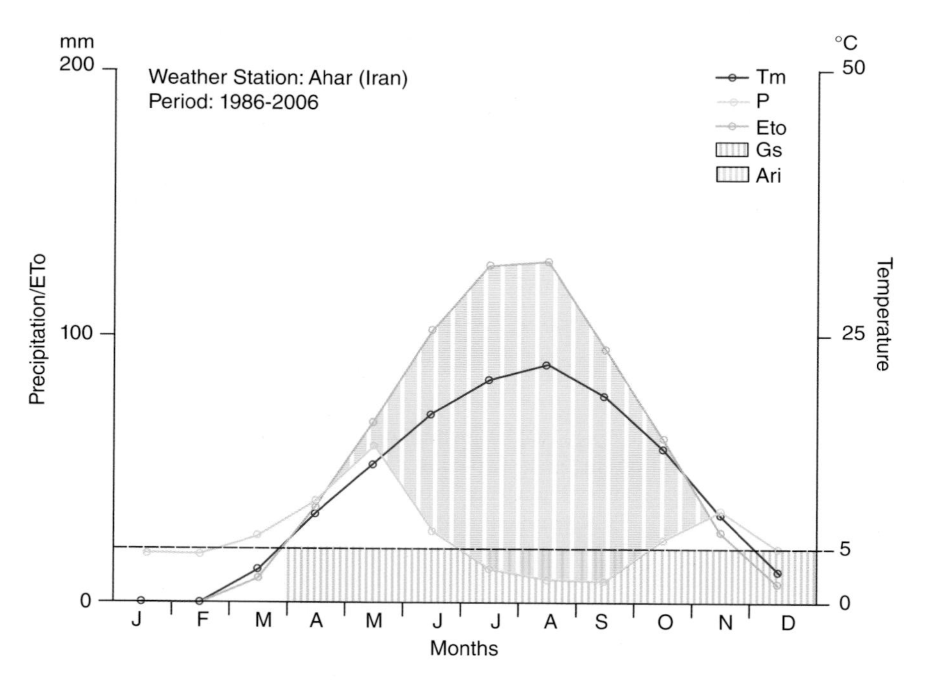

Figure 12.2 Graphical representation of the study area climate (current situation). Tm, mean temperature; P, precipitation; Gs, growing period; ETo, potential evapotranspiration; Ari, aridity index

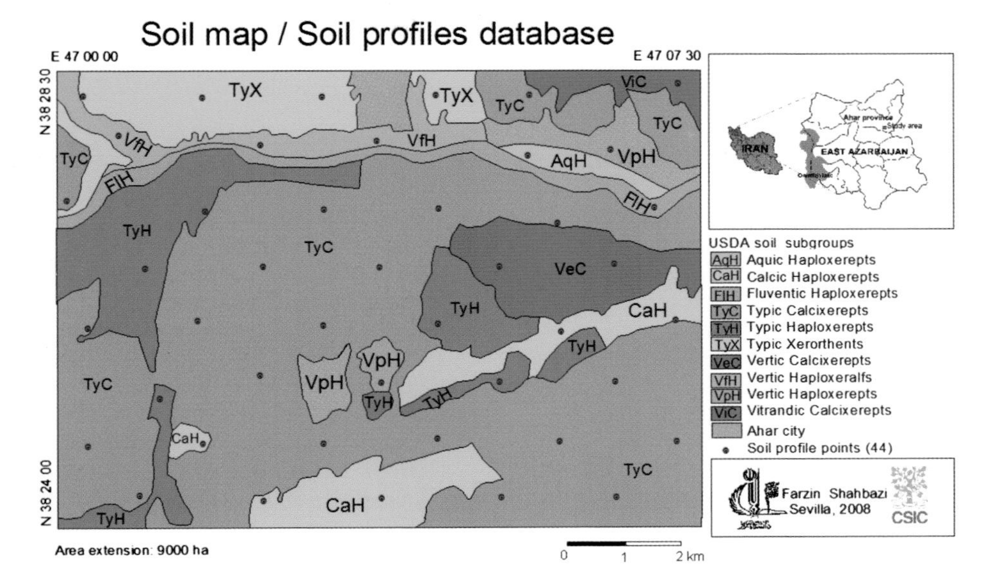

Figure 12.3 USDA soil subgroups map of the study area

The multilingual soil database SDBm plus (De la Rosa, 2004, 2009) was used to store and manipulate the large amount of soil data. The database stores the following input data: field site descriptions and soil profile characteristics; standard soil analytical data and soluble salts data; and soil physical analytical data, especially with reference to infiltration and water retention. Major facilities of the SDBm plus include input, edit, print, selection, and file generation. The 'soil layer generator' option represents a useful interface between the SDBm plus and the land evaluation and geographical information systems. The control section data for applying the models were: 0–50 cm, 25–50 cm and 0–100 cm.

Following the USDA *Soil Taxonomy* (USDA, 2006) and FAO Soil classifications (FAO, 1976) the dominant soils were classified as Inceptisols (Cambisols), Entisols (Regosols) and Alfisols (Luvisols). Additionally, 10 soil subgroups were identified, with Typic Calcixerepts (Calcaric Cambisols) as the largest subgroups (>53% area).

12.2.4 The MicroLEIS technology

The MicroLEIS DSS through its 12 evaluation models (De la Rosa *et al.*, 2004, 2009; see also Table 12.2), analyses the influence of selected soil indicators on critical soil functions including (i) land productivity (agricultural and forest soil suitability, crop growth, and natural fertility), and (ii) land degradation (runoff and leaching potential, erosion resistance, subsoil compaction, workability, and pollutant absorption and mobility).

These empirically based models were developed based on artificial intelligence techniques, using soil information and knowledge of the Mediterranean region. Input variables are physical/chemical soil parameters (e.g. useful depth, stoniness, texture, water retention, reaction, carbonate content, salinity, or cation exchange capacity) collected in standard soil

Table 12.2 MicroLEIS land evaluation models according to the soil function evaluated and the strategy supported for land use planning (adapted from De la Rosa *et al.*, 2004, 2009)

Constituent model	Land evaluation issue (modeling approach)	Specific strategy supported
Land use planning-related		
Terraza	Bioclimatic deficiency (parametric)	Quantification of crop water supply and frost risk limitation
Cervatana	General land capability (qualitative)	Segregation of best agricultural and marginal agricultural lands
Sierra	Forestry land suitability (qualitative/neural network)	Restoration of semi-natural habitats in marginal agricultural lands: selection of forest species
Almagra	Agricultural soil suitability (qualitative)	Diversification of crop rotation in best agricultural lands: for traditional crops
Albero	Agricultural soil productivity (statistical)	Quantification of crop yield: for wheat, maize, and cotton
Raizal	Soil erosion risk (expert system)	Identification of vulnerability areas with soil erosion problems
Marisma	Natural soil fertility (qualitative)	Identification of areas with soil fertility problems and accommodation of fertilizer needs
Soil management-related		
ImpelERO	Erosion/impact/mitigation (expert system/neural network)	Formulation of management practices: row spacing, residue treatment, operation sequence, number of implements, and implement type
Aljarafe	Soil plasticity and soil workability (statistical)	Identification of soil workability timing
Alcor	Subsoil compaction and soil trafficability (statistical)	Site-adjusted soil tillage machinery: implement type, wheel load, and tire inflation
Arenal	General soil contamination (expert system)	Rationalization of total soil input application
Pantanal	Specific soil contamination (expert system)	Rationalization of specific soil input application: N and P fertilizers, urban wastes, and pesticides

surveys, monthly agro-climatic parameters for a long-term period, and agricultural crop and management characteristics. Since the late 1980s, the MicroLEIS DSS has evolved significantly towards a user-friendly agro-ecological decision support system for environmentally sustainable soil use and management.

The design philosophy is a toolkit approach, integrating many software instruments: databases, statistics, expert systems, neural networks, Web and GIS applications, and other information technology. Input data warehousing, land evaluation modeling, model application software and output result presentation are the main development modules of this system.

MicroLEIS DSS helps in converting knowledge on land use and management systems, as estimated by research scientists, into information that is readily comprehensible to policy-makers and farmers. Its land evaluation models allow a site-specific application, providing an effective tool for assessing the suitability of land for a specified use and management at a particular location. All the information needed to select the suitable land use and management can be entered separately, hence it is possible to establish the exact soil, climate, and farming conditions. Combining MicroLEIS DSS results with Geographic Information System (GIS) helps to extract information from the evaluation models to be used and displayed as thematic geo-referenced maps. It is at this level of assessment that policy decisions are usually made (Davidson *et al.*, 1994). A CD-ROM version of MicroLEIS DSS is included with the book *Evaluación Agroecologica de Suelos para un Desarrollo Rural Sostenible* (De la Rosa, 2008). A spin-off from the CSIC (named Evenor-Tech; www.evenor-tech.com) is now launched based on MicroLEIS technology.

12.3 Modelling with MicroLEIS in the Ahar region

Land use planning decisions are supported essentially by land capability and land suitability models. Land use planning is generally aimed at a regional level. In MicroLEIS DSS it is supported by Terraza, Cervatana, Sierra, Almagra, Albero, Raizal and Marisma (see descriptions of the models in Table 12.2), in order to implement strategies for segregation of arable land surfaces, restoration of semi-natural habitats, diversification of crop rotation, and identification of risk areas. In this research, however, the soil erosion risk was not evaluated (Raizal model).

Relating major land use to soil capability and soil suitability for each site is considered the first objective in sustainable land use planning. Any kind of agricultural use or management system will have a negative environmental impact when applied on land with very low suitability.

12.3.1 Arable land identification

The Terraza model gives an empirical prediction of the bioclimatic deficiency of a site for the growth of several crops. Besides the solar radiation and temperature typical of the site, the lack of water and the risk of frost are considered the main climatic deficiencies for crop growth. However, the Cervatana model forecasts the general land use capability or suitability for a broad range of possible agricultural uses.

The results of applying the Terraza (bioclimatic deficiency) model and the Cervatana (land capability) model to the selected 10 benchmark soil subgroups are shown in Table 12.3, where dominant classes are presented in each soil. Eight soil subgroups are classified as excellent to moderately suitable, and another two as marginally suitable. Typic Calcixerepts, Typic Haploxerepts, Vertic Calcixerepts, Vertic Haploxeralfs, Calcic Haploxerepts and Vertic Haploxerepts showed the highest capability for most agricultural crops (S1 class) in 22.8%, 7%, 5.6%, 3.1%, 1.83% and 1.43% of the land area, respectively. Soil and topography are the major limiting factors for the Fluventic Haploxerept and Vitrandic Calcixerept subgroups and to some extent for parts of the Typic Calcixerepts (2.42%) and Typic Xerorthents (4.84%). Some 11.75% of the area was identified as marginal land, but currently dedicated to agricultural use. Unsuitable conversions of natural vegetations to

Table 12.3 Land capability evaluation results from application of the Terraza and Cervatana qualitative models[a]

		Land capability classes[b]	
USDA soil subgroups	Approx. extent (ha)	Highly/moderately suitable	Marginally suitable
Aquic Haploxerepts	89.1	S2l	
Calcic Haploxerepts	669.8	S2l	
Fluventic Haploxerepts	262		S3l
Typic Calcixerepts	4793.5	S2l	
Typic Haploxerepts	1131.4	S1	
Vertic Calcixerepts	504.2	S1	
Vertic Haploxeralfs	278.5	S1	
Vertic Haploxerepts	326.5	S2l	
Vitrandic Calcixerepts	141.7		S3t
Typic Xerorthents	693	S2lr	

[a]Development, inputs and validity of these models are described in De la Rosa *et al.* (2004, 2009).
[b]Land capability classes: S1, Excellent; S2, Good; S3, Moderate; N, Not suitable.
Limitation factors: t, topography: slope type and slope gradient; l, soil: useful depth, texture, stoniness/rockiness, drainage, and salinity; r, erosion risk: soil erodibility, slope, vegetation cover, and rainfall erosivity; b, bioclimatic deficiency (length of growing period) without considering the frost risk.

intensively tilled agricultural cultivation is one of the primary reasons for soil degradation on marginal lands (Figure 12.4).

12.3.2 Semi-natural habitats

A mismatch between current land use and the agro-ecological potential and limitations of that land require a fundamental change in land use. The Sierra (forestry land suitability) model comprises two sub-modules: a 'Description of forest species', detailing the edaphoclimatic requirements of 22 typical Mediterranean forest species; and an evaluation module, 'Selection of appropriate species', for selecting the best species for the land-unit evaluated. This is a first approach to the ecological requirements of these forest species, after selecting the biophysical parameters apparently determining their vegetative development, grouped by site, soil and climate. The input variables considered for modeling and application analyses can be grouped in three categories: soil, climate and site data (Heredia, 2007). The Sierra 2 model application results are summarized in Table 12.4.

The mastic tree (*Pistacia lentiscus*) is the most viable species for reforestation in the area. It is interesting to note the different number of viable tree species in comparison with the number of viable shrub species predicted for whole soil subgroups. This is due to the difference in influence of the soil factor useful depth and humidity.

In order to adopt an agroforestry farming system, the land evaluation results of the Sierra model can be combined with those predicted by the Almagra model for selecting the best combination of trees and crops to produce maximum environmental benefits in each soil unit.

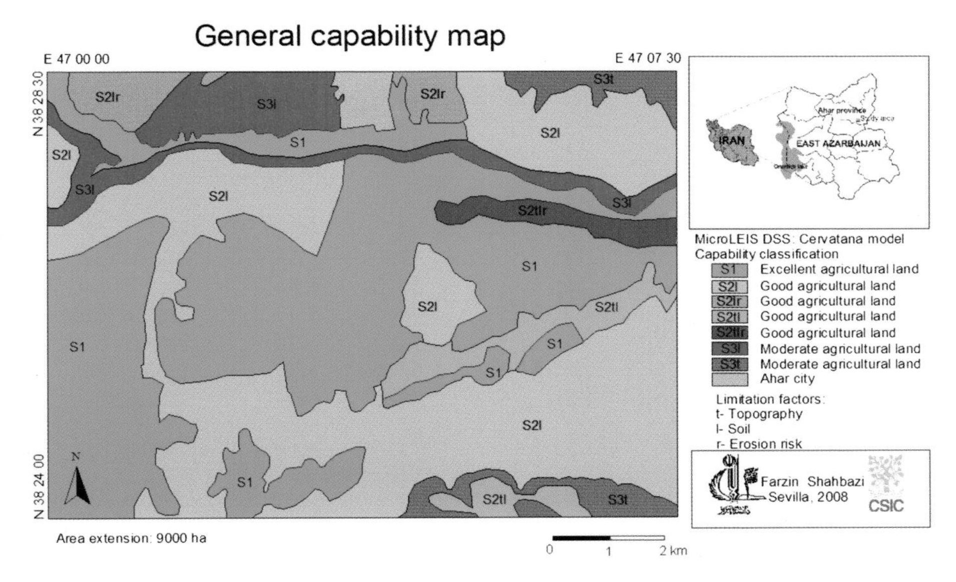

Figure 12.4 General capability map of the study area

12.3.3 Crop diversification

The Almagra model fits the types of biophysical evaluation that use as diagnostic criteria those soil characteristics or conditions favorable for crop development in terms of productivity (De la Rosa *et al.*, 2008). It is based on an analysis of the edaphic characteristics most directly affecting productivity development under different agricultural uses.

The results of applying the Almagra (agricultural soil suitability) model in the eight benchmark soil units previously classified as agricultural lands are shown in Table 12.5.

Table 12.4 Reforestation results from point application of the Sierra2 model[a] to the marginal agricultural lands

Benchmark soil subgroups	Viable shrub species
Typic Xerorthents	Esparto (Stipa tenacissima), broom-like kindery-vetch (*Anthyllis cytisoides*), dentate lavender (*Lavandula dentata* L.), nastic tree (*Pistacia lentiscus* L.), lygos (*Retama sphaerocarpa*), rock rose (Cistus albidus L.)
Vitrandic Calcixerepts	Esparto (*Stipa tenacissima*), broom-like kindery-vetch (*Anthyllis cytisoides*), dentate lavender (*Lavandula dentata* L.)
Fluventic Haploxerepts	Dentate lavender (*Lavandula dentata* L.)
Typic Calcixerepts	Esparto (*Stipa tenacissima*), broom-like kindery-vetch (*Anthyllis cytisoides*), rock rose (*Cistus albidus* L.)

[a]Development, inputs and validity of this model are described in Heredia (2006).

Table 12.5 Soil suitability evaluation results from point application of the Almagra model[a] to the best agricultural lands

	Soil suitability classes						
Benchmark soil subgroups	Wheat	Maize	Potato	Soybean	Sugar beet	Alfalfa	Peach
Aquic Haploxerepts	S1[b]	S2c	S2tcs	S2s	S2a	S2sa	S2tdcsag
Calcic Haploxerepts	S2t	S2tc	S2tc	S1	S2t	S1	S2tdcg
Typic Calcixerepts	S2t	S2tc	S2tc	S2t	S2ta	S2t	S4t
Typic Haploxerepts	S2t	S2tc	S2tc	S2t	S2ta	S2t	S4t
Typic Xerorthents	S3t	S3t	S3t	S3t	S3t	S2c	S2pt
Vertic Calcixerepts	S2t	S2tc	S2tc	S2t	S2ta	S2t	S4t
Vertic Haploxeralfs	S2ta	S2tca	S2tsa	S2tsa	S2t	S2tsa	S4t
Vertic Haploxerepts	S2t	S2tc	S3t	S2ts	S2ta	S2tsa	S4t

[a] Development, inputs and validity of this model are described in De la Rosa *et al.* (2004, 2009).
[b] Soil suitability classes: S1, Optimum; S2, High; S3, Moderate; S4, Marginal; S5, Not suitable.
Soil limitation factors: p, useful depth; t, texture; d, drainage; c, carbonate content; s, salinity; a, sodium saturation; g, profile development.

For this qualitative model, matching tables following the principle of maximum limitation for soil factors are used to express soil suitability classes for 12 Mediterranean crops. In this research, only seven typical and traditional crops were selected. The control, or vertical section for measuring texture, carbonates, salinity and sodium character was established by adapting the criteria developed for the differentiation of families and series in the *Soil Taxonomy*. For annual crops, the control section is between the surface and 50 cm depth, or between the surface and the limit of useful depth when the latter is less than 50 cm. For semi-annual and perennial crops, the control section range is between the surface and 100 cm depth.

Aquic Haploxerepts had high suitability for all the selected crops except wheat. Carbonates and salinity were the major limiting factors in the cultivation of maize and soybean. Optimum soil suitability for the cultivation of wheat, soybean and alfalfa was found in 48.89% of the Typic Calcixerepts area, whereas 10.83% of the Calcic Haploxerepts area had the same suitability. Generally, the excessive carbonate content in soils is the main limiting factor found at the evaluation. Wheat (*Triticum aestivum*), soybean (*Glycine max*) and alfalfa (*Medicago sativa*) are the most suitable crops for most of the units. In none of the subclasses were wheat, maize or potato affected by useful depth, drainage and profile development, whereas for some parts of the study area these were the main limiting factors in the development of a peach crop. So, in 30.62% of the total area, cultivation of peach (a perennial crop) can be recommended. Land suitability classification maps of wheat, alfalfa and peach are represented as examples of an annual, semi-annual and perennial, respectively, for the study area (Figures 12.5 to 12.7).

Results from the Almagra model were combined with GIS for improved visualization. This showed that 26.43%, 55.78% and 5.04% of the total area has optimum, high and moderate suitability, respectively, for the cultivation of wheat. Soybean and alfalfa have very similar suitability classes to wheat. Potato, sugar beet and peach do not have any optimum suitability classes. Therefore, the best crop rotation in the study area involves wheat, soybean and alfalfa. Considering the maize suitability classification from the Almagra model, it can be added to the crop rotation cycle. The final results of crop diversification are shown in Table 12.6.

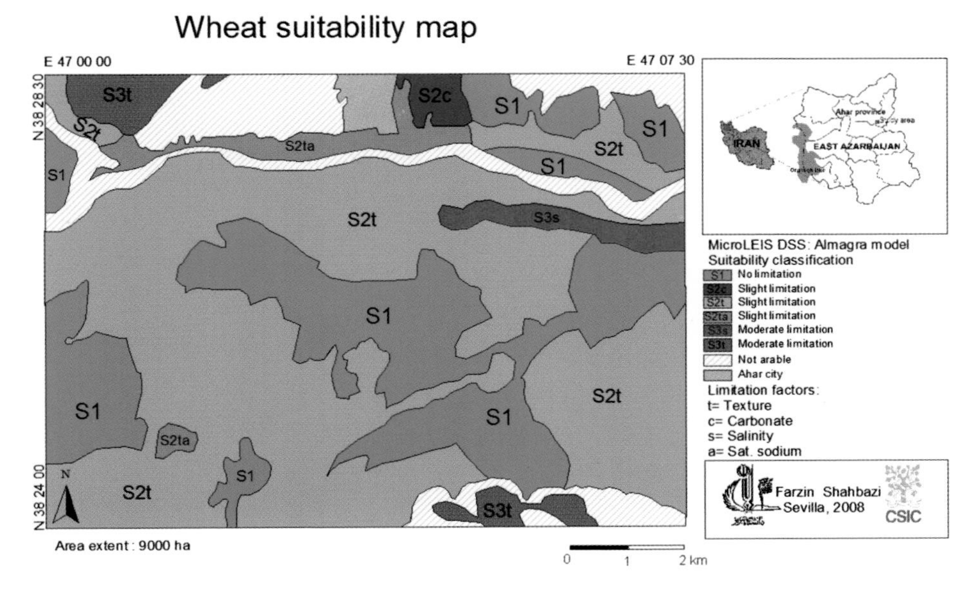

Figure 12.5 Land suitability map of wheat for Ahar province

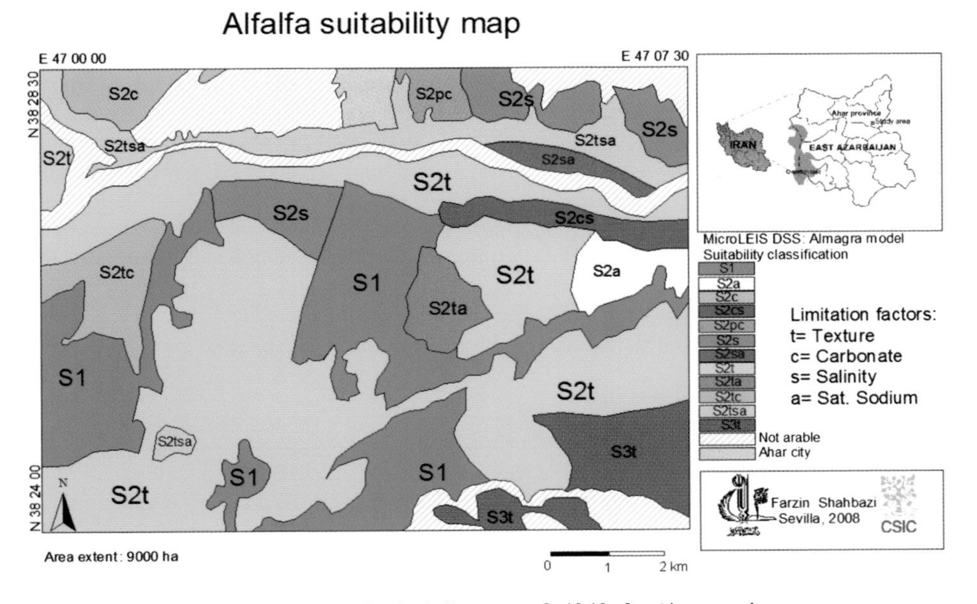

Figure 12.6 Land suitability map of alfalfa for Ahar province

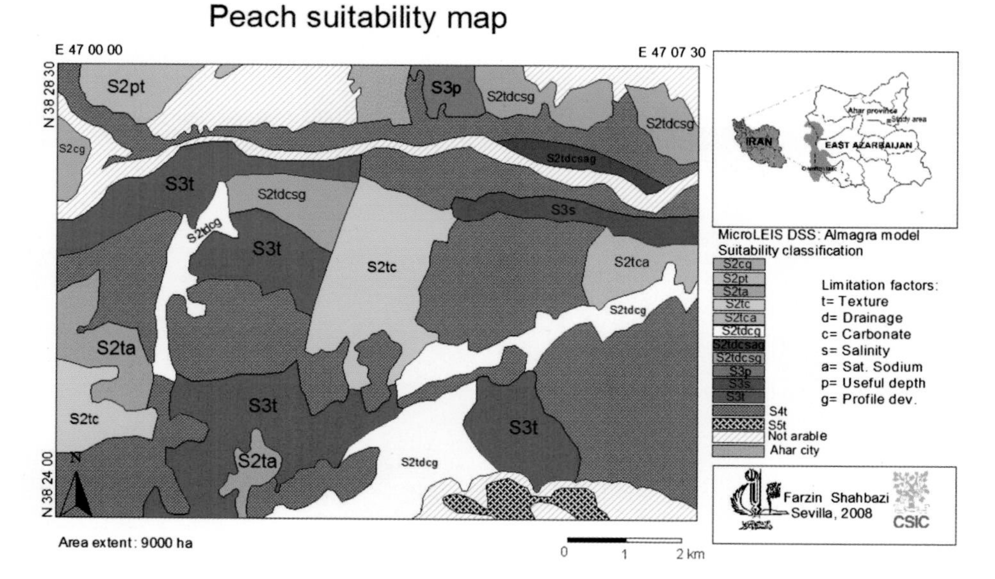

Figure 12.7 Land suitability map of peach for Ahar province

12.3.4 Soil productivity capability evaluation

The Albero model deals with the characteristics of a quantitative system of evaluating soil productivity capability, making use of computerized multiple regression techniques (De la Rosa *et al.*, 1981). It is a first approach to predicting the productivity of maize, wheat and cotton, based on a limited number of soil properties. Without analyzing the effect of climatic characteristics, and considering a high level of agricultural management practices, it attempts to explain the variability in productivity due exclusively to edaphic characteristics.

The predicted yields calculated by application of the Albero statistical regression model (Table 12.7) demonstrate the optimum soil physical/chemical quality of the

Table 12.6 Summary of land suitability classification results (% of the total area) using the Almagra model

Suitability classes[a]	Wheat	Maize	Potato	Soybean	Sugar beet	Alfalfa	Peach
S1	26.43	2.03	0	22.36	0	19.59	0
S2	55.78	80.18	66.71	59.85	84.21	62.36	31.61
S3	5.04	5.04	20.54	5.04	3.04	5.3	16.57
S4	0	0	0	0	0	0	37.85
S5	0	0	0	0	0	0	1.22

[a]Soil suitability classes: S1, Optimum; S2, High; S3, Moderate; S4, Marginal; S5, Not suitable.
1% of total area is occupied by Ahar city.
11.75% of total area is recommended for reforestation (see Table 12.4.)

Table 12.7 Agricultural soil productivity evaluation results from point application of the Albero model[a] to the arable land area

	Wheat		Maize	
Benchmark soil subgroups	Predicted yield (t/ha)	Approx. extent of area (ha)	Predicted yield (t/ha)	Approx. extent of area (ha)
Aquic Haploxerepts	4–4.5	89.1	6.5–7	89.1
Calcic Haploxerepts	4.5–5	237.3	6.5–7	393.8
Typic Calcixerepts	3.5–4	3000	6.5–7	2058
Typic Haploxerepts	3.5–4	800	7–7.5	784
Typic Xerorthents	2.5–3	257	6–6.5	164
Vertic Calcixerepts	3.5–4	324.2	7.5–8	324
Vertic Haploxeralfs	4.5–5	280	8–8.5	280
Vertic Haploxerepts	5–5.5	198	7.5–8	198

[a] Development, inputs and validity of this model are described in De la Rosa *et al.* (2004, 2009).

Vertic Haploxeralfs and Vertic Haploxerepts (Vertic properties) for maize and wheat production.

12.3.5 Soil fertility capability evaluation

The Marisma model was designed to group the soils according to edaphic criteria that have a direct influence on the interactions of applied fertilizers and other agricultural practices (De la Rosa, 2008). For this purpose, soils were classified at an initial level according to soil and subsoil textures. Then, 13 modifiers were taken into consideration, connected specifically with similar aspects of the soil fertility.

Table 12.8 shows the results of applying the Marisma (soil fertility capability) model in the eight agricultural benchmark units. This model gives special emphasis to the soil's chemical qualities, but also considers several physical soil parameters related to textural class.

Typic Calcixerepts showed considerable variation in fertility classes and certain management practices such as protecting against loss of surface soil and flushing of nitrogen at the beginning of the rainy season, not applying rock phosphate, etc. The fertility capability classification map is shown in Figure 12.8. The greatest management difficulties were found in 4.93% of Typic Haploxerept and 3.41% of Calcic Haploxerept areas, where care must be taken not to work the soils when wet.

12.4 Conclusions

Land evaluation appears to be a useful way to identify and develop the potential and/or general capability of the best agricultural land, resulting from dynamic interactions between the soil, climate and management variables. In Ahar province, 45% of the total area was classified as suitable for agricultural uses given the current climate. However, almost 12% of the total area must be reforested by suitable shrub species to minimize land degradation. We also extended the evaluation procedures to predict the inherent suitability of each soil unit for supporting a specific crop over a long period of time. Soils with vertic properties have an excellent capability for most traditional crops, with

Table 12.8 Soil fertility capability evaluation results from the Marisma model

Benchmark soil subgroups	FCC classes	Diagnostic report
Aquic Haploxerepts	LCdbn	Surface crusting risk; protect against soil loss; free carbonate material in soil surface; water deficit in the growing period
Calcic Haploxerepts	CCdbn	Care must be taken not to work when wet; protect against soil loss; possible leaching of N; free carbonate material in soil surface; alkaline conditions
Typic Calcixerepts	LLdb	Surface crusting risk; good subsoil texture; possible leaching of N; free carbonate material in soil surface
Typic Haploxerepts	LLdb	Surface crusting risk; good subsoil texture; possible leaching of N; free carbonate material in soil surface
Typic Xerorthents	SSdb	Surface leaching of nitrates; low subsoil water-holding capacity; possible leaching of N; free carbonate material in soil
Vertic Calcixerepts	CCdbn	Care must be taken not to work when wet; protect against soil loss; possible leaching of N; free carbonate material in soil surface; alkaline conditions; leach with Ca salts to prevent dispersion
Vertic Haploxeralfs	CCdbn	Care must be taken not to work when wet; protect against soil loss; possible leaching of N; free carbonate material in soil surface; alkaline conditions; leach with Ca salts to prevent dispersion
Vertic Haploxerepts	SCdbsn	Surface leaching of nitrates; protect against soil loss; possible leaching of N; free carbonate material in soil; leach with Ca salts to prevent dispersion; leaching with drainage is recommended

wheat-alfalfa-soybean selected as the best crop rotation. However, great care must be taken to apply an appropriate agricultural management system on these soil types.

Land use identification and allocation according to ecological potential and limitations is the first major objective of land use planning. The second objective is to preserve the soil and its ecosystem services under the different uses. We have demonstrated in this chapter that both complex tasks can be performed by agro-ecological land evaluation analysis using MicroLEIS DSS.

The land evaluation process developed in the Ahar study area can in principle be extrapolated to others parts of Asia, thereby permitting the sharing of knowledge amongst researchers and other stakeholders. The application of MicroLEIS Technology in Asia leads to upscaling, which not only implies a larger dimension for each project but also implies sharing knowledge, experience and resources for soil use and protection.

The present research is integrated in a global project of Sustainable Land Use Planning in West Asia using MicroLEIS DSS. Further results regarding climate change impact can be found in Shahbazi *et al.* (2009). MicroLEIS Technology is an open project under continuous development and it is hoped that aspects such as biological parameters and carbon sequestration will be incorporated in the near future.

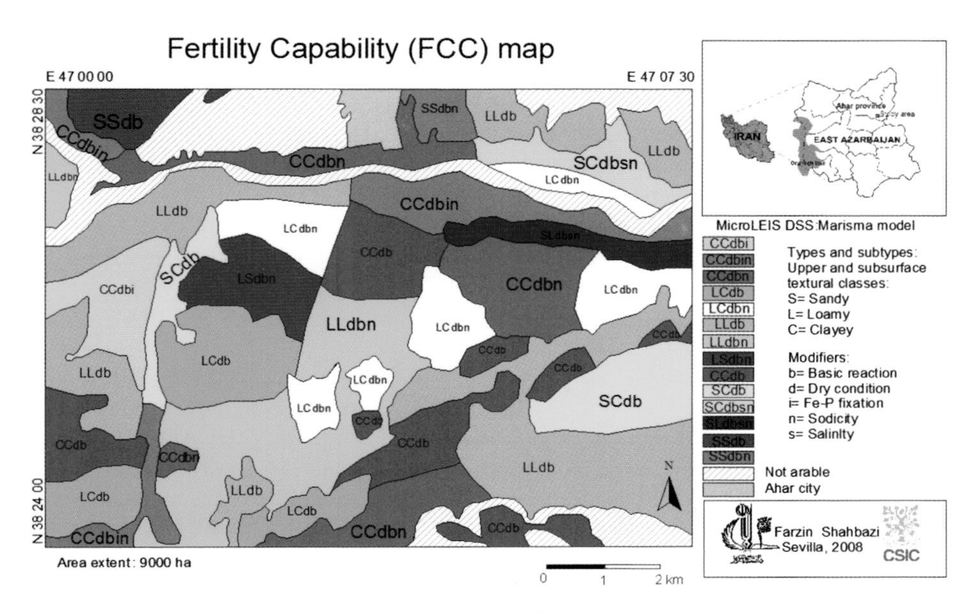

Figure 12.8 Fertility capability classification map of Ahar province

Acknowledgements

The authors wish to thank Tabriz University for its support during the course of the PhD thesis work of Farzin Shahbazi. Also, we acknowledge the support of the Spanish Science Research Council (CSIC), Institute for Natural Resources and Agrobiology (IRNAS), Seville, Spain.

References

Ball, A., De la Rosa, D. (2006) Modelling possibilities for the assessment of soil systems. In: Uphoff, N., Ball, A., Fernandes, E., *et al.* (eds), *Biological Approaches to Sustainable Soil Systems*. Boca Raton, FL: Taylor & Francis/CRC Press, pp. 683–92.

Davidson, D., Theocharopoulos, S.P., Bloksma, R.J. (1994) A land evaluation project in Greece using GIS and based on Boolean and Fuzzy Set Methodologies. *International Journal of Geographical Information Systems* **8**: 369.

Darwish, K.M., Wahba, M.M., Awad, F. (2006) Agricultural soil suitability of Haplo-soils for some crops in newly reclaimed areas of Egypt. *Journal of Applied Sciences Research* **2**: 1235–43.

De la Rosa, D. (2008) *Agro-ecological Land Evaluation for a Sustainable Rural Development*. Madrid: Mundi-Prensa [in Spanish].

De la Rosa, D., Sobral, R. (2007) Soil quality and methods for its assessment. In: Braimoh, A.K., Vlek, P.L.G. (eds), *Land Use and Soil Resources*, XXII. Springer, pp. 167–200.

De la Rosa, D., Cardona, F., Almorza, J. (1981) Crop yield predictions based on properties of soils in Sevilla, Spain. *Geoderma* **25**: 267–74.

De la Rosa, D., Mayol, F., Diaz-Pereira, E., Fernandez, M., De la Rosa, D. Jr (2004) A land evaluation decision support system (MicroLEIS DSS) for agricultural soil protection. *Environmental Modelling and Software* **19**: 929–42.

De la Rosa, D., Anaya-Romero, M., Diaz-Pereira, E., Heredia, N., Shahbazi, F. (2009) Soil-specific agro-ecological strategies for sustainable land use – A case study by using MicroLEIS DSS in Sevilla Province (Spain). *Land Use Policy* **26**: 1055–65.

FAO (1976) A framework for land evaluation. *FAO Soil Bulletin 32*. Rome: FAO.

FAO (1978) Report on the agro-ecological zones project. *World Soil Resources Report 48*. Rome: FAO.

Global Land Project (2009) International Workshop on Vulnerability and Resilience of Land Systems in Asia. Book of abstracts.

Heredia, N. (2006) Medidas especificas de proteccion del suelo haciendo uso del sistema agro-ecologico de ayuda a la decision MicroLEIS. Desarrollo de un modelo de red neuronal para la selección de especies arbustivas (Sierra^2). Zaragoza: CIHEAM-IAMZ Pub.

Heredia, N. (2007) Desarrolo de un modelo de evaluación de tierras en red neuronal (Sierra 2) para la selección de species arbustivas en la reforestación de zonas mediterráneas. Un nuevo componente del sistema MicroLEIS. Master of Science thesis, Mediterranean Agronomic Institute of Zaragoza (IAMZ), Zaragoza, Spain [in Spanish].

Karlen, D.L., Mausbach, M.J., Doran, J.W., Cline, R.G., Harris, R.F., Schuman, G.E. (1997) Soil quality: A concept, definition and framework for evaluation. *Soil Science Society of America Journal* **61**: 4–10.

Moscatelli, G., Sobral, R. (2005) Avances en la selecion de indicadores de calidad para las series de suelos representativas de la region Pampeana, Argentina. Buenos Aires: INTA [in Spanish].

Pierce, S.M., Cowling, R.M., Knight, A.T., Lombard, A.T., Rouget, M., Wolf, T. (2005) Systematic conservation planning products for land-use planning: Interpretation for implementation. *Biological Conservation* **125**: 441–58.

Shahbazi, F., Jafarzadeh, A., Sarmadian, F., *et al.* (2009) Climate change impact on land capability using MicroLEIS DSS. *International Agrophysics* **23**: 277–86.

Shim, J.P., Warkentin, M., Courtney, J.F., Power, D.J., Sharda, R., Carlsson, C. (2002) Past, present and future of decision support technology. *Decision Support Systems* **33**: 111–26.

United Nations Conference on Environment and Development (1992) Convention on biological diversity. In: *Proceedings of the United Nations Conference on Environment and Development, Rio de Janeiro, Brazil, June 1992*. New York: UN Department of Public Information.

USDA (2006) *Keys to Soil Taxonomy*, 10th edn. United States Department of Agriculture.

Vrscaj, B., Poggioa, L., Marsana, F.A. (2008) A method for soil environmental quality evaluation for management and planning in urban areas. *Landscape and Urban Planning* **88**: 81–94.

13 Impacts of Agricultural Land Change on Biodiversity and Ecosystem Services in Kahayan Watershed, Central Kalimantan

J.S. Rahajoe[1], L. Alhamd[1], E.B. Walujo[1], H.S. Limin[2], M.S. Suneetha[3], A.K. Braimoh[4] and T. Kohyama[5]

[1]Research Center for Biology, Indonesian Institute of Sciences, Bogor, Indonesia
[2]CIMTROP (Center for International Management of Tropical Peatland), University of Palangka Raya, Indonesia
[3]United Nations University Institute of Advanced Studies, Yokohama, Japan
[4]The World Bank, Washington DC, USA
[5]Faculty of Environmental Earth Science, Hokkaido University, Hokkaido, Japan

13.1 Introduction

Ecosystem services are the conditions and processes through which natural ecosystems and the species they comprise sustain and fulfill human life. They represent the multiple benefits humans can obtain, either directly or indirectly, from ecosystem functions (Daily, 1967). Many of these are crucial to human survival (e.g., food and fiber, watershed protection, climate modulation, nutrient cycling, and habitat for plants and animals). Economic valuation of ecosystem services is becoming increasingly important as a means of

Vulnerability of Land Systems in Asia, First Edition. Edited by Ademola K. Braimoh and He Qing Huang.

understanding the multiple benefits provided by ecosystems (Guo *et al.*, 2001). A study on socio-economic impacts of plantation projects in Kalimantan showed that they led to the loss of agricultural land, disappearance of traditional lifestyles and social disturbance within local communities (Potter & Lee, 1998).

Land-use change is among the most important factors that significantly affect ecosystem processes and services, since it potentially alters, either positively or negatively, the available net primary production area that is appropriated. But monitoring and projecting the impacts of such land-use changes are difficult because of the large volume of data and interpretation required and the lack of information about the contribution of alternate landscapes to these services. It has been predicted that, in the future, land-use change is likely to occur predominantly in the tropics, associated with decreases in net primary productivity and warming in surface temperature (DeFries & Bounoua, 2004). Land change in the tropics is mainly driven by agricultural expansion and deforestation (DeFries *et al.*, 1999). An example is the conversion of peat swamp forest to rice fields in Central Kalimantan in 1996. Called the One Million Hectare Mega Rice Project (MRP), it led to the establishment of big canals that resulted in the reduction of water levels in the peat swamp forest, and in the flow of water from the peat swamp forest to the Sebangau and Kahayan rivers. This resulted in decreased primary production and decreased water levels. As such conditions predispose to fires, frequent forest fires broke out in Central Kalimantan (Boehm *et al.*, 2005), consequently reducing biodiversity in the forest and attendant ecosytem services. Now, the MRP is a major fire hotspot region, especially in the dry season (Boehm, 2004). There was also a rapid conversion of peat swamp forest primarily into unused fallow land in 1999–2003. If the situation continues, there is a high risk that most of the peat swamp forest resource in Central Kalimantan will be destroyed within a few years, with grave consequences for the local hydrology, climate, biodiversity and livelihoods of the local people (Boehm, 2004).

As was mentioned earlier, forest fire is one of the factors causing forest degradation in Kalimantan. At the end of the extreme dry season in 1997 (caused by El Niño-Southern Oscillation (ENSO)), the biggest fires broke out over almost all forest types in Kalimantan and Sumatra Island. Forest fires have enormous impacts on the tropical forest ecosystems and biodiversity (Barber & Schweithelm, 2000). The estimated extents of spatial damage by fire during 1997–98 in Kalimantan were 75,000 ha of peat swamp forest, 2,375,000 ha of lowland forest, 2,829,000 ha of agricultural land, 116,000 ha of timber plantation, 55,000 ha of estate crops and 375,000 ha of dry scrub and grassland, with a total affected area of around 6,500,000 ha (Bapenas, 1999). Frequent forest fires have occurred over the past 10 years, and repeated cycles of burning have completely transformed large tracts of forest into grassland or scrubland. In a study on the effect of forest fires on biodiversity loss, carried out in the mixed dipterocarp forest in East Kalimantan, about 90% of 240 trees in a 1.6-ha permanent plot died due to forest fire (Whitmore, 1984).

A reduction in the quality of ecosystem services in the Kahayan watershed was also observed, as conversion to rubber plantation consequently led to soil erosion, sedimentation and decreased water quality in the Kahayan River. Our study site is Bawan village, which is located in the middle stream of Kahayan watershed. Bawan village is experiencing rapid deforestation and soil degradation. The changing land-use patterns at the expense of ecological services have become an issue of increasing concern. In this context, this study seeks to achieve two objectives: (i) to evaluate the impact of land-use changes on the biodiversity; and (ii) to examine how land-use changes affect the ecosystem. We therefore expect that this study can offer policy-makers some preliminary recommendations to ensure the sustainable use and management of similar ecosystems.

13.2 Study locations and methods

13.2.1 Study sites: Bawan village, Central Kalimantan

Central Kalimantan is the biggest province on the island of Kalimantan, occupying about 153,800 km^2. It lies between latitudes 0°45N and 3°30S, and stretches between longitudes 111° and 116°E. The capital city of this province is Palangkaraya, which is located on the Upper stream of the Kahayan River. The town occupies an area of about 2400 km^2. Plantations cover 3,139,000 ha of Central Kalimantan, including oil palm, rubber, rattan, coffee, cocoa and coconut. Food crops cover 5,980,750 ha, including paddy, cassava, pineapples, corn, bananas, rambutan and cempedak (local fruit).

The topography is flat in 32.97% of the area, hilly areas make up 9.83% and the area of extreme slope accounts for 40%. Almost four-fifths of Central Kalimantan is made up of tropical forests, producing valuable commodities such as rattan, resin and timber. The biggest river in Central Kalimantan is the Kahayan River, which runs from north to south through Kualakulun and Palangkaraya. The upper stream and downstream of the Kahayan River are located in Kalukung Mountain and Pulangpisau, respectively. The pH of the Kahayan River is around 5.5–7.5 in the rainy season, and varies between 4.0 and 7.0 in the dry season. The lower pH during the dry season is due to the strong effects of sulphuric acid from pyrite-containing peat that is discharged during the rainy season (Haraguchi, 2005). One of the reasons for low water quality is gold mining. Some 932 gold mining machines were observed in 2003, and 999 machines in June 2004, from the upper reaches to the downstream sections of the Kahayan River. Mercury (Hg) content in the Kayahan River was 0.18, 0.39 and 0.23 ppb (parts per billoin) in the upper, middle and lower reaches (Kido *et al.*, 2009). A positive relationship between the Hg content in the water or sediment and the number of the mining machines in Kahayan was observed, suggesting that Hg contamination was directly related to the gold mining activities (Yamada *et al.*, 2005).

The annual precipitation in Palangkaraya was 2731 mm (average from 1989 to 2008). Average monthly rainfall was in the range 153.5–303.1 mm, and fell below 100 mm for a few months of the dry season. The annual mean temperature varied between 26.8 and 28.1°C (Figure 13.1). The lowest annual rainfall was recorded for 1996, 2001 and 2004, while the highest annual temperature was recorded in 1998, a year after the biggest forest fire broke out in Central Kalimantan.

Bawan village is in the Pulang Pisau District, Banama Tingang sub-district. about 86 km from Palangkaraya, and lies at an elevation of 25 m above sea level (Figure 13.2). Bawan village is about 87 km^2 in area. Most of Bawan is covered by heath forest, with scattered patches of peat swamp forest. Bawan village is located about 15 m from the River Kahayan. The banks of the Kahayan River are mainly covered by rubber plantations (Figure 13.2). The number of households in Bawan village is 211, and the total population is 869. The population in the study area consists mostly of indigenous Dayak groups. About 39% of the respondents were working with the government of Banama Tingang sub-district, 23% undertook farming as their main livelihood.

The study site for biomass estimation and tree species biodiversity was in the heath forests. The heath forests of Borneo occur on poor sandy soils, strongly podzolized or even seasonally waterlogged to form humus podzols, humus iron podzols, and ground water humus podzols (Specht & Womersly, 1979; Whitmore, 1998). In general, the soil in heath forests is very acidic and poor in nitrogen content (Moran *et al.*, 2000). Tropical peatlands are one of the largest near-surface reserves of terrestrial organic carbon, with 65% or more organic matter content (MacKinnon *et al.*, 1996), and are ombrogenous (rain fed)

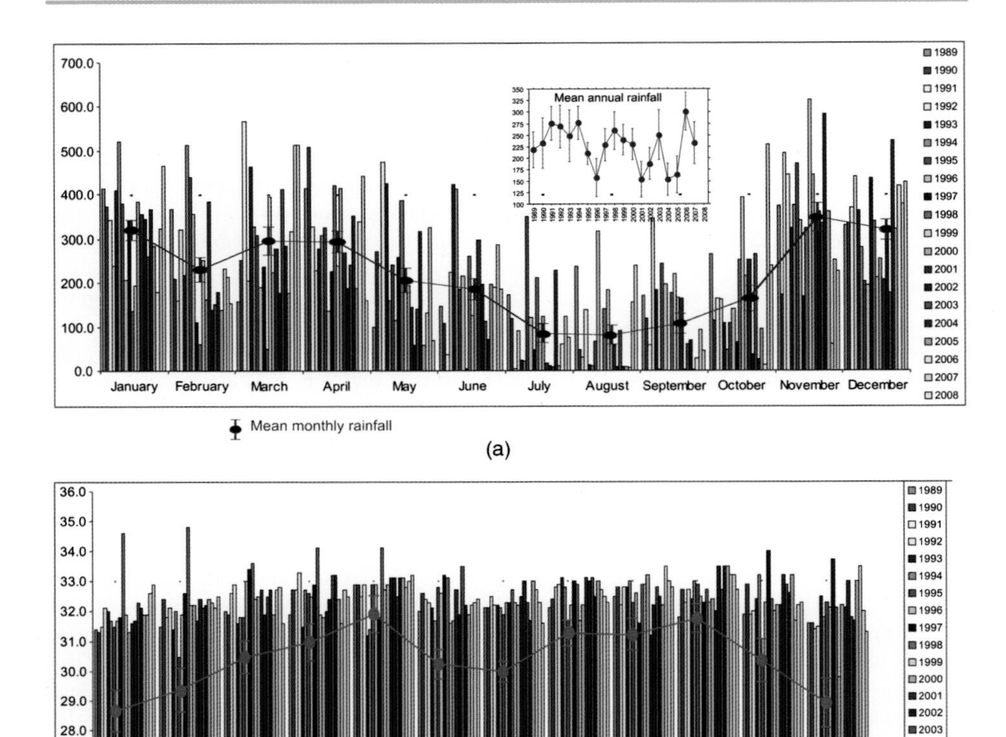

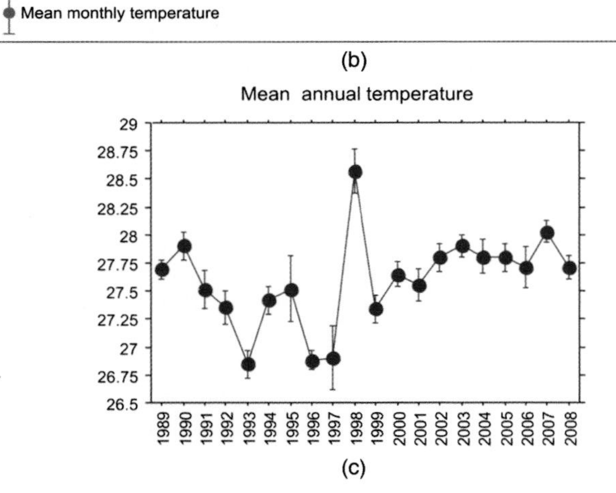

Figure 13.1 Monthly values for (a) rainfall and (b) temperature over the 20 years from 1989 to 2008 in Palangkaraya. (c) Mean annual temperatures for the same period

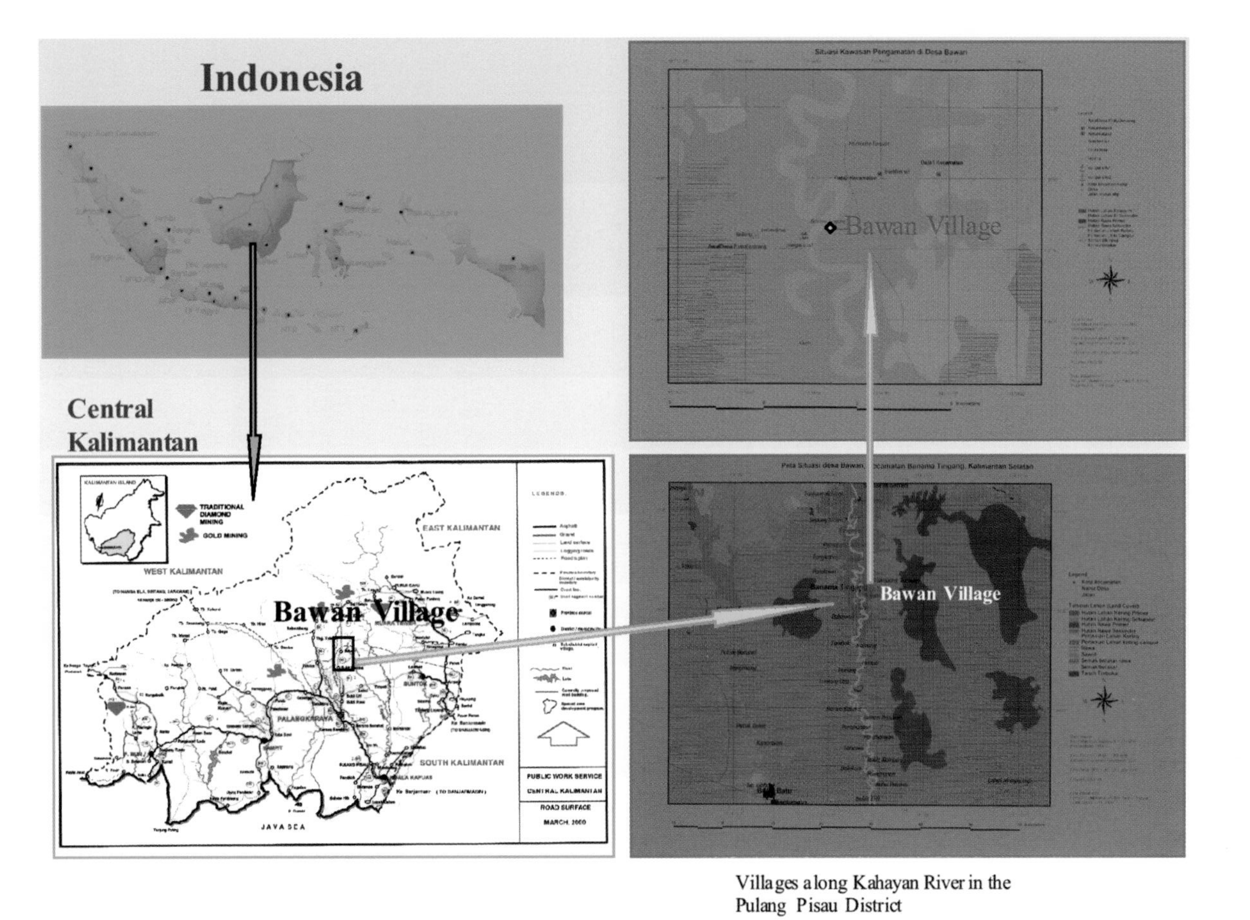

Figure 13.2 Bawan village is located on the bank of the Kahayan River, about 5–15 m distance from the river

(Driessen, 1978; Morley, 1981). The peat deposits are usually at least 50 cm thick, but they can extend up to 20 m. Because peat swamps are waterlogged they are nutrient deficient and acidic (pH usually is <4). The pH was recorded at around 3.4 in the natural peat swamp forest and 3.5–4.1 in the agricultural land, while in the rice land it was about 3.6 (J.S. Rahajoe, unpublished data).

13.2.2 Participatory rural appraisal

Stakeholder meetings and focus group discussions within a PRA (participatory rural appraisal) framework were carried out in Palangkaraya and Bawan village in December 2008 and February 2009, respectively. The stakeholder meeting was held to identify the current problems of Kahayan watershed *vis-à-vis* access to ecosystem services by the stakeholders. The stakeholder meeting was attended by participants from various institutions such as: the Provincial Center for Environmental Management, the Ministry of the Environment, the Provincial Center for Watershed Management, the Ministry of Forestry, an academic from the Faculty of Agriculture, Palangka Raya University, the Department of Forestry, the Department of Mining and Energy, an NGO, the Marine Affairs and Fisheries Department, and elders of the local people (Dayak) from the middle to the upper reaches of the Kahayan watershed.

The PRA exercise helped to identify the changes to ecosystem services and biodiversity during the last 40 years in Bawan village. The participants were divided into four groups: (1) Mapping of ecosystems group; (2) Income and expenditure group; (3) Institution and Infrastucture group; and (4) Farming activity group. Based on the information from the PRA exercise, a follow-up survey was conducted among selected respondents in the village based on a structured questionnaire.

13.3 Results and discussion

13.3.1 Current status of the Kahayan watershed

The stakeholder meeting identified areas of concern and priority for Kahayan watershed management:

1. Biophysical characterisation of the Kahayan watershed, including land use, hydrology, soil fertility, etc.
2. Assessment of the effect of gold mining on water quality along the Kahayan River as well as the effect of land conversion on soil erosion and nutrient depletion, etc.
3. The need to include traditional knowledge of the local Dayak people in land-use planning of the Kahayan watershed.

A main concern addressed in this meeting was that zonation of the Kahayan watershed should be undertaken by a combination of government and Dayak traditional rule. Other concerns discussed included: (i) land use overlapping among stakeholders; (ii) rule differences between the upper and lower reaches; (iii) soil erosion and sedimentation; (iv) use of the river as a municipal waste reservoir; (v) illegal gold mining along the river; (vi) water pollution (decrease in water quality) due to the overuse of chemical fertilizer; (vii) increase in the extent of rubber plantations along the Kahayan watershed; (viii) river flows becoming stronger; and (ix) loss of fishing resources.

The meeting also led discussions about addressing the Kahayan's watershed problems. These included:

- Coordination among stakeholders to control and manage the Kahayan watershed.
- Land-use regulation of the Kahayan watershed based on scientific results.
- Improve opportunities for employment.
- Establish an Integrated Kahayan Watershed Forum.
- Increase the land rehabilitation every year.
- Maintain the green belt alongside the river for about 2–5 km.

13.3.2 Biodiversity and forest products in Bawan village

Twelve species of timber tree and 14 medicinal plants were commonly used by the local people before the 1960s. The survey among the villagers revealed three major issues:

1. Since 2006 only six of the timber tree species and four medicinal plant species had been easy to find in the forest (Table 13.1). The populations of benuas and meranti, the

Table 13.1 Availability of major species of timber and medicinal plants in the forest

No.	Before 1960	After 2000
Timber species (local names)		
1	Benuas (*Shorea* spp.), Dypterocarpaceae	Benuas (*Shorea* spp.), Dypterocarpaceae
2	Meranti (*Shorea* spp.), Dypterocarpaceae	Meranti (*Shorea* spp.), Dypterocarpaceae
3	Krewing (*Cotylelobium burckii* Heim)	Krewing (*Cotylelobium burckii* Heim)
4	Pelepek (*Shorea materialis* Ridl.)	Pelepek (*Shorea materialis* Ridl.)
5	Madahirang	Madahirang
6	Kayu tahan	Kayu tahan
7	Ulin (*Eusideroxylon zwageri* T. & B.)	Few
8	Ramin (*Gonystilus*)	None
9	Pilau	None
10	Lanan	None
Medicinal plants		
1	Pasak Bumi (*Eurycoma longifolia*)	Pasak Bumi (*Eurycoma longifolia*)
2	Tabat Barito	Tabat Barito
3	Jaka sembung	None
4	Kelapapa	None
5	Sasenduk	None
6	Saluang belum	Saluang belum
7	Panamar gantung	Panamar gantung
8	Kumis kucing	None
9	Katipei pari	None
10	Akar kuning (*Arcangelisia flava* Merr.)	Akar kuning (*Arcangelisia flava* Merr.)
11	Memplam	Memplam
12	Tambat bumi	None
13	Tikang siou	None

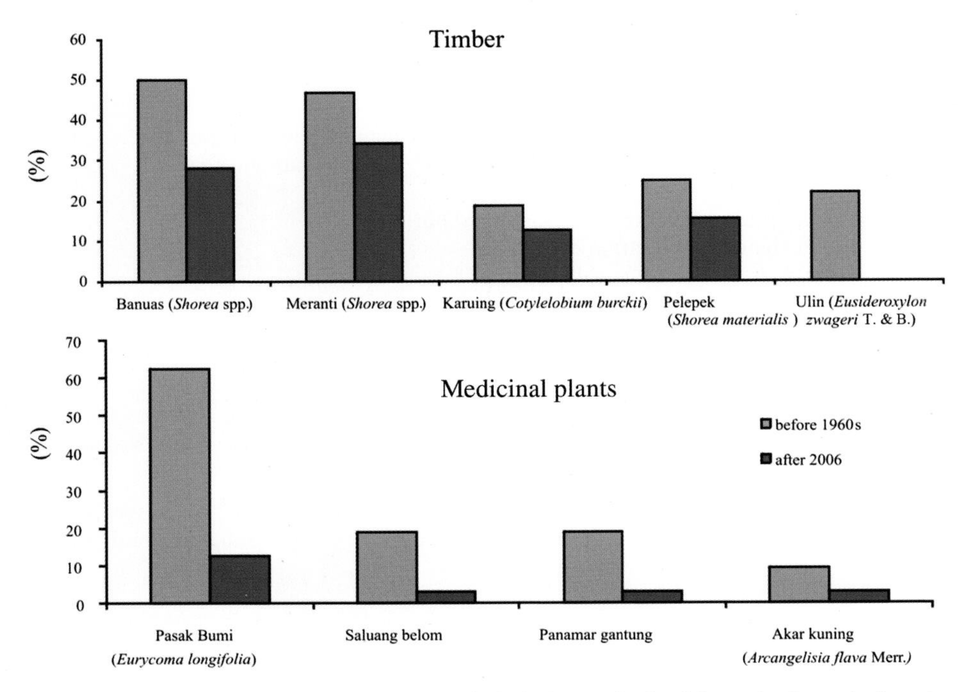

Figure 13.3 Main species of timber and medicinal plants obtained from the forest before the 1960s and after 2006

main timber species during the 1960s, had declined. This tendency was also found for medicinal plants (Figure 13.3).

2. Timber production had fallen for more than 40 years during the period 1960 until 2006, due to forest degradation and land conversion to farming or plantations (Figure 13.4).
3. From 1968 until the 1980s, three private forest concessions were in force leading to wood and rattan exploitation. To facilitate accessibility, the forest concession companies built a road for transporting forest products to the nearest river, from where they were delivered by boat downstream. This forest concession led to the degradation of the forest ecosystem and consequently affected the livelihoods of the local people. The location of the concession companies can be seen in Figure 13.4a (light green colour). These areas are now used for rubber plantations (Figure 13.4b).

To establish species diversity in the heath forest, a 0.5-hectare permanent plot was established for long-term ecological study in 2009. This permanent plot was located between Bawan and Tumbang Terusan villages. Tree density was about 1224 individuals per hectare, with a basal area of 25.6 m^2/ha. This value was lower compared to the heath forest in Lahei village about 35 km from the study site, where the tree density was 1982/ha and the basal area was 27.6 m^2/ha (Suzuki *et al.*, 1998; Miyamoto *et al.*, 2002). Forty-nine plant species were recorded, belonging to 21 families. Dipterocarpaceae is the major family in this study site, accounting for 12 of the tree species (Table 13.2). The importance value (IV) of 15 tree species from the highest to the lowest is shown in the Table 13.3. The highest IV was recorded for *Calophyllum* cf. *calcicola*. The distribution of tree class diameter is shown in

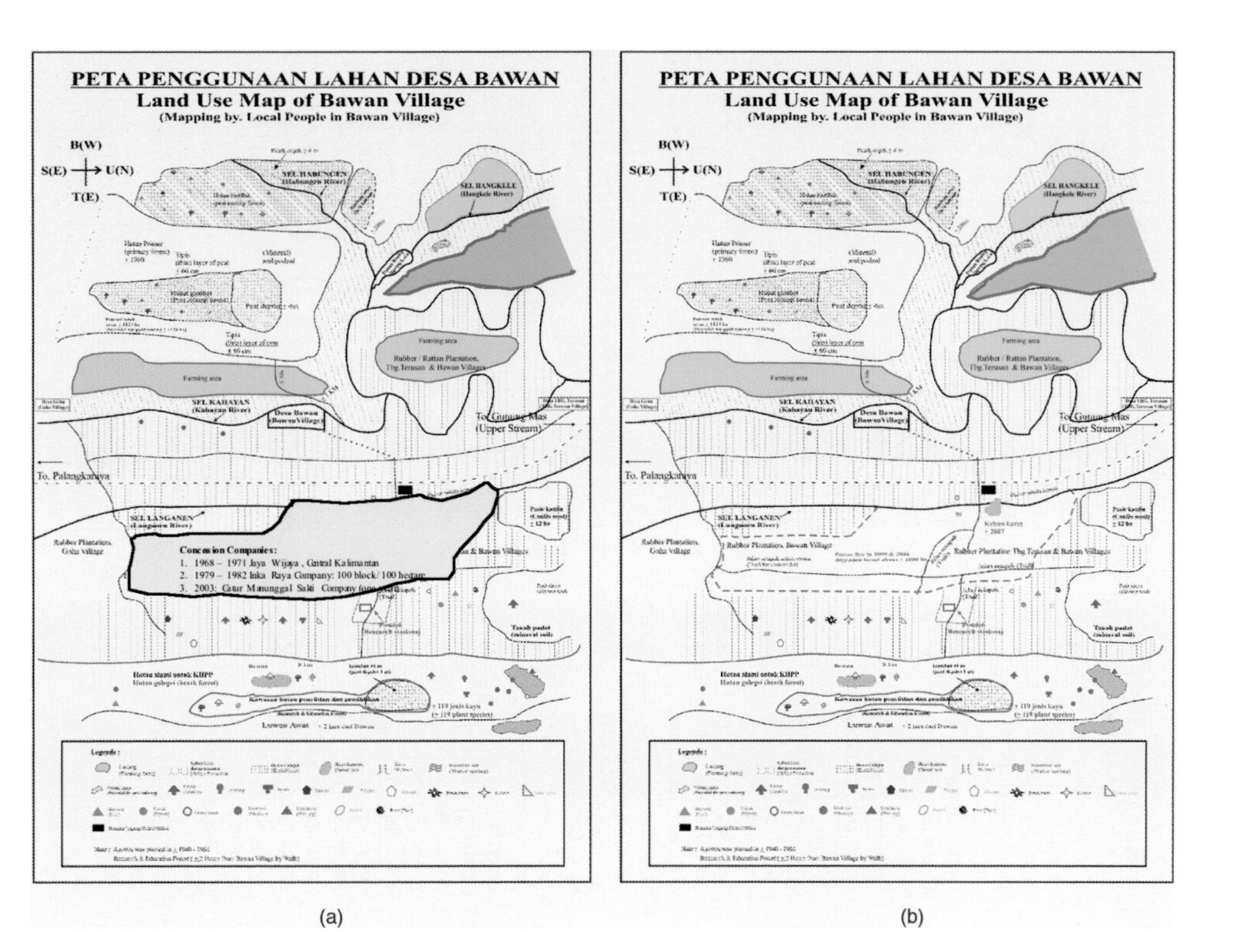

Figure 13.4 Land-use map of Bawan village (mapping by Bawan villagers during the participatory rural appraisal program)

Table 13.2 Tree species in the heath forest of Bawan village

No.	Plant species	Local name	Family
1	*Gluta renghas* L.	Rangas Manuk	Anacardiaceae
2	*Antidesma coriaceum* Tul.	Pupuh Pelanduk	Euphorbiaceae
3	*Ardisia sanguinolenta* Bl.	Pupuh Pelandok	Myrsinaceae
4	*Baccaurea javanica* (Blume) Muell. Arg.	Mahui	Euphorbiaceae
5	*Calophyllum* cf. *calcicola* P.F. Stevens	Jinjit	Clusiaceae
6	*Calophyllum gracilipes* Merr.	Tampang Gagas	Clusiaceae
7	*Calophyllum lanigerum* Miq.	–	Clusiaceae
8	*Cinnamomum javanicum* Bl.	Sintuk	Lauraceae
9	*Cotylelobium burckii* (Heim)	Krewing	Dipterocarpaceae
10	*Cratoxylum glaucum* Korth.	Jarunggang	Hypericaceae
11	*Croton oblongus* Burm. f.	Kepot Bojoko	Euphorbiaceae
12	*Dialium patens* Baker	Pupuh Palanduk	Fabaceae
13	*Diospyros curranii* Merr.	Uring Pahe	Ebenaceae
14	*Dipterocarpus elongatus* Korth.	Krewing Bayan	Dipterocarpaceae
15	*Elaeocarpus petiolatus* Wall.	Meranti	Elaeocarpaceae
16	*Elongatus* sp.	Krewing Bayan	Dipterocarpaceae
17	*Garcinia merguensis* Wight.	Enyak Berok Kuning	Clusiaceae
18	*Gluta renghas* L.	Rangas Manuk	Anacardiaceae
19	*Hopea ferruginea* Parij.	Meranti Bunga	Dipterocarpaceae
20	*Horsfieldia irya* (Gaerth.) Warb.	Meranti Kumpang	Myristicaceae
21	*Lithocarpus dasystachys* (Miq.) Red	Ampaning	Fagaceae
22	*Memecylon edule* Roxb.	Banjaris	Melastomataceae
23	*Neoscortechinia kingii* (Hook.f.) Pax	Parupuk	Euphorbiaceae
24	*Nephelium maingayi* Hiern	Ketiau	Sapindaceae
25	*Payena* cf. *khoomengiana* J.T. Pereira	Nyatu	Sapotaceae
26	*Plectronia glabra* Kurz.	Kayu Tulang	Rubiaceae
27	*Prunus grisea* (C. Muell.) Kalkm.	Meranti	Rosaceae
28	*Santiria laevigata* Bl.	Garunggang, Meranti daun besar	Burseraceae
29	*Shorea brunnescens* Ashton	Meranti Batu	Dipterocarpaceae
30	*Shorea materialis* Ridl.	Palepek	Dipterocarpaceae
31	*Shorea rugosa* Heim.	Rasak	Dipterocarpaceae
32	*Shorea scaberriana* Burck.	Rasak	Dipterocarpaceae
33	*Shorea beccariana* Burck.	Mahambung	Dipterocarpaceae
34	*Shorea teysmaniana* Dyer.	Lentang, Lentang bintik	Dipterocarpaceae
35	*Sindora leicocarpa* Backer.	Meranti Ehang	Fabaceae
36	*Stemonurus secundiflorus* Bl.	Ehang	Icacinaceae
37	*Syzygium garciniifolium* (King) Merr. & L.M. Perry	Bangaris, Kapur naga	Myrtaceae
38	*Syzygium* sp. 1.	Jambu-jambu	Myrtaceae
39	*Tristaniopsis whiteana* Grift.	Belawan Punai	Myrtaceae
40	*Ternstroemia aneura* Miq.	Ehang, Enyak Berok	Theaceae
41	*Tristaniopsis grandifolia* (Ridl.) P.G. Walsen & JT.	Meranti Emang	Myrtaceae
42	*Vatica umbonata* (Hook.f.) Burck.	Ehang Burung	Dipterocarpaceae
43	*Diospyros curranii* Merr.	Uring Pahe	Ebenaceae
44	*Diospyros bantamensis* K. & V.	Tutup Kebali	Ebenaceae
45	*Pimelodendron griffithianum* (Muell.Arg.) Benth.	Kayu Gita	Euphorbiaceae
46	*Syzygium zelanicum* (L.) D.C.	Galam Tikus	Myrtaceae
47	*Syzygium* sp. 3	Kayu Belawan	Myrtaceae

Note: This list was based on the ecological study in the heath forest, which was located between Bawan and Tumbang Terusan villages.

Table 13.3 Importance values (highest to lowest) of 15 tree species in the heath forest

Family	Species	Importance value
Clusiaceae	*Calophyllum* cf. *calcicola* P.F. Stevens	10.585
Icacinaceae	*Stemonurus secundiflorus* Bl.	9.749
Theaceae	*Ternstroemia aneura* Miq.	9.749
Dipterocarpaceae	*Hopea ferruginea* Parij.	8.914
Euphorbiaceae	*Neoscortechinia kingii* (Hook.f.) Pax.	8.914
Euphorbiaceae	*Baccaurea javanica* (Blume) Muell. Arg.	6.128
Anacardiaceae	*Gluta renghas* L.	5.014
Dipterocarpaceae	*Shorea brunnescens* Ashton	5.014
Euphorbiaceae	*Croton oblongus* Burm.f.	4.457
Elaeocarpaceae	*Elaeocarpus petiolatus* Wall.	3.621
Myrtaceae	*Syzygium* sp.	3.064
Myrsinaceae	*Ardisia sanguinolenta* Bl.	2.507
Rubiaceae	*Plectronia glabra* Kurz.	2.507
Dipterocarpaceae	*Shorea materialis* Ridl.	1.950
Euphorbiaceae	*Antidesma coriaceum* Tul.	1.393

Figure 13.5. The largest class, trees with diameters of 10 to 19.9 cm, contained 312 trees. The estimation of biomass production was 147.87 t/ha and the carbon stock was about 78.65 t/ha. The biomass in this study site was lower than that of heath forest in Lahei village – 232 t/ha with carbon stock of 127.6 t/ha (Miyamoto *et al.*, 2007). This information provides an estimate of the carbon that would potentially be released from the heath forest if the forest was converted to plantation and agricultural land in the vicinity of Bawan village. The differences in above-ground biomass and carbon stock in the heath forest relate to the species composition and to the differences in carbon content of each species (Anderson *et al.*, 1983; Moran *et al.*, 2000; Rahajoe, 2003).

Information from the Bawan villagers helped to easily locate six timber species in the forest, when the results of an ecological study in the area recorded the presence of only two species in the heath forest (suggesting a decrease in plant diversity). Other species

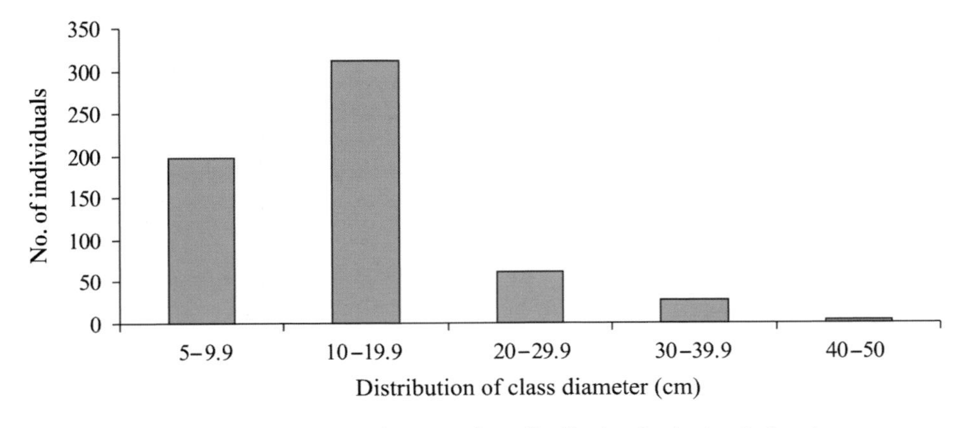

Figure 13.5 The tree diameter class distribution in the heath forest

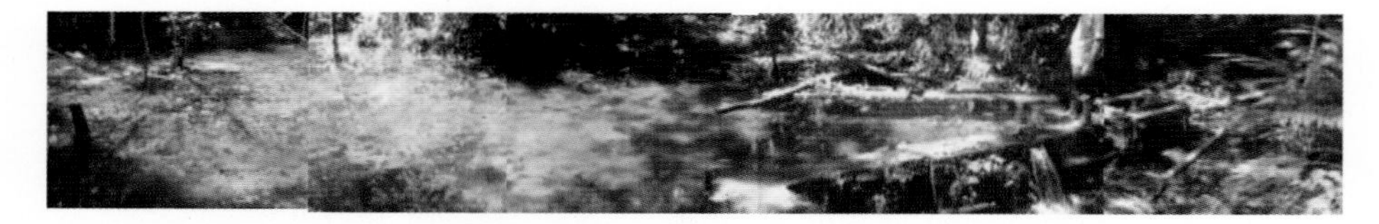

Figure 13.6 The spring used by Bawan villagers for drinking water, located about 6 km from the village

were found in the peat swamp forest or other locations in the heath forest. This is in line with the decreased pattern of plant species observed by Tyynellä *et al.* (2003) during the forest conversion program in South Kalimantan.

13.3.3 Ecosystem services in Bawan village

During the period 1949–2009, strong flows of the river resulted in erosion of the river bank in Bawan village. This led to changes to the Bawan village border, with some Bawan villagers moving their houses more inland about two to three times during the 60-year period. The area along the Kahayan watershed, from Bawan village to the other village upstream in Gunung Mas sub-district, was converted to rubber plantation (see Figure 13.2: yellow color indicates the rubber plantation). Also, illegal gold mining operates along the Kahayan River. There are about 1000 gold mining machines along the river from Palangkaraya to Gunung Mas sub-district (Yamada *et al.*, 2005). This resulted in high sedimentation, strong river water flow and poor water quality. Therefore Bawan villagers only used this water for showering their livestock and washing the equipment used for spraying insecticides. Since 1996 for household purposes they have used a spring located about 6 km from the village (Figure 13.6), in the middle of forest near an 11-year-old rubber plantation.

A landscape map of Bawan village drawn by local people from Bawan, Tumbang Terusan and Goha villages (during the focus group discussions) showed that the forest around Bawan village was covered by the peat swamp and heath forests (Figure 13.4). It also illustrated that the villagers still protect the forest for many purposes, such as to preserve sacred forest sites, to protect the spring, to safeguard forest area with a high potential for gold mining, and for research and education purposes. The research and education forest was initiated by CIMTROP (Palangkaraya University).

Based on information from Bawan villagers, more than 100 plant species were recorded in the peat swamp and heath forests, while from ecological studies, only 48 tree species in the heath forest were recorded (Table 13.2). It is predicted that species diversity will rise with increasing permanent plot area. Therefore it is recommended to expand the permanent plot in the peat swamp and the heath forest. In 2000 and 2006 forest fires broke out in Bawan village and surrounding areas following which the area was used for establishing a rubber plantation (Figure 13.4). About 1100 ha of new rubber plantations have been established in this location.

13.3.4 Rubber plantations in Bawan village

The world's largest producers of natural rubbers are three Asian countries: Thailand, Indonesia and Malaysia. These supply 80% of the world's natural rubber requirements. Rubber is Indonesia's most important agricultural export commodity, and 75% of national rubber comes from smallholder. In the first half of 2006, Indonesia contributed 2.28 million tonnes to global natural rubber exports. Before the 1960s, rubber growers and companies enjoyed good incomes from high prices. However, the prices have fluctuated in the international market and the price of natural rubber has decreased in the past decade. This has also affected the income of Bawan villagers, who have either returned to the forest products for their livelihoods or to cultivating agricultural crops for their daily needs.

The history of rubber planting in Central Kalimantan dates from the era of Dutch colonists, who in the early 20th century introduced rubber to the region. Before World War I, the price of rubber latex was high. Rubber plantations, introduced to Indonesia from 1900, entered South Kalimantan from Pagat near Barabai. By 2004 the total area planted with rubber in Central Kalimantan was about 807,254 ha, 47% of which 357,345 ha (47%) represented rubber plantations, comprising 349,152 ha of local estates (98%) and 5464 ha (2%) of large estates. The productivity of local rubber was low (500–600 kg/ha/year) while that of the large estates was 1000–1500 kg/ha/year (http://imperiumcentrebiz.blogspot.com; accessed 30 May 2014).

The government introduced an oil palm plantation program to Bawan villagers. However, the villagers refused to convert the rubber plantations to oil palm due to lack of technical expertise in cultivating oil palm, and unwillingness to cease rubber tapping, which was a major livelihood activity (Figure 13.7 highlights the three major products of the region from 1992 to 2007).

During the period 2002–2007, the area of rubber plantation increased, but rubber production fell in Pulang Pisau district (Figure 13.8). This fall was due to low rubber productivity, resulting from low fertilizer inputs, low levels of farmers' technical expertise, poor quality of rubber seeds, non-optimal tree density and advanced age of most rubber trees (at over 20 years old they had reached a less productive age). Replacement of the old rubber trees is necessary and can be done in several ways, including combining local farming culture with planting seasonal food crops. As per custom in the village, rubber trees are combined with rice or other crops (Figures 13.9 and 13.10). The integrated tree plantation approach combining an indigenous livelihood and industrial timber production is being implemented also in West Kalimantan (Tyynelä *et al.*, 2003).

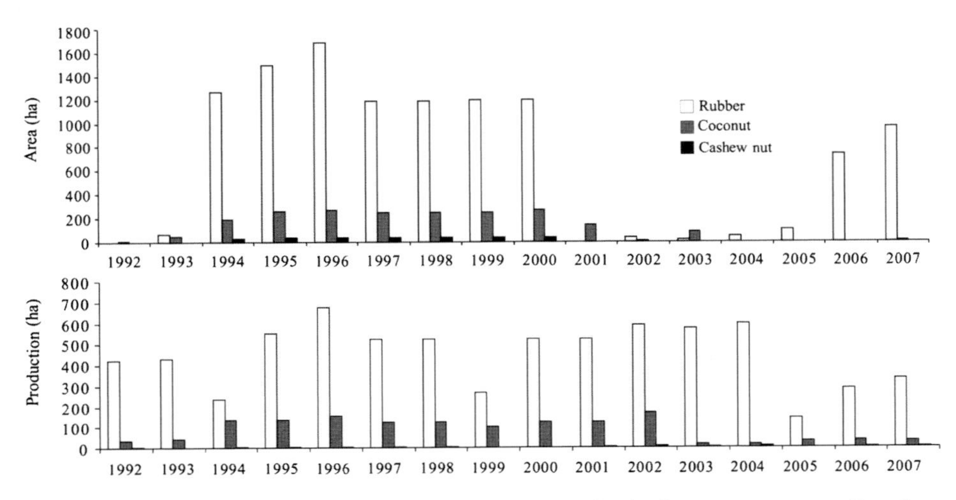

Figure 13.7 Three major plantations in the Pulang Pisau district from 1992 to 2007. Data from BPS (2002–2007)

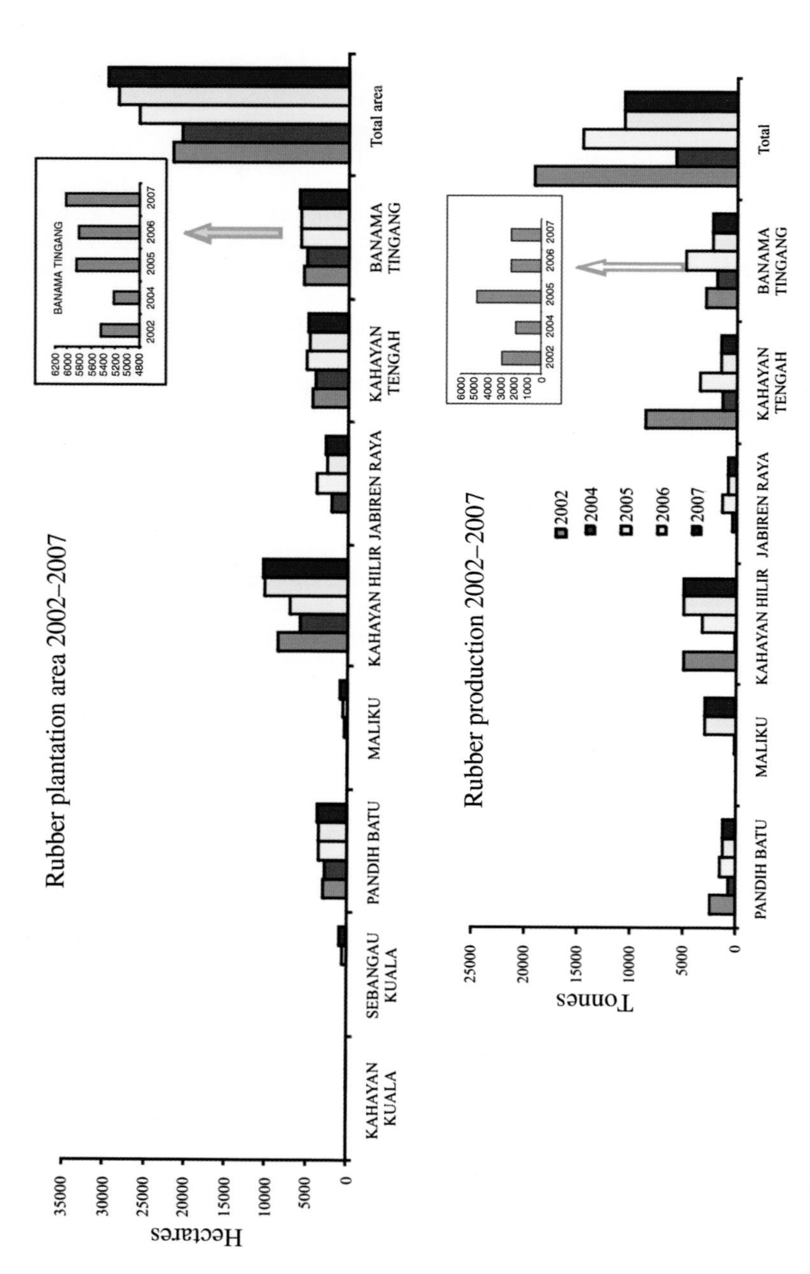

Figure 13.8 Rubber plantation area and rubber production during the period 2002–2007 in the Pulang Pisau Regency. Data from BPS (2002–2007)

Figure 13.9 A combination of rubber tree plantation (planted 2 years ago) and cassava

Figure 13.10 A combination of rubber tree plantation (planted 1 year ago) and rice

It is fortunate that rubber grows in areas with poor soil fertility, such as in the heath forest (pH 3.2–3.8; N = 0.04–0.40%; P_2O_5 = 1.97–2.71 mg/100 g; K_2O = 10.2–21.2 mg/100 g; Ca^{++} = 0.52–1.80, Mg^{++} = 0.22–0.28, Na^{+} = 0.17–0.28 and K^{+} = 0.11–0.17 meq/100 g) (Djuwansyah 2000). This is the best ecosystem in which to expand the rubber plantation in this area, where the main forest type is heath forest.

Changes in land use for the period 1968–2003 in Bawan village are illustrated in Figure 13.2. After the government stopped the operation of three concession companies in the region, the local people used the area and also areas affected by forest fire for rubber plantations by clear cutting the land. There was no limit on the extent of plantations, which was determined by the muscle power of the person who cleared the land. They could use as much land as they could clear cut. Therefore during the period of 2006–2008, rubber plantations managed by local people expanded to about 1100 ha. They received rubber seedlings from the government, or planted their own seedlings or ones from their relatives.

Before the 1980s, the main income of the village apart from the rubber plantation was from the manufacture of small boats and wooden house roofs, with supplemental revenue derived from agriculture. Nowadays the main income is from rubber plantations and jobs with the government. Most of the households in Bawan village have rubber plantations, with an area of about 1–4 ha. Almost all rubber produced in Bawan village is sold to middlemen, who sell the latex to the traders in Banjarmasin, South Kalimantan. Recently, the rehabilitation and replanting of rubber plantations increased in Bawan village, and thus was followed by decreased rubber production (see inset in Figure 13.8).

13.3.5 Changes in farming systems and agricultural produce

From 2002 to 2007, cassava was the main product of Pulang Pisau district, followed by ipomea (Figure 13.11). Peak cassava and ipomea production was recorded in 2004, while groundnut production peaked in 2007. Agricultural products such as rice, cassava, vegetables and soybeans in Bawan village are consumed only for the household or personal use of the villagers.

Based on information from the elders, there was a change in the patterns of weather and flooding frequency in their village over the time period. Before the 1960s the weather was

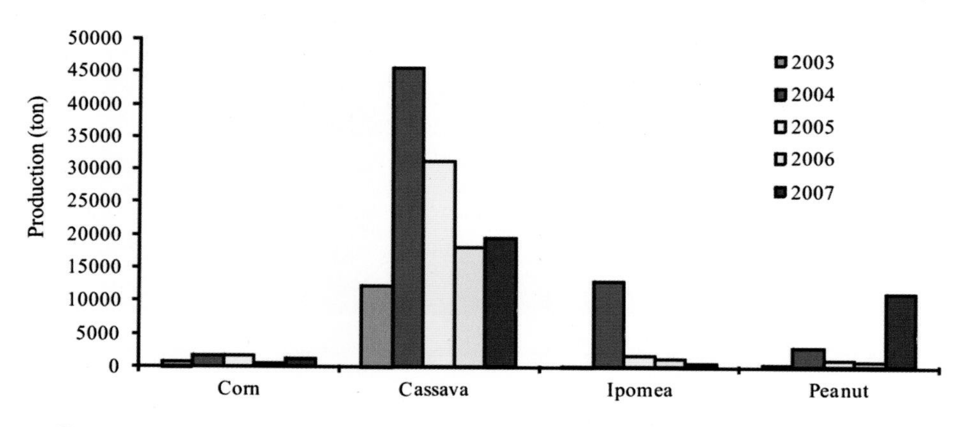

Figure 13.11 Crop production over the period 2002–2007. Data from BPS (2002–2007)

predictable. Then flooding frequency increased in the rice field, from one to three or four times a year. This resulted in failed harvests and reduced rice production.

In Bawan village, the farmers used to cultivate rice in September during the rainy season, using local rice varieties (Siung rice, pelita rice or amuntai rice). Siung rice is adaptable to the highland/upland area and amuntai rice grows in the lowlands under waterlogged conditions. Harvest time is in April or May, depending on the weather. But today, as it is difficult to predict the weather, cultivation timing has to be frequently changed. Some farmers told how soon after they sow the rice, the rice field floods, forcing them to recultivate. This increases the cost of rice cultivation and sometimes results in crop failure.

In the 1960s a farmer needed four to six cans (1 can = 10 kg) of seeds to cultivate rice in one hectare, and the rice production was about 200 cans. But today rice production is only about 50 cans, because of unpredictable weather. Bawan villagers face many problems in cultivating rice, notably flooding of the river over the lowland rice field, submerging the crop. The flood frequency is reported to have increased after 2000, to four to eight times a year, compared with just once a year before. Therefore, it has become almost impossible to cultivate rice in the lowlands. Cultivation in the highlands is hampered by limited water for growth during the dry season. Nonetheless, there has been a reduction in the lowland rice area and an increase in the upland rice area in the Pulang Pisau Regency (Figure 13.12). This pattern of decreased rice cultivation in lowland areas and decreased total rice production during the 5 years from 2002 to 2007 is illustrated in Figure 13.13. The pattern is the opposite for upland rice, which showed an increase in area and increased rice production. This indicates that there is a change from lowland rice to upland rice, and conversion of farming area to plantation lands (Figure 13.8), which is corroborated by the data in Figure 13.2 (the yellow area along the Kahayan watershed is plantation area). Due to decreasing rice production in the village, five years ago the villagers started to purchase rice for consumption to supplement their own rice production.

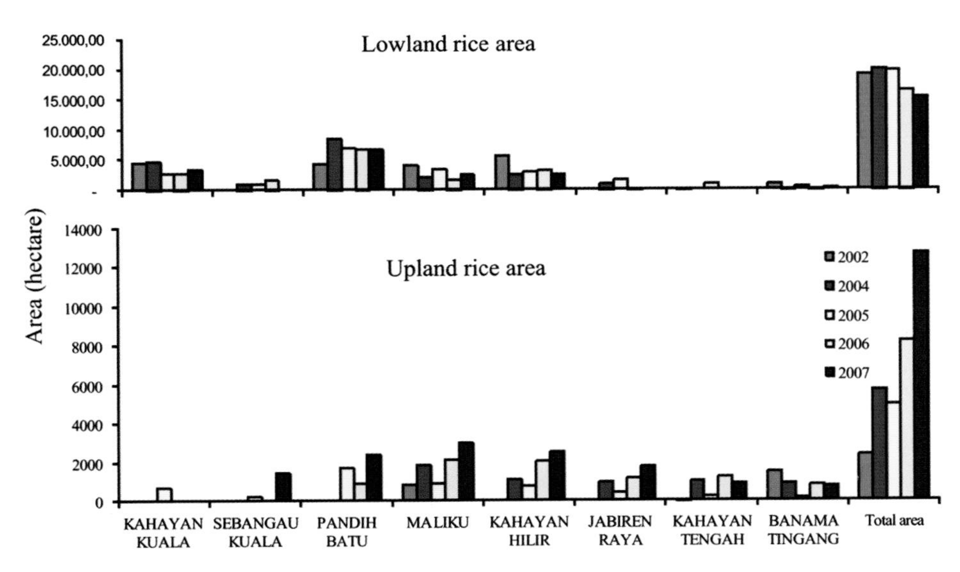

Figure 13.12 Lowland and upland rice areas in the period 2002–2007 in Pulang Pisau Regency. Data from BPS (2002–2007)

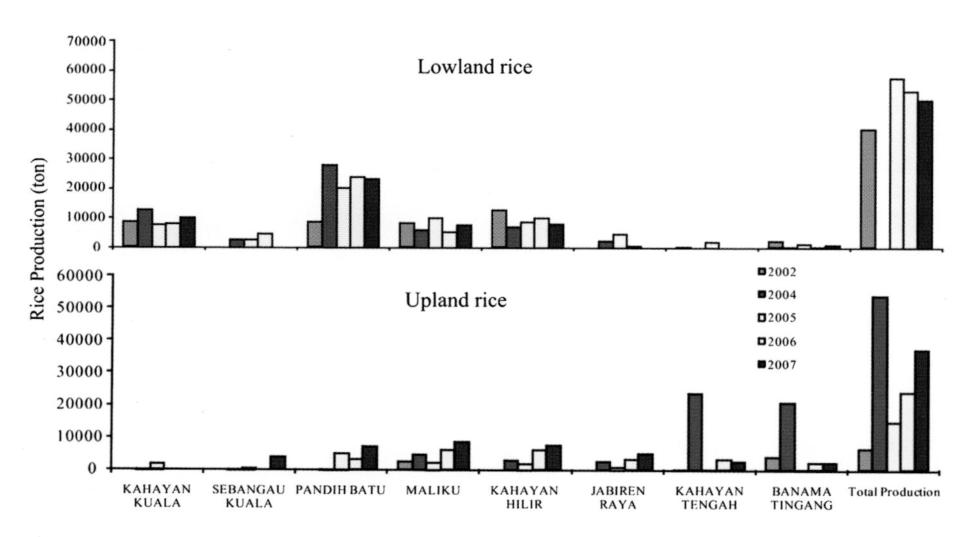

Figure 13.13 Lowland and upland rice production in the period 2002–2007 in Pulang Pisau Regency. Data from BPS (2002–2007)

13.4 Conclusion

The impact of changes in agricultural land use on biodiversity and ecosystem services is described for Bawan village. Unpredictable weather over a 40-year period has led to failure of rice cultivation and products (thereby affecting food security), increased flooding frequency in the village, a change in their potable water source from the Kahayan River to a spring, and a decrease in plant diversity and products from the forest. One hectare of heath forest conversion results in 73.78 t of carbon released from the above-ground biomass. This reflects the enormous carbon sequestration potential from land-use systems that conserve forest. There is a need to develop such land-use systems that not only benefit the environment, but also improve the livelihood of the local community.

Acknowledgements

We acknowledge the Bawan villagers, who supported the PRA and survey. We are grateful to the CIMTROP staff for supporting the field study, especially to Darma, Yuda and Pati. The research was supported by Asia Pacific Network for Global Change Research for the project entitled 'Managing ecosystem services in Southeast Asia: a critical review of experiences in montane upper tributary watersheds' (ARCP2009-18NMY-Braimoh)

References

Anderson, J.M., Proctor, J., Vallack, H.W. (1983) Ecological studies in four contrasting lowland rain forests in Gunung Mulu National Park, Serawak. III. Decomposition processes and nutrient losses from leaf litter. *Journal of Ecology* **71**: 503–27.

BAPENAS/The National Development and Planning Agency (1999) Planning for fire prevention and drought management project. Vol. 2. Cause, extent, impact, and cost of 1997/1998 fires and drought. Jakarta: BAPPENAS.

Barber, C.V., Schweithelm, J. (2000) Trial by fire: Forest fires and forestry policy in Indonesia's era of crisis and reform. World Resources Institute. Forest Frontiers Initiative. in collaboration with WWF Indonesia and Telapak Indonesia Foundation. 76 pp.

Boehm, H.D.V. (2004) Land cover change on wetland in Kalimantan, Indonesia, between 1999 and 2003. In Wise Use of Peatlands. *Proceedings of the 12th Peat Congress*, Paivanen, J. (ed.), **1**: 614–26.

Boehm, H.D.V., Ramírez, O.I.S., Bustillo, D. (2005) Environmental field trials and GIS image analysis in the Tangkiling District along River Rungan in Central Kalimantan, Indonesia. Presented at the International Peatland Symposium on 23.9.2005 in Palangkaraya.

BPS (2002) Kecamatan Banama Tingan Dalam Angka, 2000-2002. Catalog no. 1403.6203.140. Kabupaten Kapuas, Kalimantan Tengah: Biro Pusat Statistik (Center for Statistical Bureau).

BPS (2003) Banama Tingan Dalam Angka, 2003. Kalimantan Tengah: Biro Pusat Statistik.

BPS (2004) Banama Tingan Dalam Angka, 2004. Kalimantan Tengah: Biro Pusat Statistik.

BPS (2005) Banama Tingan Dalam Angka, 2005. Kalimantan Tengah: Biro Pusat Statistik.

BPS (2006) Banama Tingan Dalam Angka, 2006. Kalimantan Tengah: Biro Pusat Statistik.

BPS (2007) Banama Tingan Dalam Angka, 2007. Kalimantan Tengah: Biro Pusat Statistik.

Daily, G.C. (1967) Introduction: What are ecosystem services. In: Daily, G.C. (ed.), *Nature's Service: Societal Dependence on Natural Ecosystems*. Washington, DC: Island Press, pp. 1–10.

DeFries, R., Field, C., Fung, I., Collatz, G., Bounoua, L. (1999) Combining satellite data and biogeochemical models to estimate global effects of human-induced land cover change on carbon emissions and primary productivity. *Global Biogeochemical Cycles* **13**: 803–15.

DeFries, R., Bounoua, L. (2004) Consequences of land use change for ecosystem services: A future unlike the past. *GeoJournal* **61**: 345–51.

Djuwansyah, R. (2000) Some characteristics of tropical Podzol in Kalimantan. In: Iwakuma, T., Inoe, T., Kohyama, T., *et al.* (eds), *Proceedings of the International Symposium on Tropical Peat Lands*. Graduate School of Environmental Earth Science, Hokkaido University and Research Center for Biology, Sapporo, pp. 33–37.

Driessen, P.M. (1978) Peat soils. In: *Soil and Rice*. Los Banos, Philippines: IRRI, pp. 768–79.

Guo, Z., Xiao, X., Gan, Y., Zheng, Y. (2001) Ecosystem functions, services and their values. A case study in Xingshan County of China. *Ecological Economics* **38**: 141–54.

Haraguchi, A. (2005) Effect of sulfuric acid discharge from acid sulfate soil on the limnological environment in Central Kalimantan, Indonesia. *Limnology* **8**: 175–82.

Kido, M., Yustiawati, Syawal, M.S., *et al.* (2009) Comparison of general water quality of rivers in Indonesia and Japan. *Environmental Monitoring Assessment* **156**: 317–29.

MacKinnon, K., Hatta, G., Halim, H., Mangalik, A. (1996) *The Ecology of Kalimantan*. Periplus.

Miyamoto, K., Suzuki, E., Kohyama, T., Seino, T., Mirmanto, E., Simbolon, H. (2002) Habitat differentiation among tree species with small-scale variation of humus depth and topography in tropical heath forest of Central Kalimantan, Indonesia. *Journal of Tropical Ecology* **19**: 1–13.

Miyamoto, K., Rahajoe, J.S., Kohyama, T., Mirmanto, E. (2007) Forest structure and primary productivity in a Bornean heath forest. *Biotropica* **39**: 35–42.

Moran, J.A., Barker, M.G., Moran, A.J., Becker, P. (2000) A comparison of the soil water, nutrient status, and litterfall characteristics of tropical heath and mixed-dipterocarp forest sites in Brunei. *Biotropica* **32**: 2–13.

Morley, R.J. (1981) Development and vegetation dynamics of a lowland ombrogenous peat swamp in Kalimantan Tengah, Indonesia. *Journal of Biogeography* **83**: 383–404.

Potter, L., Lee, J. (1998) Tree planting in Indonesia: trends, impact and directions. Final Report of Consultancy for the Center for International Forestry Research (CIFOR). Department of Geography, University of Adelaide, 125 pp.

Rahajoe, J.S. (2003) The role of litter production and decomposition of dominant tree species on the nutrient cycles in natural forests with various substrate quality. Doctoral dissertation, Graduate School of Environmental Earth Science, Hokkaido University.

Specht, R.L., Womersly, J.S. (1979) Heathlands and related shrublands of Malesia (with particular reference to Borneo and New Guinea). In: Specht, R.L. (ed.), *Heathlands and Related Shrublands*. New York: Elsevier, pp. 321–38.

Suzuki, E., Kohyama, T., Simbolon, Haraguchi, A., Tsuyuzaki, S., Nishimura, T.B. (1998) Vegetation of kerangas (heath) and peat swamp forests in Lahei, Central Kalimantan. In: *Annual report of Environmental Conservation and Land Use Management of Wetland Ecosystem in South Asia*. Core University Program between Hokkaido University, Japan, and Research Center of Biology, LIPI Indonesia, Sapporo, pp. 3–4.

Tyynelä, T., Otsamo, R., Otsamo, A. (2003) Indigenous livelihood system in industrial tree plantation in West Kalimantan, Indonesia: Economics and plant species-richness. *Agroforestry Systems* **57**: 87–100.

Whitmore, T.C. (1984) *Tropical Rain Forests of the Far East*, 2nd edn. Oxford University Press.

Whitmore, T.C. (1998) *An Introduction to Tropical Rain Forests*, 2nd edn. Oxford University Press.

Yamada, T., Inoue, T., Dohong, S., Darung, U. (2005) Mercury contamination in river water and sediment in Central Kalimantan, Indonesia. In: *Environmental Conservation and Land Use Management of Wetland Ecosystem in Southeast Asia*. Annual Report. Hokkaido University–Indonesian Institute of Sciences. Japan Society for the Promotion of Science, pp. 115–20.

14 Spatio-Temporal Evolution of Urban Structure in Shanghai

Wenze Yue[1], Peilei Fan[2] and Jiaguo Qi[3]

[1]*Department of Land Management, Zhejiang University, Hangzhou, China*
[2]*School of Planning, Design and Construction, Center for Global Change and Earth Observations, Michigan State University, East Lansing, USA*
[3]*Department of Geography, Center for Global Change and Earth Observations, Michigan State University, East Lansing, USA*

14.1 Introduction

As global cities are important centers for economic growth and population concentration, a deep understanding of the evolution of coupled human-environment systems in global cities has significant implications for assessing the vulnerability of global cities to climate hazards and designing adaptation and mitigation strategies. Most related studies have treated cities as homogenous units and rarely have they delved under the city level and investigated its subunits, which have distinct social, economic, institutional, and political structures, mainly due to the limitation of data.

Urban socio-spatial differentiation – the arrangement of individuals into divisions of power and wealth occupying locations in different geographic space within a city – is a relatively nascent phenomenon in China. Historically, most large cities in China did not show any marked trend for socio-spatial differentiation as the urban form was characterized by mixed social areas built upon different land uses. However, during the period of economic reform, along with rapid urban expansion, the internal structures of Chinese cities experienced tremendous transformation. Socio-spatial differentiation emerged and became pronounced between different areas within cities, due to relocation of residents and factories, infrastructure provision, and improvement of the urban environment, enabled by local governments and triggered by China's urban land and housing reforms. Will Chinese

Vulnerability of Land Systems in Asia, First Edition. Edited by Ademola K. Braimoh and He Qing Huang.

cities follow the trend of socio-spatial differentiation seen in cities of industrialized countries? What are the policy implications of socio-spatial differentiation to adaptation and mitigation to climate hazards? This chapter hopes to contribute to the debate by studying Shanghai's socio-spatial differentiation. It also serves as a socio-economic basis for a vulnerability analysis of Shanghai.

Studies on socio-spatial differentiation of Shanghai have focused on specific aspects, such as housing and real estate (Wu, 2001) and relocation of residents (F. Wu, 2004; He & Wu, 2005; Li & Wu, 2006b). In order to have a comprehensive assessment, integrating different aspects of sustainable urban development, including land use, economic growth, population change, and environment quality, this chapter studies the coupled human-environment systems of 13 urban districts and investigates how they evolved differently due to their different locations, historical experiences, and development policies from 1949, especially during the reform era that started from 1978. While some districts exhibited a positive urbanization trend, indicated by rapid economic growth, better environmental conditions, and more appropriate land use, such as Pudong New District, Minghang District, and Xuhui District, others showed a negative urbanization trend, illustrated by slow economic growth, worse environmental conditions, and less appropriate land use, such as Zhabei District and Yangpu District. In this chapter we use a multidisciplinary approach based on theories and methods in urban studies, human geography, and environmental sciences. We employ socio-economic data analysis, remote sensing, and GIS spatial analysis to characterize and assess the evolution of urban districts in Shanghai.

The rest of the chapter is organized as follows. Section 14.2 introduces the theoretical framework that we apply to this research and describes the data and method. Section 14.3 presents findings on co-evolution of land use, economic conditions, and environmental conditions at both the city and district levels. Section 14.4 identifies a set of factors driving the evolution of urban structure and spatial differentiation in Shanghai and the implications for the city to mitigate and adapt to climate hazards. Section 14.5 concludes and offers policy recommendations to mitigate the increasing disparity between urban districts.

14.2 Theoretical framework, study area, data, and methodology

14.2.1 Theoretical framework

14.2.1.1 Vulnerability analysis and difference of social units Vulnerability is defined as 'the degree to which a system, subsystem, or system component is likely to experience harm due to exposure to a hazard, either a perturbation or stress/stressor' (Turner *et al.*, 2001). Vulnerability analysis has drawn mainly from three major concepts: entitlement, coping through diversity, and resilience. All three concepts emphasize that the differences between social units help to explain why they face different vulnerabilities, have different coping capacities, and respond to disturbance at different speeds. Vulnerability analysis therefore must account for differences between social units as impacts and strategies to respond to perturbation and stressors are strongly linked with the social, economic, institutional, and political structure of the social units.

Most studies on coupled human-environment systems of global cities have primarily focused on the city level due to the unavailability of data – for instance, studies on Shanghai's vulnerability to climate hazards conducted by Serbinin *et al.* (2007) and WWF (2009). In this chapter, adopting the urban socio-spatial differentiation framework widely used in urban social geography, we examine the coupled human-environment system of Shanghai at a finer scale to provide a socio-economic basis for the vulnerability analysis of Shanghai for future studies. We briefly introduce the concepts of socio-spatial differentiation and its Chinese characteristics in the following paragraphs.

14.2.1.2 Socio-spatial differentiation Under the labels of 'divided cities' (Fainstein *et al.*, 1992; Marcuse, 1993), 'dual cities' (Mollenkopf & Castells 1991; Kempen, 1994), 'polarized cities' (Burgers, 1996; Hamnett, 1994; Wessel, 2000; Walks, 2001), and 'partitioned cities' (Marcuse & Van Kempen, 2002), aggravated urban socio-spatial differentiation has attracted increasing attention from the academic field in industrialized countries in recent years. Macro socio-economic changes, i.e., economic restructuring and globalization, segmented labor market and ethnic segregation, and the retreat of the welfare state, are said to directly cause the increasing inequality of urban space in post-Fordist cities (Li & Wu, 2006b). First, economic restructuring in industrialized countries, characterized by the decreasing importance of traditional manufacturing and the rise of the high-skill sectors (high-tech manufacturing and producer services), especially in global cities, has resulted in social polarization, leading to the re-sorting of urban space for production (spaces for offices and factories) and consumption (housing and entertainment) (Sassen, 1991; Burgers, 1996; Walks, 2001; Marcuse & Van Kempen, 2002). Second, the consequent segmented labor market and the discriminatory housing market based on ethnicity standards resulted in a segregated residential structure, with the extreme case of black underclass ghettos in US cities (Clark, 1986; Wilson, 1987; Van Kempen & Ozuekren, 1998). Third, the transition from the Keynesian welfare state to the Schumpeterian workfare state leads to further differentiation of urban spatial spaces (Jessop, 1994; Musterd & Ostendorf, 1998). While the Keynesian welfare state shelters disadvantaged groups from marginalization thus countering the residential disparity, the Schumpeterian workfare state aims to strengthen structural competitiveness by intervening on the supply side and to subordinate social policy to the needs of labor market flexibility. In addition to the macro factors, micro-level factors such as life cycle, social and employment status, and household composition also contribute to spatial differentiation of residential urban space (Knox & Pinch, 2000).

Despite the usefulness of the above-mentioned three dimensions as an analytical framework, the underlying forces and their consequent impact on urban spatial differentiation in Chinese cities differ significantly from those in advanced countries. First, unlike economic restructuring in cities of most industrialized countries, Chinese cities have experienced rapid industrialization in the reform period; only a handful of large cities, such as Shanghai and Beijing, have just started to experience deindustrialization, illustrated by the decline of traditional manufacturing and the rise of high-tech manufacturing and producer service sectors. Further, while the impact of globalization is mostly limited to the production sphere in Western cities, global capital has directly maneuvered the spatial differentiation and changed the urban landscape of Chinese cities through heavy investment not only in the manufacturing sector (usually located at the urban fringe) but also in the real estate sector in both central and peripheral areas of some large cities. Second, ethnicity discrimination does not appear to be such a serious challenge for most Chinese cities; rather, discriminatory policies based on the education and skill of migrants has been more pronounced in creating social stratification of residential spaces in urban China. Third, although the Chinese state has dramatically changed its role, especially because the central government withdrew their direct control of cities' urban environment, the role of the state is much more complicated than the transformation from the Keynesian welfare state to the Schumpeterian workfare state in the industrialized countries. It is more appropriate to say

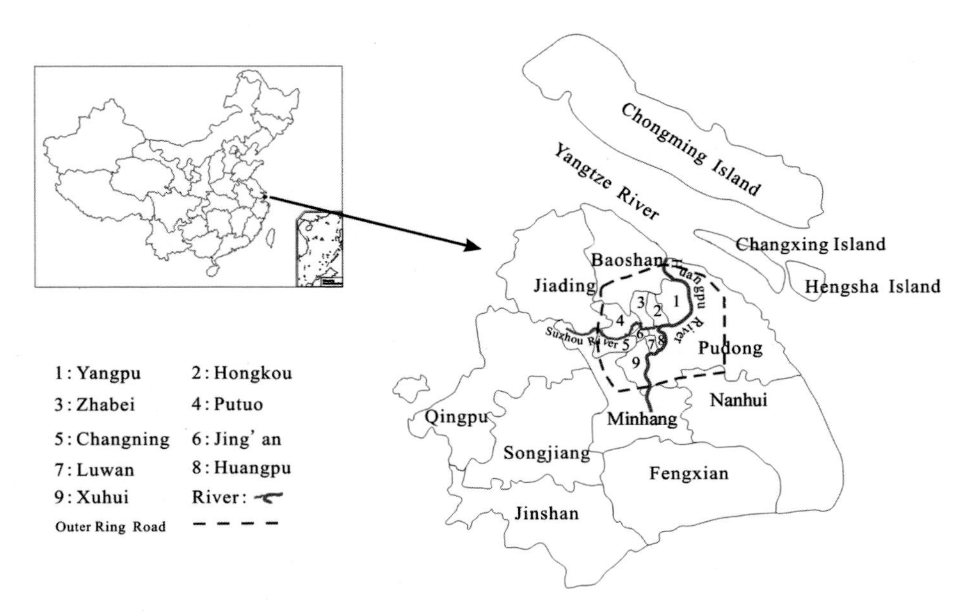

Figure 14.1 Location and administrative divisions of Shanghai

that the Chinese state – from central to municipal to district governments – constitutes a multi-scaled state system. Despite the central government's withdrawal of direct control through fiscal decentralization and land reform policies, local governments at both municipal and district levels gained more autonomy in changing urban spaces by working closely with global capital and domestic private sectors, resulting in the increasing socio-spatial differentiation not only between cities but also within cities.

14.2.2 Study area

Located on the east coast of China at latitude 31°14′N and longitude 121°29′E, Shanghai has a registered population of 18.6 million and total area of 6340.5 km^2 (Figure 14.1). Shanghai has remained China's largest economic center since 1850. This primacy reached its peak in the 1920s and 1930s, when Shanghai was an important international city in East Asia made prosperous by global and local trades. During the socialist planning period (1949–1978), Shanghai was rapidly transformed into a manufacturing center (Cai & Victor, 2003), with the secondary industries contributing over 70% of its total output and a low urbanization ratio. Shanghai's urban development has progressed at an unprecedented pace since the economic reforms began in 1978 (Xu *et al.*, 2004); the urbanization ratio increased significantly, from 59% in 1978 to 86% in 2007.

With most of its area flat, Shanghai has an average elevation of 4 m above sea level. Situated on the soft mud of the Yangtze River delta, its natural subsidence is estimated at 2 to 2.6 m. The city has a subtropical monsoon climate with four distinct seasons, and frequent typhoons in summer and autumn. Shanghai's major climate hazards are identified as sea-level rise, storm surge, coastal erosion, and salt water intrusion; these will significantly impact its economy in the foreseeable future (WWF, 2009).

While this chapter does not conduct an analysis of Shanghai's vulnerability to climate directly, we analyze the evolution of coupled human-environment systems at a district level, focusing on urban land use, economics, population, and environmental evolution, as a socio-economic basis for vulnerability analysis. Although Shanghai has a total of 18 districts and one county, we selected only the central districts and their immediate outer districts as our study objects due to data limitation caused by administrative changes. We divided the samples into two parts. The first collection contained 10 central districts of Shanghai, including Huangpu, Luwan, Xuhui, Changning, Jing'an, Yangpu, Hongkou, Zhabei, Putuo, and the part of Pudong New Area along the Huangu River. The second collection included the first collection and three outer districts, namely Baoshan, Jiading, and Minhang, and the rest of Pudong New Area.

14.2.3 Data and methodology

We relied on remote sensing imageries and used a geographic information system (GIS) to extract district-level data, in addition to statistical data collected from the Statistical Bureau of Shanghai and multiple district governments. We used the land use data of Shanghai from 1947 to 2003 (1947, 1958, 1964, 1979, 1984, 1988, 1993, 1996, and 2003). We derived the urban built-up area boundaries from the aerial remote-sensing surveys and Landsat5 TM images, and divided land use into different categories according to the standard land use classification (GBJ 137-90) issued in 1991 by China's Ministry of Construction. We employed an overlay method in GIS analysis to extract the growth area and ratio for the urban built-up area. In order to extract land use conversion information at a Shanghai and district scale, respectively, we converted vector data of land use to raster data, used the 'Grid Calculator' to detect the land use change, and obtained a conversion matrix for every 2 years; we then derived this data of each land type by district vector data.

Statistical data for the population, economy, and environment were collected from various years of the *Shanghai Statistical Yearbook*, *Shanghai Population Statistical Yearbook*, *Shanghai Economic Statistical Yearbook*, the historical books of districts or counties (Quxian Zhi), statistical yearbooks and bulletins for each district, as well as from web pages of statistics bureaus of Shanghai and districts.

14.3 Findings

14.3.1 Urban evolution of Shanghai

14.3.1.1 Expansion of the urban built-up area: 1947 to 2008 The urban built-up area of Shanghai has increased rapidly since the 1940s, expanding 18 times from 76 km^2 in 1947 to 1462 km^2 in 2008 (Figure 14.2). However, there are significant differences in the pace of growth between different periods. While annual land expansion ratios were 4.6 km^2, 3.2 km^2, 2.7 km^2, and 6.0 km^2 for the periods of 1947–58, 1958–64, 1964–79, and 1979–84, respectively, they dramatically spiked to 26.1 km^2, 43.5 km^2, and 123.2 km^2 for the periods of 1984–88, 1988–2002, and 2002–2008, respectively. The fastest expansion occurred during the period from 2002 to 2008, when the city expanded by 751.38 km^2 within a short span of 6 years, due to the connection of the main urban core with the nearby satellite towns. Whereas before 1996 the city expanded mainly along the northeast-southwest axis, Shanghai has mainly developed towards the east and west since 1996.

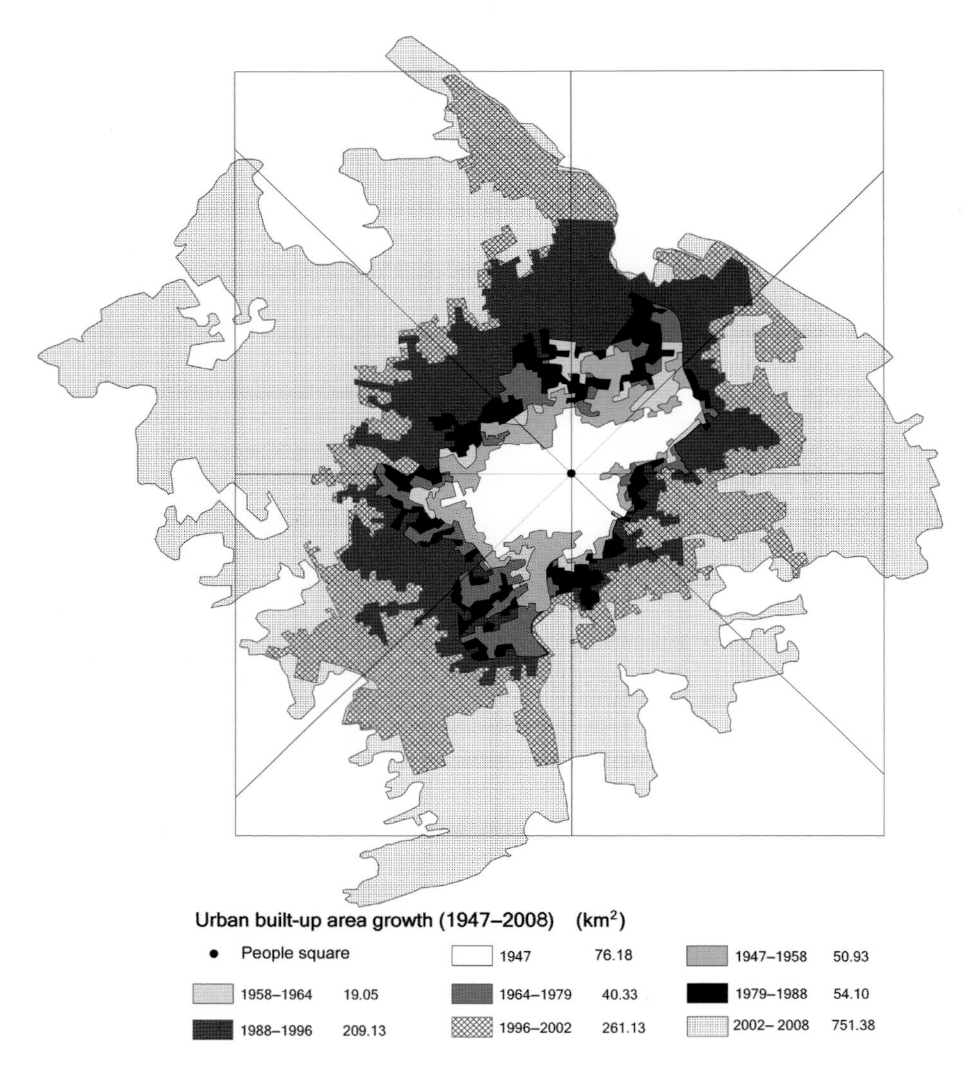

Figure 14.2 Expansion of urban built-up area, 1947–2008

14.3.1.2 Land use change in the central urban area Besides the expansion of the urban built-up area, Shanghai's urban land also experienced dramatic conversion. First, its urban residential land increased continuously: the area of urban residential land in 2003 was 2.62 times (105.25 km^2) that in 1947 (40.2 km^2). It is worth noting that the phase of fastest increase was from 1988 to 1993, when the policy of commodity housing was started in Chinese cities. Most residential land was converted from agriculture and reserved land.

Second, Shanghai's industrial land experienced first an increasing then a decreasing pattern from 1947 to 2003. During the Maoist period, the overall goal of urban planning in China was to create decentralized and self-sufficient urban forms by encouraging rapid industrialization. Shanghai focused on building itself as a production-oriented industrial city from 1947to 1988. The urban transformation during this period was mainly characterized by rapid industrialization, especially by the formation of state-owned enterprises that spread throughout the city and were mixed with traditional communities. However, when land reform and housing reform started in the 1980s, factories started to be relocated in the urban periphery or in other cities, and industrial land was converted to commercial land for greater profit. Shanghai's industrial land use history vividly reflects the industrialization path taken by the city: it first increased rapidly from 19.10 km^2 in 1958 to 53.02 km^2 in 1988, remained stable until 1996, then decreased slightly from 52.65 km^2 in 1996 to 44.21 km^2 in 2003.

Third, like urban residential land, Shanghai's urban commercial and public land demonstrates an overall growth. It increased steadily from 27 km^2 in 1947 to 52.84 km^2 in 1988, remained stagnant from 1988 to 1996, and then experienced a sharp rise after 1996, from 51.10 km^2 in 1996 to 63.01 km^2 in 2003, due to the increased public infrastructure investment.

14.3.1.3 Economic development Shanghai's economic development can be divided into three phases. During the first phase of the socialist planning system (1949–78), the growth rate remained slow. During the second phase (1978–92), after the economic reforms, Shanghai started to slightly hasten its pace of growth. Shanghai's economy really took off during the third phase (1992 to present), after the establishment of Pudong New Area, reflected by a much higher growth rate of GDP and GDP per capita after 1992 (Figure 14.3).

14.3.1.4 Population Due to the strict household registration system implemented in urban China, in contrast to the dramatic changes in economic growth, Shanghai's population growth remains steady, except for a spike in 1958 when there was an administrative adjustment of Shanghai's municipality (Figure 14.4). The total population even declined during the Cultural Revolution as a large proportion of the intellectual youth (high school and college students) were sent to receive re-education in the countryside in a different part

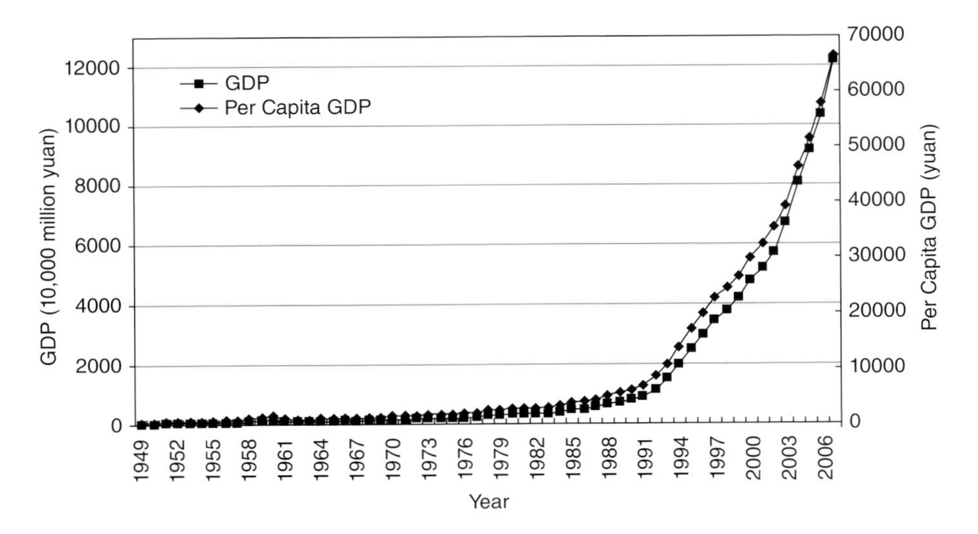

Figure 14.3 Economic development of Shanghai city, 1949–2007

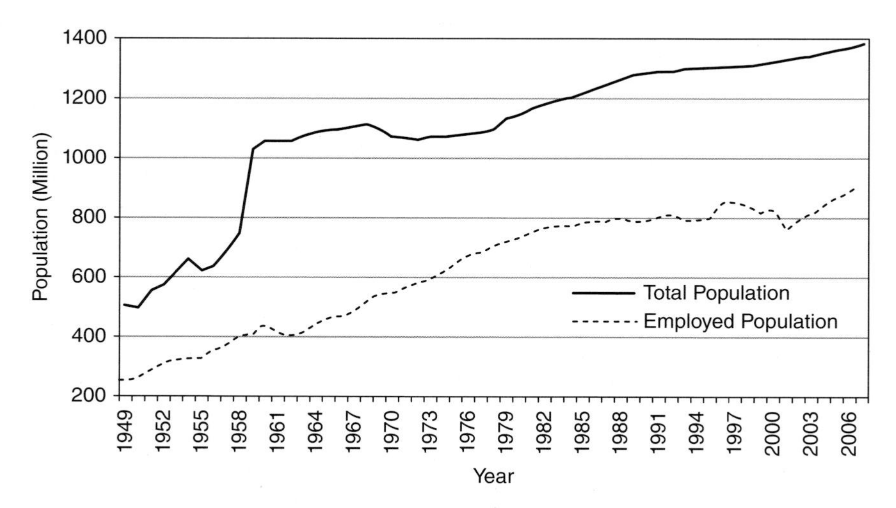

Figure 14.4 Population of Shanghai, 1949–2007

of China. Shanghai's employed population showed a steady increase until 1992 when economic restructuring occurred in Shanghai. It should be noted that Figure 14.4 only reflects the population with household registration and does not include migrants who do not have formal household registration but account for a large proportion of Shanghai residents. We will discuss this in detail in Section 14.4.

14.3.1.5 Environment Measured by PM_{10}, SO_2, NO_2, and days when air quality is greater than or equal to Grade II, Shanghai's urban environment shows significant improvement since the mid-1990s. For instance, the density of PM_{10} fell from 0.239 mg/m^3 in 1996 to only 0.088 mg/m^3 in 2007, whereas the density of NO_2 decreased from 0.089 mg/m^3 in 1996 to only 0.054 mg/m^3 in 2007. Further, the days on which air quality was greater than or equal to Grade II grew from 295 to 328 (Table 14.1).

14.3.2 Urban transformation at the district level

14.3.2.1 Land use change in the central urban area Different districts in Shanghai's central urban area have exhibited different trends in residential land use change from 1947 to 2003. Overall, all districts except Huangpu, Jing'an, and Luwan gained residential land during the period. It is worth noting that residential land in Hongkou and Pudong grew rapidly before 1996 but decreased from 1996 to 2003.

Industrial land in most central districts followed the general trend of Shanghai, i.e., industrial land first experienced fast expansion until 1988, remained stagnant until 1996, and then decreased sharply afterwards. For instance, Yangpu district remained as the district with the largest industrial land area until 1996 but lost over 30% of its industrial land from 1996 to 2003. The only exceptions were Xuhui, Pudong, and Putuo, whose industrial land kept growing from 1947 to 2003.

Corresponding to the rise of tertiary industry in the 1990s, urban commercial and public land started to grow rapidly for most districts, especially from 1996 to 2003, except for Xuhui, Changning, and Putuo.

Table 14.1 Shanghai's urban environment

Indicator	1996	1997	1998	1999	2000	2001	2002	2003	2004	2005	2006	2007
SO_2 (mg/m^3)	0.059	0.068	0.053	0.044	0.045	0.043	0.035	0.043	0.055	0.061	0.055	0.055
NO_2 (mg/m^3)	0.089	0.105	0.100	0.099	0.090	0.063	0.058	0.057	0.062	0.061	0.051	0.054
PM10 (mg/m^3)	0.239	0.231	0.215	0.168	0.156	0.100	0.108	0.097	0.099	0.088	0.086	0.088
Average pH level of rain	–	–	–	–	5.19	5.2	5.4	5.2	4.92	4.93	4.73	4.55
Frequency of acid rain (%)	–	–	–	–	26.0	25.2	10.9	16.7	32.7	40.0	56.4	75.6
Air quality greater or equal to Grade II (days)	–	–	–	–	295	309	281	325	311	322	324	328
Ratio of good air quality (%)	–	–	–	–	80.8	84.7	77.0	89.0	85.2	88.2	88.8	89.9

Source: Shanghai Statistical Yearbook, various years.

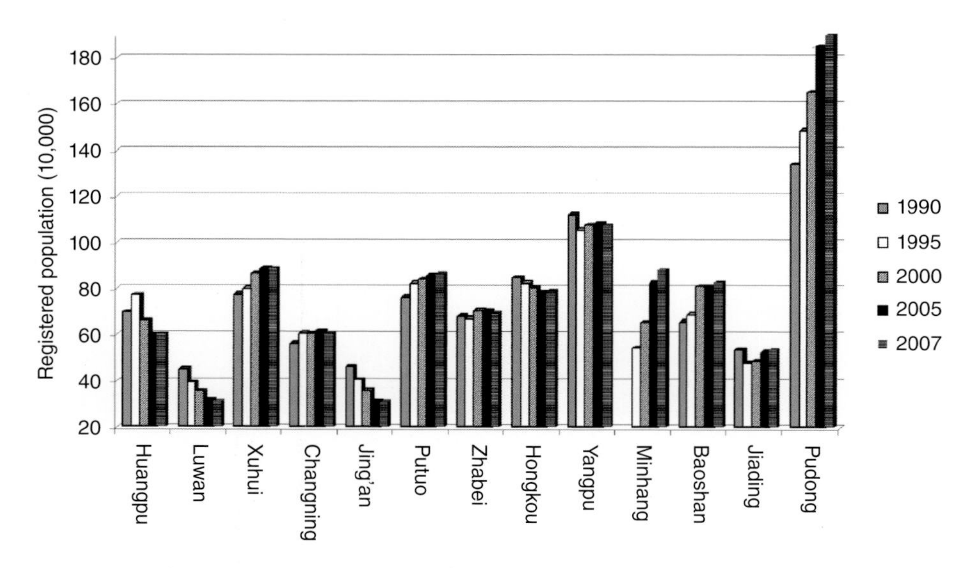

Figure 14.5 Registered population at the district level, 1990–2007

14.3.2.2 Population Some central districts such as Huangpu, Luwan, Jing'an, and Hongkou experienced a population decline, while others, like Xuhui, Changning, and Putuo, increased their population (Figure 14.5). Meanwhile, all outer districts significantly increased their population (Table 14.2).

Corresponding to the pattern of industrial land, most migrants live in the urban periphery, as 50% of the migrants are located in the four outer districts, i.e., Pudong, Baoshan,

Table 14.2 Population of districts in Shanghai, 2007

District	Residential population (000s people)	Migrant population (000s people)	Proportion of migrant population
City	18,580.8	4992.2	27%
Pudong	3053.6	983.4	32%
Huangpu	521.9	79.9	15%
Luwan	268.0	35.0	13%
Xuhui	965.9	135.1	14%
Changning	650.2	96.6	15%
Jing'an	252.1	29.1	12%
Putuo	1134.0	237.7	21%
Zhabei	743.9	99.7	13%
Hongkou	781.7	100.8	13%
Yangpu	1175.1	120.2	10%
Baoshan	1332.1	345.2	26%
Minghang	1895.6	791.7	42%
Jiading	999.0	435.7	44%

Source: SSB, 2008.

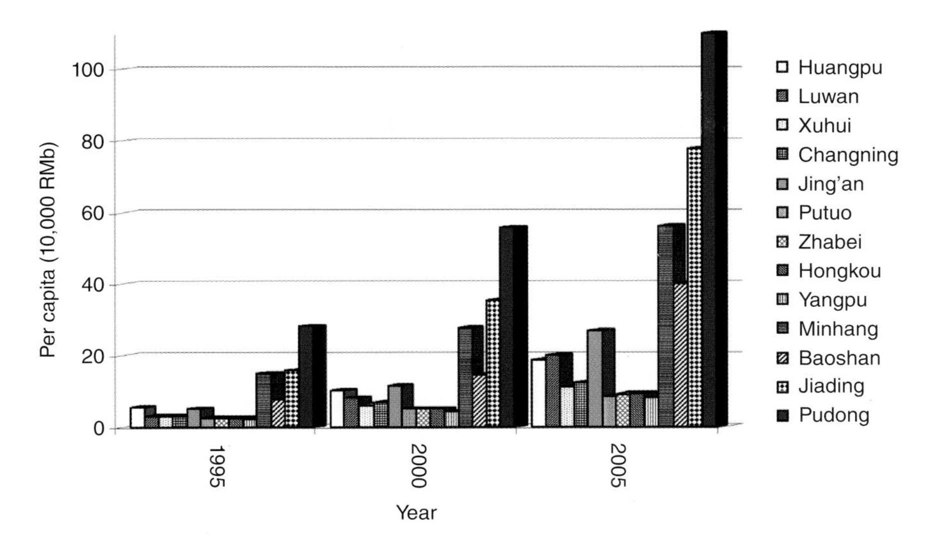

Figure 14.6 Shanghai's GDP per capita at the district level, 1995, 2000, and 2005. RMB, Renminbi

Minghang, and Jiading. For instance, 32%, 42%, and 44% of the residents in Pudong, Minhang, and Jiading, respectively, are migrants. In contrast, in some central districts, such as Yangpu, Jing'an, and Luwan, migrants represent only 10–13% of the population.

14.3.2.3 Economic development Although Shanghai experienced rapid economic growth, the districts varied considerably in their abilities to develop their economies (Figure 14.6). All four outer districts and some central districts such as Jing'an, Huangpu, and Luwan had a higher GDP per capita than other districts. In particular, Pudong New Area significantly topped all other districts in GDP per capita.

Although manufacturing has decreased its contribution to GDP from 60% in 1990 to 43.5% in 2007, it has remained an important sector of the economy for most districts. Ranking of gross output value of manufacturing (Table 14.3) corresponds well with the overall economic performance in Figure 14.6. For instance, while the districts of Pudong, Minhang, Baoshan, and Jiading moved from the bottom four in 1978 to the top four in 2007, other districts such as Yangpu and Huangpu dropped from first and second in 1978 to fifth and ninth in 2007.

14.3.2.4 Ecology and environment Using public green area per capita as an indicator, we measured change of the urban environment at the district level and found out that most districts have improved values of the indicator. Although some central districts, such as Jing'an, Luwan, and Huangpu, have very low values for public green area per capita, they have improved their situations as public green area grew more rapidly than in other districts. Outer districts such as Baoshan, Jiading, Minhang, and Pudong have both higher values and a higher growth rate of public green areas than the city average (Table 14.4).

Table 14.3 Districts' ranking of gross output value of the manufacturing sector, 1978, 1990, and 2007

District	1978	1990	2007
Huangpu	2	4	9
Luwan	9	13	10
Xuhui	4	3	6
Changning	6	11	11
Jing'an	8	12	13
Putuo	3	6	7
Zhabei	7	7	8
Hongkou	5	9	12
Yangpu	1	1	5
Pudong	13	8	1
Minhang	10	5	2
Baoshan	12	2	4
Jiading	11	10	3

Source: Data for 1978 and 1990 taken from Chen Xialin (2004) A study on competitiveness of urban district in Shanghai. Doctoral dissertation, Fudan University, p. 57. Data for 2007 from *Shanghai Statistical Yearbook*.

Table 14.4 Public green space per capita

District	1999 (m^2)	2003 (m^2)	2006 (m^2)	Annual growth (1999–2006)
Huangpu	1.11	1.22	1.37	3%
Luwan	0.73	1.37	1.56	11%
Xuhui	2.87	3.94	4.63	7%
Changning	4.14	5.58	6.37	6%
Jing'an	0.40	0.82	0.94	13%
Putuo	2.39	4.22	5.15	12%
Zhabei	1.64	2.31	2.81	8%
Hongkou	1.21	1.62	1.81	6%
Yangpu	2.61	2.86	3.74	5%
Baoshan	9.09	13.62	18.90	11%
Minhang	4.40	22.78	24.71	28%
Jiading	8.05	18.94	23.59	17%
Pudong	8.69	23.06	24.20	16%

Source: Shanghai Statistical Yearbook, various years.

14.4 Discussion

As shown in Section 14.3, although Shanghai as a city experienced profound urban change after the 1990s, reflected by transformation of the urban landscape, fast economic development, and substantial improvement of the urban environment, there is a noticeable disparity in terms of the evolution of the coupled human-environment between districts. While some districts, such as Pudong New District, Minghang District, and Xuhui District, exhibited a positive urbanization trend, indicated by rapid economic growth, better environmental conditions, and more appropriate land use, others, such as Zhabei District and Yangpu District, exhibited a negative urbanization trend, illustrated by slow economic growth, worse environmental conditions, and less appropriate land use. What has caused such an uneven landscape within a common city boundary? We argue that although the same forces shape Shanghai's overall trajectory of urban development, namely the interaction of the market and the state at a multi-scaled level and the influx of private capital, especially foreign direct investment (FDI), they have varied impacts on different districts. We therefore focus on the influence of the three main aspects mentioned in Section 14.2: economic restructuring and globalization, inflow of migrants, and the role of the multi-scaled state, on the spatial differentiation of Shanghai and discuss briefly the policy implications of socio-spatial differentiation for vulnerability analysis of Shanghai to climate hazards.

14.4.1 Economic restructuring and globalization

The most obvious driving force for socio-spatial differentiation in Shanghai is the impact of economic restructuring, characterized by the relocation of traditional manufacturing to urban periphery, the rise of high-tech manufacturing in development zones, and the growth of producer services in both new and old centers of the city, partially driven by the inflow of global capital. Shanghai gradually transformed itself from a traditional manufacturing base to a city focusing on high-tech manufacturing and tertiary industry, promoting itself as the economic and financial center of East Asia and attempting to restore its former image as the 'Paris of the East' from the 1930s. For instance, while in the 1960s and 1970s, secondary and tertiary industries contributed to 70% and 20%, respectively, of Shanghai's GDP, the figures changed to 46% and 53% in 2007. A review of the data at district level shows that while in 1978 most central districts led in industrial outputs, in 2007 the outer districts and counties on the urban periphery, i.e., Pudong New Area, Nanhui, Minghang, Jiading, Baoshan, and Jinshan, topped in industrial output. Most central districts had very little industrial output, and tertiary industry contributed more than 80% of their GDP, especially in Huangpu, Luwan, Hongkou, Jing'an, and Changning.

The participation of global capital has greatly facilitated the speed and upgraded the scale of Shanghai's economic restructuring. Shanghai's foreign direct investment (FDI) grew exponentially from US$0.003 billion in 1981 to US$7.92 billion in 2007. While a large portion of FDI has been poured into the manufacturing sector, such as automotive, electronics, telecommunication equipment, and chemical manufacturing, FDI has increasingly flown into tertiary sectors such as finance, banking, and other producer services (Wei *et al.*, 2006). Moreover, global capital has not only been involved in the production sphere, it also has directly supplied huge amounts of capital to change the built environment of Shanghai, especially in the 1990s. In fact, the real estate sector has become one of the largest sectors to absorb FDI, following closely the secondary and tertiary industries. For instance, 12.15% of Shanghai's FDI in 2007 was invested in the real estate sector.

The spatial distribution of FDI is very uneven. While Pudong New Area, Jiading, and Minghang led in FDI due to manufacturing investment, central districts such as Jing'an, Luwan, and Xuhui also scored highly in attracting FDI in producer services. In contrast, districts such as Putuo and Zhabei seemed unable to attract as much FDI. Compared with domestic projects, FDI is less bound by intra-government politics, social responsibilities, and even planning control, therefore is able to move Shanghai's landscape more towards a highly differentiated pattern that follows market principles (Wu, 2000). For instance, financial firms, foreign firms, and hotels show clear clustering patterns in Shanghai. The Bund re-emerged as the locus of foreign financial firms whereas Lujiazui became a concentrated location for foreign high-profile companies (Rose, 1997). The uneven distribution of Shanghai's FDI can be attributed to the following: the establishment of Pudong New Area and other investment zones, the location of pre-existing industrial and commercial districts, and the layout of the new transportation systems (Wei *et al.*, 2006).

14.4.2 Changing population profile and impact on the housing market

Housing provision for the changing population profile, mainly due to global and domestic immigration, is another important driving force of spatial differentiation in Shanghai. Shanghai has a large inflow of migrants, both international and domestic, due to the opportunities provided by economic restructuring and globalization. In 2007, Shanghai's residential population reached 18.6 million, of which almost one-third (5 million) were migrants in residence for more than half a year residence. A further 1.6 million stayed for less than half a year, bringing the total floating population in Shanghai to 6.6 million (SSB, 2008). At the high end, foreign employees working for foreign companies and joint ventures constituted a new social group, which can be termed the 'global elite', enjoying the highest quality of urban life in Shanghai. In 2007, there were about 133,000 foreigners living in Shanghai, with 51% of them coming from Japan, South Korea, and the USA, and 65% of them working in foreign companies and joint ventures (SSB, 2008). Another group was composed of highly skilled and highly educated white collar workers, which can be labeled the 'domestic elite', the result of Shanghai's municipal government's development strategy to build Shanghai into an 'intellectual highland'. To attract people with the required intellectual abilities, in 2002 a new residential card registration system was enacted, requiring migrants to hold a post-baccalaureate degree or possess a special talent, and an employment permit requiring the 'graduate to hold a degree above or equal to a Master's degree, or a Bachelor's degree from colleges/universities developed by the State Council, or from those local colleges within "211 projects"...' (Li & Wu, 2006a). This domestic elite was needed for jobs in the growing high-tech manufacturing and tertiary sectors in Shanghai. In contrast, rural migrants constitute the opposite end of the spectrum, and have to go through a long process to legally remain in Shanghai. Rural migrants are largely excluded from formal employment; most have a level of education below 'junior high school' and work in '3D' jobs, i.e., dirty, dangerous, and demeaning (Solinger, 1999). While the urban periphery absorbed a large number of migrants (mostly rural migrants), the central districts also had a significant migrant population (Table 14.2). However, it would be more interesting to reveal what kind of migrants different districts attract. Due to the data limitations, we are only able to provide a rough estimate based on other scholars' work on the housing market of Shanghai.

Migrants in Shanghai have not only boosted demand for housing but also further aggravated the highly differentiated urban spaces. Foreign residents mostly target high-end commodity housing, which is located mainly in the western part of Shanghai. Among about 300 listed projects of overseas housing in 1995, 89 were in the Changning District, 73 in the Xuhui District, 32 in the Jing'an District, and 61 in Pudong New Area (Wu, 2000). Domestic elites (white collar) generally chose medium-high priced housing areas and concentrated in Yangpu, north of Hongkou, south of Xuhui, and part of Changning (Li & Wu, 2006c). Most rural migrants clustered in the urban peripheries, such as Baoshan, the outer part of Zhabei and Putuo, and southeast of Xuhui (Li & Wu, 2006c).

Due to their disadvantage in accessing urban housing because they do not have Hukou (household registration) in Shanghai, most migrants are unable to gain ownership of affordable housing and are largely excluded from the mainstream housing distribution system. Most migrants rent private housing or live in enterprise dormitories, which usually are less well built and less well equipped, occupying far less space per person (W. Wu, 2004). Further, due to high mobility, the level of social connectedness and civil engagement remains low in areas predominantly settled by migrants. The poor quality of housing and the low level of social connectedness in areas mainly settled by migrants will pose great challenges for district governments in attempting to involve these communities in responding to potential disaster.

14.4.3 The role of the multi-scaled state

In a way uniquely characteristic of a transitional economy, the state plays a complex role in urban transformation and spatial differentiation in cities such as Shanghai, and the capacity of different districts to cope with climate hazards.

First, the heavy involvement of central government is reflected in the development of Pudong New Area, Shanghai's most notable example of urban transformation. Although Shanghai was designated by the central government in 1984 as one of 14 open coastal cities, until 1990, the city did not enjoy similar preferential treatment for taxation, foreign trade, and fiscal autonomy, due to its strategic importance to the national economy as it contributed one sixth of the state's revenue. In 1990, the State Council announced Shanghai's new development plan, under which Pudong New Area would enjoy a greater degree of autonomy and preferential treatment. The development of Pudong, an area of 522 km^2, east of the Huangpu River, implied a determination by central government to make Shanghai into a global city. Various development zones, such as Shanghai Minhang Economic and Technological Development Zone, Hongqiao Export and Trade Development Zone, Shanghai Caohejing Hi-tech Park, Lujiazui Finance and Trade Zone, Jingqiao Export Processing Zone, Waigaoqiao Free Trade Zone, Zhangjiang Hi-tech Park, and Xinghuo Development Area, are designated by the central government and offer preferential treatment to foreign investors with additional incentives, including concessions on income tax for foreign investors, custom duties, and tax for equipment, vehicles, and building materials related to foreign investment.

Second, Shanghai's municipal government has been actively involved in urban transformation and building a 'world city', which has led to spatial differentiation through land-leasing instruments, local partnership with private sectors, and infrastructure provision. The property reform of the 1980s allowed the municipal government to control state-owned land by authorizing leases for the right to use the land. The Shanghai municipal government used this instrument to manage the urban land use changes according to its desired direction (Wu, 2003). For instance, the Bund, the most famous tourist attraction and symbol of the city, was once the financial center of the Far East. Intending to redevelop the Bund into the Central Business District (CBD) of Shanghai, the municipal government enacted a new regulation to control land use and set up a specific development company to relocate public offices and redevelop the area for international insurance companies, commercial banks, and other international service companies. Through the efforts of the government, the Bund is now an area where financial institutions are concentrated, and is considered as China's Wall Street. Coupled with Lujiazui, Shanghai is preparing to reclaim its position as financial center of the Far East.

The growth machine theory implies that the public and private sectors collaborate to secure land-based revenue. After being granted fiscal autonomy and the ability to set local taxes by the new tax system introduced in 1994, local governments generally used concessions or waivers of development-related tax to attract more real estate development. Because value-added tax and business income tax have become the main sources of local revenue, local governments are keen to promote tertiary industry, especially high-density commercial development (Wu, 2000). Shanghai municipal government and various district governments, for instance, have been actively involved in site clearance, involving relocation of residents and paying them compensation, to provide land needed for desired investment and projects. District governments can also resort to mayoral decisions when complicated conflicts of interests arise.

Moreover, local government took on a major responsibility for infrastructure provision as a result of the fiscal reforms, which enabled it to retain revenue obtained from locally managed projects, such as premiums from land-leasing. For instance, Shanghai's infrastructure investment increased from 0.45 billion Renminbi (RMB) in 1978 to RMB 37.9 billion in 1996 and further leapt to 146.6 billion in 2007 (SSB, 2008).

Third, the different resources and capabilities of the district governments has further aggravated the uneven landscape of Shanghai. China's changing fiscal system, which led to fiscal autonomy, has forced local governments to take responsibility for levying and collecting revenue, which in turn has led to 'subcontracting' of the responsibility to county and district governments (Wu, 2000), and Shanghai is no exception. Although Shanghai's municipal government gained a preferential fiscal contract package in 1988 to retain certain local revenues, it also faced mounting pressure from its decreasing revenue base due to the deterioration of state enterprises. Therefore, since 1990, the municipal government has gradually entrusted district and county governments with a range of administrative powers, including planning, financial management, maintenance and public works, pricing of staple commodities, and foreign trade and industrial and commercial administration. As districts are endowed with different land, resources, and industrial structures, they have different resources and capabilities for exploiting this fiscal autonomy. While some district governments, such as Jing'an, Pudong, and Luwan, enjoyed rapid growth and high revenue per capita, others, like Yangpu, Putuo, Hongkou, and Zhabei, had very limited revenue gains in the past decade (Figure 14.7). While strong revenue bases have enabled some district governments to invest in infrastructure provision and environmental improvements, weak revenue bases have constrained others, leading to increased disparity between districts. This disparity between district governments has direct implications for their coping capacity in the event of climate hazards. Districts with insufficient resources won't have the resources to build adequate infrastructure or respond to disasters in an efficient way; this calls for intervention from the municipal government.

14.5 Conclusion

This chapter studies the coupled human-environment evolution of Shanghai as a unit and as a system composed of different social units, namely its 13 urban districts, from 1949 to 2008. It is our first attempt to explore the spatial differentiation issue at the district level of Chinese cities, which has significant implications for vulnerability to climate hazards. Although this is a preliminary assessment, we argue that studies of vulnerability to climate hazards should also consider the coupled human-environment evolution at finer scales, such as district level, as different entitlements, endowments, and coping capacities to deal with perturbation and stress are rested within the different social, economic, institutional, and political structures.

In this chapter, we first confirmed the rapid expansion of the urban built-up area of Shanghai, especially after 1988. The expansion rate reached an astonishing annual growth

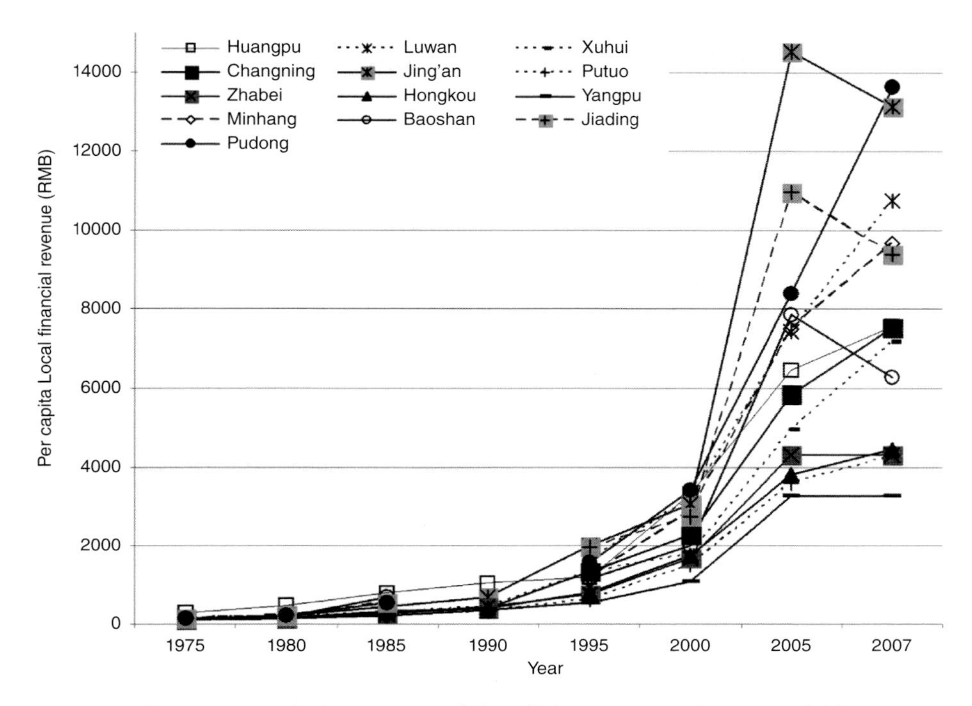

Figure 14.7 District revenues of Shanghai, 1975–2007. RMB, Renminbi

rate of 123.2 km^2 from 2002 to 2008. While residential, public, and commercial land all follow this increasing trend, Shanghai's industrial land experienced an initially increasing then decreasing pattern from 1947 to 2003, corresponding well to the city's industrialization and subsequent deindustrialization. However, due to the strict hukou (household registration) system, Shanghai's official population lagged far behind its economic growth. Shanghai has also improved its urban environment significantly in recent years.

However, when we examine at the district level and assess the coupled human-environment evolution that includes land use, economic conditions, and environmental conditions, we found that the districts have exhibited considerable heterogeneity. Overall, some districts exhibited a positive urbanization trend, indicated by rapid economic growth, better environmental conditions, and more appropriate land use, while others showed a negative urbanization trend, characterized by slow economic growth, worsening environmental conditions, and less appropriate land use. For instance, all four outer districts (Pudong, Minghang, Jiading, and Baoshan) as well as central districts such as Jing'an, Huangpu, and Luwan, have a higher GDP per capita and a higher growth rate in public green area per capita than the rest.

In our discussion, we focused on economic restructuring and globalization, the inflow of migrants, and the role of the multi-scaled state to explain the spatial differentiation of Shanghai and its implications for analysis of vulnerability to climate hazards. First, economic restructuring, involving relocation of traditional manufacturing to the urban periphery, the rise of high-tech manufacturing in development zones, the growth of producer service sectors in both new and old centers of the city, and the inflow of global capital, have all shaped Shanghai's spatial differentiation. Little attention, however, has been paid to whether or not Shanghai's natural environment and resources can handle such rapid urbanization in a sustainable way, thus leaving Shanghai in an overall vulnerable position for future climate hazards. Second, housing provision for migrants, including the 'global elite', 'domestic elite', and rural migrants, has facilitated spatial differentiation in Shanghai, especially through price differentiation of the residential spaces. In areas predominantly settled by migrants where social connectedness is weak, local governments will face great challenges in involving citizens in preparing for and responding to disasters. Third, the state has played a complex role in contributing to spatial differentiation of Shanghai, reflected in the heavy involvement of central government in setting up Pudong New Area, the hands-on direct control from municipal government via land-leasing instruments, collaboration with private developers, infrastructure provision, and the different resources and capabilities of district governments due to fiscal decentralization.

Although we do not propose a return to the more homogenous spatial structure found before the period of economic reforms, we feel that the increasing disparity between urban districts poses great challenges to achieving a more sustainable urban future that encompasses efficiency, equality, and environment in Shanghai. Those districts that have inadequate financial resources and political power, unless supported by a higher-level authority, are particular vulnerable to the threats posed by climate hazards.

Acknowledgements

We thank Professor Mei Anxin at the Department of Geography of East China Normal University for providing the land use data from 1947 to 2003. We appreciate two anonymous reviewers for their constructive comments. We also would like to acknowledge the following for their support: Center for Advanced Study of International Development (CASID) of Michigan State University and NASA's Land Cover and Land Use program through a grant to Michigan State University (NNX09AI32G).

References

Burgers, J. (1996) No polarisation in Dutch cities? Inequality in a corporatist country. *Urban Studies* **33**: 99–105.

Cai, J., Victor, F.S. (2003) Measuring world city formation – The case of Shanghai. *The Annals of Regional Science* **37**: 435–46.

Clark, W.A.V. (1986) Residential segregation in American cities: a review and interpretation. *Population Research and Policy Reviews* **5**: 95–127.

Fainstein, S.S., Gordon, I., Harloe, M. (eds) (1992) *Divided Cities: New York and London in the Contemporary World*. Oxford: Blackwell.

Hamnett, C. (1994) Social polarisation in global cities: theory and evidence. *Urban Studies* **31**: 401–24.

He, S., Wu, F. (2005) Property-led redevelopment in post-reform China: a case study of Xintiandi redevelopment project in Shanghai. *Journal of Urban Affairs* **27**: 1–23.

Jessop, B. (1994) Post-Fordism and the state. In: Amin, A. (ed.) *Post-Fordism: A Reader*. Oxford: Blackwell, pp. 251–79.

Kempen, E.T.V. (1994) The dual city and the poor: social polarization, social segregation and life chances. *Urban Studies* **31**: 995–1015.

Knox, P., Pinch, S. (2000) *Urban Social Geography: An Introduction*. London: Prentice Hall.

Li, Z., Wu, F. (2006a) Socioeconomic transformations in Shanghai (1990-2000): Policy impacts in global-national-local contexts. *Cities* **23**: 250–68.

Li, Z., Wu, F. (2006b) Residential disparity in urban China: a case study of three neighborhoods in Shanghai. *Housing Studies* **21**: 695–717.

Li, Z., Wu, F. (2006c) Sociospatial differentiation in transitional Shanghai. *Acta Geographica Sinica* **61**: 199–211 [in Chinese].

Marcuse, P. (1993) What's so new about divided cities? *International Journal of Urban and Regional Research* **17**: 355–65.

Marcuse, P., Van Kempen, R. (2002) *Of States and Cities: The Partitioning of Urban Space*. Oxford: Oxford University Press.

Mollenkopf, J., Castells, M. (1991) *Dual City: Restructuring New York*. New York: Russell Sage Foundation.

Musterd, S., Ostendorf, W. (1998) *Urban Segregation and the Welfare State: Inequality and Exclusion in Western Cities*. London and New York: Routledge.

Rose, F. (1997) Shanghai: Exception or rule? Consideration of urban development paths and process in China since 1978. Paper presented at the 5th Asian Urbanization Conference, London.

Sassen, S. (1991) *The Global City*. Princeton, NJ: Princeton University Press.

Sherbinin, A. de, Schiller, A., Pulsipher, A. (2007) The vulnerability of global cities to climate change. *Environment and Urbanization* **19**: 39–64.

Solinger, D.J. (1999) *Contesting Citizenship in Urban China: Peasant Migrants, the State, and the Logic of the Market*. Berkeley, CA: University of California Press.

SSB (Shanghai Statistical Bureau) (2008) *Shanghai Statistical Yearbook, 2008*. Shanghai: Shanghai Statistical Bureau.

Turner, B.L. II, Kasperson, R.E., Matson, P.A., *et al.* (2001) A framework for vulnerability analysis in sustainability science. *Proceedings of the National Academy of Sciences of the U. S. A.* **100**: 8074–9.

Van Kempen, R., Ozuekren, A.S. (1998) Ethnic segregation in cities: new forms and explanations in a dynamic world. *Urban Studies* **35**: 1631–56.

Walks, R.A. (2001) The social ecology of the post-Fordist/global city? Economic restructuring and socio-spatial polarisation in the Toronto urban region. *Urban Studies* **38**: 407–47.

Wei, Y.D., Leung, C.K., Luo, J. (2006) Globalizing Shanghai: foreign investment and urban restructuring. *Habitat International* **30**: 231–44.

Wessel, T. (2000) Social polarisation and socio-economic segregation in a welfare state: the case of Oslo. *Urban Studies* **37**: 1947–67.

Wilson, W.J. (1987) *The Truly Disadvantaged: The Inner City, the Underclass, and Public Policy*. Chicago: University of Chicago Press.

Wu, F. (2000) Global and local dimensions of place-making: remaking Shanghai as a world city. *Urban Studies* **37**: 1359–77.

Wu, F. (2001) Housing provision under globalisation: a case study of Shanghai. *Environment and Planning A* **33**: 1741–61.

Wu, F. (2003) Globalization, place promotion, and urban development in Shanghai. *Journal of Urban Affairs* **25**: 55–78.

Wu, F. (2004) Intraurban residential relocation in Shanghai: modes and stratification. *Environment and Planning A* **36**: 7–25.

Wu, W. (2004) Sources of migrant housing disadvantage in urban China. *Environment and Planning A* **36**: 1285–1304.

WWF (World Wide Fund for Nature) (2009) *Mega-Stress for Mega-Cities: A Climate Vulnerability Ranking of Major Coastal Cities in Asia*. Gland, Switzerland: WWF.

Xu, J., Yue, W., Tan, W. (2004) A statistical study on spatial scaling effects of urban landscape pattern: a case study of the central area of the external circle highway in Shanghai. *Acta Geographica Sinica* **59**: 1058–67 [in Chinese].

III
Institutions

15
Governing Ecosystem Services from Upland Watersheds in Southeast Asia

Louis Lebel and Rajesh Daniel

Unit for Social and Environmental Research, Chiang Mai University, Chiang Mai, Thailand

15.1 Introduction

The ecosystem services derived from upland watersheds are important to the well-being of people living in them, others living downstream and to society more widely (Millennium Ecosystem Assessment, 2005; Braumann *et al.*, 2007; Lebel *et al.*, 2008). Perceived or realized services often include: providing food, timber, fuel-wood and non-timber products; pollination and pest control for crops; water for irrigation or hydropower; sites for cultural activities; flood protection and buffered base flows; carbon sequestration; and water filtration. The specific benefits people obtain from a watershed are highly dependent on the mixture of ecosystems present, landscape structure and social contexts.

As a consequence of this variety of valued services, pursuing multiple management objectives is a practical reality for most upland watersheds in Southeast Asia[1]. It is also a source of contestation and conflict. Managing a watershed for one particular service or user may result in trade-offs in provision of other services and for other actors. Local communities and governments have frequently tried to prioritize, eliminate or integrate use of

[1] In this chapter, Southeast Asia is taken to include the following countries: Brunei, Cambodia, Indonesia, Lao PDR, Malaysia, Myanmar/Burma, the Philippines, Singapore, Thailand, Timor-Leste and Vietnam – as well as Yunnan province in southwest China, and northeast India.

Vulnerability of Land Systems in Asia, First Edition. Edited by Ademola K. Braimoh and He Qing Huang.

different services with combinations of plans, rules, incentives and information (Lebel & Daniel, 2009).

Spatial planning has been the favoured approach. Governments have devised classifications for land, forests and watersheds and used these to restrict or encourage particular activities (Laungaramsri, 2000). Upland communities have also made spatial plans, but with typically more flexible and overlapping systems of rights for using different resources – that is, with a less strictly territorial and a more socio-cultural perspective (Daniel & Ratanawilailak, 2011).

An important adjunct of plans is to associate landscape units with rules of use and responsibilities. Rule-making can be by, or in consultation with, users or it can be dictated by more remote authorities. Co-management models have often been promoted because they provide opportunities to consider services valued at different levels.

Although many rules are do's and don'ts, alternatives that create incentives may be more effective in some situations (Wunder, 2007). Markets for ecosystem services have been established in various parts of the world as an alternative to regulations to encourage conservation of valued services. Their performance depends on institutional design details and socio-political contexts (Wunder *et al.*, 2008b).

The quality of information about services and impacts of use and management is crucial to most efforts at governing them but often receives insufficient attention (Carpenter *et al.*, 2009). Payments for ecosystem services, for example, require a good understanding of which actions actually secure provisioning, and indicators that can be monitored (Engel *et al.*, 2008). In many cases unambiguous, place-specific evidence that particular land-covers provide a service is lacking – for example, flood protection benefits (Bruijnzeel, 2004; Locatelli & Vignola, 2009; van Dijk *et al.*, 2009). Local, experience-based knowledge and scientific knowledge are not as frequently integrated as needed (Berkes, 2009). Building awareness and understanding through integrated assessments and monitoring is crucial (Lele, 2009) but will never be a substitute for politics around which services and users should be prioritized.

In this chapter we focus on the institutional and political dimensions of governing ecosystem services from upland watersheds in Southeast Asia. We build on an earlier, much shorter, review of experiences from the global tropics (Lebel & Daniel, 2009) by considering in much more detail empirical evidence from Southeast Asia. The chapter is organized around the four themes: plans, rules, incentives and information.

15.2 Plans

A common approach to managing the complex set of services from upland watersheds is through spatial land-use planning; zoning some areas for biodiversity conservation, watershed protection, forestry, agriculture, tourism or multiple uses. Most governments in Southeast Asia have adopted policies for controlling land use in upland watersheds. The extent to which users and residents are involved in planning varies as does the influence of plans on practices.

15.2.1 Protected areas

Most conservation policies and strategies of governments are founded on the idea of separating people from their environments in systems of protected areas (Chopra *et al.*, 2005). A discourse around the benefits of ecosystem services has usually been added to early justifications based on biodiversity and recreational or cultural values, focusing on either downstream communities or more recently on global environmental benefits of reduced deforestation. The protected area approach argues that conserving natural ecosystems in a near-intact state will maintain a full suite of ecosystem processes, and thus the full range of services, that forests or other forms of native vegetation provide.

Vast areas of the tropics were declared as protected areas between 1980 and 2005 (Naughton-Treves *et al.*, 2005). Indonesia has set aside about 12.5% of its land area in protected areas for nature conservation; Malaysia about 31% (EarthTrends, 2003; UNEP-WCMC, 2003); Thailand about 19% for conservation (ICEM, 2003c; IUCN, 2007). The practice of allocating large areas as conservation areas and parks spread in the 1990s to other countries in the Mekong region. Lao PDR has established an extensive set of protected areas covering more than 21% (ICEM, 2003a) and Cambodia 18% (ICEM, 2003b). While most protected areas are located in the uplands (as lowland areas have already been cleared of native vegetation) the extent to which use of forest goods and services is restricted varies among countries (Thomas *et al.*, 2008). The effectiveness of management also varies widely, with many parks existing only on paper, promised local benefits to residents from tourism often smaller than expected, and conflicts created over access to land, resources and services (Roth, 2004a; Naughton-Treves *et al.*, 2005).

As of 2008, Thailand's Royal Forestry Department had established more than 200 protected area units covering approximately 19% of the country's land area. The government of Thailand intends to increase protected area systems to 30% of the country by 2016 (Trisurat & Pattanavibool, 2008). The implications for the upland communities, particularly ethnic peoples, living and farming in protected areas declared as national parks, wildlife sanctuaries and watershed areas, are huge and tensions involving farming communities fighting state forest land classification have become increasingly frequent (Wittayapak, 1996; Poffenberger, 1999). At larger scales, the outstanding challenge is that benefits from the existence value of biodiversity frequently do not align with the value of habitat conversion to agriculture for local poor communities (Fisher & Christopher, 2007).

However, parks have failed at forest conservation; special deals are possible to convert land for tourism and even personal use. In Thailand's Khao Yai National Park, a proposed golf course and resort that began construction was halted only after protests (Laungaramsri, 2002a; Ross, 2003). Powerful military or political figures can acquire 'protected' areas or land designated for resettlement of displaced villagers but typically remain uninvestigated (Phongpaichit, 1999).

Studies of biodiversity conservation in Ruteng Park on Flores Island in Indonesia provide a closer look at the links between the conservation of biodiversity and the livelihoods of rural people who live on the fringes of the parks and protected areas. An early study demonstrated the economic value of drought mitigation services to farmers downstream of forest in upland watersheds in line with policies of the Indonesian government and park (Pattanayak & Kramer, 2001). A subsequent study linked forest cover to prevalence of diarrhoea, presumably through impacts on drinking water quality, illustrating another under-appreciated service of the park (Pattanayak &Wendland, 2007). Parks can provide multiple ecosystem services and benefits.

15.2.2 Forest and watershed classifications

Control and authority over forest areas of most countries in Southeast Asia rests with the national or state government. In Indonesia, the Philippines and Thailand, state control over forests has grown since the establishment of forest service agencies during the late 19th and early 20th centuries. Governments have exerted authority over forests through land classification and zoning schemes, for example, by attaching regulations prohibiting local access or use depending on whether a parcel of land is classified as 'forest' or 'upland' (Lebel *et al.*, 2004).

In Thailand, forests are defined by the 1941 Forestry Act as 'land without occupants' and as 'land with no right-holders' (Forestry Act, 1941). As land areas become classified as 'forest reserves' there is a great deal of ambiguity of ownership of agricultural lands in rural areas particularly in collectively used lands such as community forests, sacred forests and fallow farmlands (Sato, 2003). Agricultural land in forest reserves makes up the majority of agricultural land in Thailand in areas typically classified as degraded forests. Only a minority of private land holdings used for agriculture have full title deeds. Often agricultural land holdings do not have the concept of ownership in the modern sense and are limited to either usufruct, or *sithi krobkrong*, and squatter's rights, or *sitthi japjong* (Sato, 2003). The changes made to legal categories and procedures for land over the past several decades are an important source of contemporary conflicts over land. The current legal categories of rural and forest land in Thailand illustrate some of the governance problems created by categories and definitions driven by interests in a narrowly framed set of services.

The National Forest Reserve Act of 1964 attempted to centralize forest control; by 1985, the Royal Forestry Department (RFD) declared approximately 45% of the country's total area as forest reserves. But lands designated as state national forest reserve often had no trees or already had people residing in those areas (Hirsch & Lohmann, 1989; Flaherty & Jengjalern, 1995).

The RFD's designation of forests as being driven by the expansion or maintenance of its own power and control over territory (Vandergeest & Peluso, 1995) does not completely explain official motives but may also include departmental factionalism. For example, designation of reserved forests before 1938 that was undertaken by the authority of local administrative sections could not realistically expand the authority of RFD. Some regional forest officers enthusiastically urged the designation of reserved forests at that time because they thought if they did not do so, the forests would disappear. Thailand's foresters were also conscious of the need for spatial enclosure of forest lands for so-called scientific forestry (Wataru, 2003).

Other Mekong region countries follow systems similar to that of Thailand with the difficulty that the state spatial classification system finds it unable to incorporate other types of land uses, such as swidden rice farming systems. Swidden systems are a mosaic of different-year fallows and secondary forest areas, some of which subsequently also are transformed into upland rice fields and then back again into fallows. Fallow forests that are part of the swidden rice cultivation cycle of upland communities cover large parts of montane Southeast Asia. Secondary forests, which regenerate on the fallow swiddens, are rich in tree species and complex with respect to stand structure. The farm-forest fallow-swidden ecosystem (including the trees and wildlife species) is part of an extensive indigenous knowledge system. But as swidden cultivation is actively discouraged by officials, the system is undergoing many changes and also increasingly being replaced by permanent rice farming. The land use changes are causing a reduction in the area covered by fallow forest ecosystem with subsequent negative impacts on biodiversity (Rerkasem *et al.*, 1994, 2009; Schmidt-Vogt, 1998; Laungaramsri, 2002b; Walker, 2004; van Vliet *et al.*, 2012; Sturgeon *et al.*, 2013).

State management uses positionality to make strictly bounded static spatial categories such as 'conservation forest' and 'village land'. Categories of this type can be easily delimited on a map and made legible to future officials. Conversely, swidden space and village management does not fit since it does not take the shape of straight lines but instead follows streams and mountain ridges, contains rough edges, and often defines location in relation to another's field or a landscape marker (Roth, 2004b).

Concerns around landslides, floods, soil erosion and sedimentation have driven much research and policy on agriculture on sloping lands (Blaikie & Muldavin, 2004; Forsyth & Walker, 2008). A recurrent rationale for policies and projects has been that maintaining or increasing forest cover will secure key ecosystem services – often without much specific attention to tree species involved or impacts of alternative land uses. The scientific evidence base for many services, like flood protection, however, often remains modest and controversial (Bruijnzeel, 2004, Locatelli & Vignola, 2009; van Dijk *et al.*, 2009). Scientific knowledge about watershed services is frequently used selectively or misrepresented in justifying upland policies (Forsyth, 1996, 1998; Walker, 2003, Blaikie & Muldavin, 2004).

In Thailand, watershed classification (Chankaew, 1996), like the definitions of forest lands, was also used as an instrument to strengthen state control of upland resources, restrict expansion of farmlands in upland catchments and threaten highland farmers with resettlement (Vandergeest & Peluso, 1995; Laungaramsri, 2000). This is illustrated on the ground where watershed classification can lead to entire provinces or districts in northern Thailand coming under strict conservation status. For instance, most of Mae Hong Son province falls under the highest order of protection, Watershed 1A, thus prohibiting all settlement and agriculture, placing huge stress on the communities who live and farm in the province. In practice implementation of the classification has been left incomplete as the Thai state does not have the capacity, political support or available land to resettle all upland farmers into lowland areas (Walker & Farrelly, 2008).

Watershed classifications across the region are grossly similar with classes of high to lower restrictions on uses. Most areas with restricted classifications are in the uplands (Thomas *et al.*, 2008). The Watershed Classification Project carried out by the Mekong River Commission Secretariat between 1989 and 2001 elaborated a basin-wide classification indicating the sensitivity of watersheds with regard to resource degradation (mainly by soil erosion). It aimed to develop a decision-support tool. Along with the classification, the project produced general recommendations for sustainable land use in each watershed class. For instance, for 'Watershed Class 1: Protection Forest Areas' with very steep slopes and rugged landforms, commonly uplands and headwater areas, the project said that as a rule, these areas should be under permanent forest cover. Notably, the project also added a caveat that 'account needs to be taken of traditional rights and land use practices' (Heinimann *et al.*, 2005).

State schemes for classifying and planning land uses do not correspond closely with actual provision of ecosystem services; swidden, multi-species orchards, and agroforestry may yield more services than monocrop plantations labelled as forests and assumed to be service-rich (Cairns, 2007; Bhagwat *et al.*, 2008; Xu *et al.*, 2009). In southwestern China, indigenous land-use practices may be more beneficial to long-term conservation objectives than protected areas (Sharma & Xu, 2007; Xu & Melick, 2007). In the typical, dynamic and mosaic landscapes of much of upland Southeast Asia, various ecosystem services are not coincident in space or time. Hydrological services like base stream flows at the end of the dry season and flood protection services during the wet season may both be valued even when how to secure them is not fully understood (Forsyth & Walker, 2008; Neef & Thomas, 2009).

15.2.3 Participatory land-use planning

A recurrent challenge for planning is getting adequate information about ecosystem services at scales relevant to decision-making (Turner & Daily, 2008). Local knowledge is often crucial but only available if planning agencies allow space for meaningful local participation (Thomas, 2006).

A good example is the work of ICRAF, CARE and local government and non-government organizations in the Mae Chaem watershed of Chiang Mai province (Thomas, 2005). Such activities are also a helpful background to more bottom-up processes of basin management in areas beyond the rural periphery and merging into peri-urban or desakota landscapes. When multiple stakeholders including scientific and local experts are involved, underappreciated services can be better understood, as shown by the example from Ruteng Park described earlier.

Institutional frameworks such as watershed management committees, organizations or networks, which may be mandated by government or emerge independently, can help solve local resource allocation problems. However, such organizations may not have much formal decision-making authority or budget but take on basic planning, conflict resolution and negotiation functions (Thomas, 2006). In the Upper Ping River basin, the Thai government established several river sub-basin committees that each adopted slightly different committee structures and activity plans to deal with the range of stakeholders and issues important in their sub-basin (Thomas, 2006). The challenge was introducing new organizations with recognized cross-sectoral planning mandates in a context where individual agencies and water user groups already had well-organized networks and coalitions (Thomas, 2006; Mollinga *et al.*, 2007). Local communities also make plans to manage their watersheds (Wanishpradist, 2005). Overall, however, it is rare to find direct involvement of stakeholders in analysis of ecosystem services as a basis for informed negotiations and decision-making (Fisher *et al.*, 2008).

A range of accounting techniques is available to help people understand dependencies on ecosystem services from generalized ecological footprints through to valuation of specific services (Jenerrete *et al.*, 2006; Patterson & Coelho, 2009). The ecological basis of many exercises, however, remains tenuous as context-specific evidence is frequently lacking or inadequate. The importance of participatory and deliberative methods for accounting and evaluation is likely to grow (Spash, 2007) especially where knowledge about hydrological services is strongly contested (Forsyth & Walker, 2008).

Classifications reflect the interests of those who build them. There is a need to reconceptualize land-use planning for conserving ecosystem services as a process of joint assessment and negotiation. Current land classifications produce tensions and conflict; they need to be adjusted to fit 'prior use' of areas currently classified as 'forest land' irrespective of the quality of forest now, and reward good management rather than penalize it. Validation is possible with aerial photography and satellite-based remote sensing.

15.3 Rules

Rules and regulations underpin plans, helping to define rights of access and use of ecosystem services and responsibilities for their management. Institutional instruments are diverse including quotas, licences, concessions and seasonal bans as well as other customary rules, taboos and norms.

15.3.1 Property rights and land tenure

Whether an ecosystem service is a private, public, club or common good makes a difference to how it might be governed (Engel *et al.*, 2008; Patterson & Coelho, 2009). Individual property rights for services that are excludable and rival are particularly useful to farmers as they encourage investment in land – for example to grow trees that may not provide returns for many years. Formal title deeds are also useful as collateral in obtaining loans (Walker, 2006).

As public goods are used more intensively they can become rival goods that need other institutions to be managed sustainably (Fisher *et al.*, 2008). Pre-existing private property rights may hinder efforts to manage services that are common pool resources – rival, but non-excludable – and coordination mechanisms are needed (Patterson & Coelho, 2009).

Some hydrological services, like drinking water supplies from springs or streams, are managed as club goods (Engel *et al.*, 2008). Others are treated as common property of a sub-watershed, village or even a group of neighbours. Rules of use in community forests typically specify amounts or seasons during which valued but scarce forest resources can be collected (Cairns, 2007; Kerr, 2007). Rules for common pool resources are often flexibly bundled, so that allocation of scarce bamboo clumps or trees with resin or others needed for spiritual ceremonies might be allocated to individuals or households whereas access to regrowth might be open to all for grazing (Lebel, 2005). Hydrological services like flood protection or dry season base flows require coordination between upstream and downstream users, and finding ways to secure mutual benefit (Kerr, 2007).

The conventional logic that formal land tenure enhances sustainable land and forest management is challenged by experiences in Thailand (Daniel & Lebel, 2006). For example, although ethnic minority groups in the northern uplands commonly do not have permanent land use rights, long-term investment in land resources is common practice (Neef *et al.*, 2000). For ethnic people in more remote uplands, formal access to land may often be less important or desirable than access to the Thai citizenship cards and the capacity and flexibility of household members in exploiting new income sources both on- and off-farm (Knupfer, 2002; Thomas *et al.*, 2008).

A common strategy of farmers in trying to prevent land claims of state forestry officials, where local rights are not otherwise recognized, is to plant fruit trees or tea shrubs, as it was believed that forestry officials would not claim land that has already been planted with perennial crops. Another response was to convert rain-fed swidden rice fields or fallows to permanent paddy fields or other cropping systems (Neef, 2001). Additionally, some communities try to avoid land losses through 'appeasing' of the forest officers by being active in 'forest protection' through building firebreaks and reforestation (Knupfer, 2002).

Lao PDR, Vietnam and China went through periods of collectivization and then reallocation of agricultural and forest land, so some of the tenure issues are different than those in Thailand (Thomas *et al.*, 2008). Land has been allocated to both individuals and villages. The Lao government has had a particularly forceful policy of ending swidden cultivation.

Overall, it is hard to draw firm conclusions on the relative performance of formal and informal arrangements. In remote areas, a lot depends on local institutions and relations with local officials rather than formal land certificates and regulations (Walker & Farrelly, 2008; George *et al.*, 2009). Both formal and informal tenure can matter, and their interplay can be positive with respect to livelihoods and environmental outcomes. With proximity or when more profitable opportunities arise – for instance related to eco-tourism or logging concessions – ambiguities in land tenure often become more problematic (Wunder *et al.*, 2008a). Clarifying property rights can be an important aspect of governance of upland watershed services, but formal land tenure is not necessarily a prerequisite to establishing sustainable management systems for ecosystem services, especially where informal use rights are locally recognized and respected.

15.3.2 Community-based management

Property rights may be vested in communities rather than households and individuals. Community forests have become an arena for spatial negotiation of land and forestland use between the state, timber companies and local communities as well as a means of promoting local participation in forest management. But many community forest programmes are failing because external factors promoting forest degradation are stronger.

In Cambodia, a ministerial order recognizing forest sites as potential community forestry areas is the first step towards formal recognition of community management of forest areas. In December 2008, more than 100 villages in several provinces were granted formal management of about 127,000 ha of forest in 87 forest sites by Cambodia's Minister for Agriculture, Forestry and Fisheries. Another 37 potential community forest areas covering 18,000 ha were already recognized in Siem Reap in 2007, bringing a total of 145,000 ha of forest under recognized local management. The next step, of formal registration through signing a CF Agreement with the Forestry Administration, gives communities full legal access and management rights over local forest areas for 15 years, protecting those areas from commercial and other outside interests. It also enables some of Cambodia's poorest people to benefit economically, with rights to use forest resources including timber (RECOFTC, 2009).

In their comprehensive review of community-based forest management in the Philippines, Lasco and Pulhin (2006) conclude that the strategies of planting trees on farms and in landscapes has had largely positive environmental effects, for example, for soil and water conservation, carbon sequestration (also see Section 15.4.1) and biomass production. Other studies of complex agro-ecosystems like benzoin and rattan gardens in Indonesia highlight how reduced intervention in the system can lead to ecological succession processes increasingly similar to those found in native forests after disturbance, and thus, important to biodiversity conservation as well as productive use (Garcia-Fernandez & Casado, 2005). Agroforestry practices using trees and cover crops greatly reduce soil erosion compared to monoculture plantations without groundcover (Sidle *et al.*, 2006).

Management practices cannot be understood separately from their social context. For example, *miang* forest areas in Nan province of northern Thailand, important for growing tea, are not only maintained as a watershed forest comprising part of the *muang fai* traditional irrigation system, but also serve as an agro-ecotourism centre bringing in cash income and supported by the local government (Wittayapak & Dearden, 1999).

Co-management for resource use can produce differing tensions depending on the policy emphasis on village, households or individuals. In Vietnam, for instance, during the 1950s and the 1960s, the national government encouraged villages to form cooperatives, though the membership of the cooperatives and decision-making processes may have continued to reflect traditional modes of operation. By the 1960s and 1970s, the government tried to establish multi-village cooperative units or communes. However, the 1980s brought policies that shifted away from collectivization, returning authority to the village, and, most recently, to households. This may make it possible for villages to regain greater autonomy, creating opportunities for traditional institutions to re-establish their role in resource-use decision-making. Yet government policies and programmes, while de-emphasizing collectives, give little recognition to the role of the traditional villages. Instead, these new policies and programmes emphasize empowering the household or individual (Sowerine *et al.*, 1998; Sowerine, 2004).

Co-management arrangements with state agencies, firms and other actors are institutionally diverse. One of the recurrent factors important to success is networks of trust that support stable social relations (Lele, 2004; Armitage *et al.*, 2009; Berkes, 2009). In the context of upland watersheds with their dynamic and complex mix of ecosystem services valued at multiple scales (Lebel *et al.*, 2008) the challenges and rewards are particularly high.

15.3.3 Logging concessions

European forestry influences are evident in state forest management practices that follow scientific forestry norms developed in the 18th century and that were focused on supplying colonial powers with the raw materials to industrialize (Bryant *et al.*, 1993; Lang, 2000; Contreras, 2003). In mainland Southeast Asia, the colonial British Empire dominated and controlled the teak trade in India, Burma (Myanmar) and Thailand. In Vietnam, Laos and Cambodia it was the French. Colonial forestry centralized authority in national capitals, using licences, concessions and military force as needed to gain control of forest resources.

Southeast Asian governments and their corporate counterparts have viewed logging as an important source of power and revenue (Pasong & Lebel, 2000; Dauvergne, 2001; Butler & Laurance, 2008). Revenues from timber exports in the Philippines during 1950–69 were used for rebuilding the country from the devastation of wars. In Malaysia, timber rents make a significant contribution to growth, exports, savings, investment, government revenue and fiscal capacity. Forestry is a dominant sector in the Laotian economy: despite restrictions on logging and high export taxes implemented in 1989 – which decreased its share of total exports by 36% – timber and wood products remained the major export (replacing hydroelectricity). In 2000, the forestry sector contributed 5% to GDP, increasing from 3.4% in 1990 (FAO, 2002). In Cambodia, both the Khmer Rouge and afterwards the elected government exported timber to Japan, Thailand and Vietnam (Billon, 2000).

The Indonesian government took control of forest resources in 1967, distributing over 60 million hectares in timber concessions to private companies often connected to military leaders (Pasong & Lebel, 2000; Dauvergne, 2001). The industry is controlled by only a handful of players; Barito Pacific, for example, in the early 1990s held more than 10% of the concessions and controlled over 6 million hectares.

The management of forests in Indonesia is based on a land-use classification that distinguishes protected forests, limited and general production forests, and conversion forests and areas for parks and reserves (Dick, 1991). Clear felling is allowed in conversion forests, and transmigration settlements are developed from contiguous logged areas. In production forests, concession periods by timber and logging companies was extended to a 35-year harvesting cycle to induce replanting and payment of the appropriate fees to the government from forest exploitation.

The need to generate foreign exchange for debt servicing and the increased demand for industrial raw materials, both in the domestic and international markets, have helped induce the replacement of complex forest ecosystems by monoculture plantations. The pulp and paper industry, for instance, in Indonesia has expanded rapidly since the 1990s. Some estimates suggest that as little as 10% was actually harvested from plantation timber with the rest coming from illegal cuts in natural forests. Forests were also cleared to plant fast-growing species (Barr, 2001). This conversion to tree plantations has resulted in forest loss and internal displacement, increased social conflicts over land, and worsened small-landholder tenure insecurity (Lang, 2002).

In Lao PDR, the rising price of natural rubber products due to the demand of the Chinese market over the last decade has attracted rubber plantation investors from China, Thailand and Vietnam to seek land concessions all over the country (Manivong & Cramb, 2008). Throughout the region one of the biggest concerns is the impact on water use as rubber is a water-demanding crop, and many of the areas where it is expanding are highly seasonal with dry season water scarcities already a constraint on agriculture (Xu, 2006; Mann, 2009; Ziegler *et al.*, 2009).

15.3.4 Logging bans

Several countries have invoked logging bans in native forests often in response to serious flood events (Daniel, 2005; Xu *et al.*, 2007). Until the 1989 logging ban, the Thai government gave concessions to companies to log large parts of the forest area that lie outside national parks and wildlife sanctuaries. Indiscriminate logging practices usually led to continuing deforestation where 30–40% of the residual younger trees were destroyed. Hence, most logged-over forest areas became quickly degraded, and were further exploited by human activities as rural communities and land speculators obtained access through the logging roads (Kashio, 1995a, 1995b).

After the 1989 logging ban, the Thai government promoted a wood import policy and Thai logging companies expanded logging concessions into the neighbouring countries of Laos, Cambodia and Myanmar (Daniel, 2005). The logging of neighbouring country forests has also resulted in illegal logging operations in Thailand's forests, particularly along the borders (Cooper & Palmer, 1992). Illegal logging in the forests of the Salween National Park and the Salween Wildlife Sanctuary along the Burma/Myanmar and Thai border was one of the best-known timber scandals in the post-logging ban period.

China's logging ban was introduced in 1988 alongside several other major policies, like the sloping land conversion programme of 1999, that combined enforcement and incentives to increase forest cover on sloping lands (Bennett, 2008). These national policies have had major impacts on livelihoods and land uses in the ethnically diverse subtropical watersheds of Yunnan Province (Xu & Melick, 2007; Xu *et al.*, 2007; Xu *et al.*, 2009) leading to, for example, the almost complete elimination of swidden cultivation by the Hani and its replacement by rubber (Xu *et al.*, 2009).

15.4 Incentives

Market-based instruments or other forms of incentives that reward good management practices are an alternative to spatial planning and regulations. Here we look specifically at payments for ecosystem services and certification schemes.

15.4.1 Payments for ecosystem services

Payments for environmental or ecosystem services (PES) have emerged as an alternative or complement to spatial planning and regulatory approaches to conservation (Wunder, 2007; Engel *et al.*, 2008). PES schemes are voluntary transactions in which an environmental service is bought by a buyer from a provider if and only if the provider secures service provision (Wunder *et al.*, 2008b). Such schemes share similarities to eco-certification of products and other incentive-based mechanisms, like environmental taxes or subsidies (Engel *et al.*, 2008; Jack *et al.*, 2008). Common challenges include clarifying property rights, getting prices right and linking actions to compensation (Fisher *et al.*, 2008). PES appear to be most relevant when an ecosystem service is under threat in marginal lands where opportunity costs are modest and land claims clear (Wunder, 2007).

For PES, the distinction of whether the ES provided are public goods or not is important. Not all ES are pure public goods, i.e., consumption by one user does not affect consumption by another; many other ES are, in fact, either excludable or rival in consumption. In particular, many water services are 'club goods' where only those holding water rights or those located in a well-delineated watershed benefit. This has implications both for identifying the users and arranging for them to pay for service provision, as well for directing the benefits to the providers. This can raise questions of equity as well (Corbera *et al.*, 2007).

To date only a few such schemes have been operating for a significant period in Southeast Asia; quite a few of these deal with watershed protection and related services. Many are related to the RUPES programme (Reward the Upland Poor for Environmental Services) (Swallow *et al.*, 2007; Van Noordwijk *et al.*, 2007; Leimona *et al.*, 2009).

In the Philippines, concern over loss of biodiversity ranked very high among stakeholders in designing a PES programme in the Peñablanca Protected Landscape. The PES programme thus was initiated with high conservation, cash payments and investments in carbon crediting as the most beneficial options. The design of the PES programme showed that the linkages between land use and the level of environmental services were crucial to the programme's sustainability; financial, economic, social and environmental factors were of equal importance in designing the PES programme (Bennagen *et al.*, 2006).

In northern Thailand, upstream, ethnic minority communities are expected to conserve upland watersheds, stream flows and biodiversity while also simultaneously being widely perceived by lowland communities and policy-makers as a threat to, rather than providers of, ecosystem services (George *et al.*, 2009; Sangkapitux *et al.*, 2009). A study in the Mae Sa watershed found that payments for water resources by downstream resources users was possible with upstream farmers willing to adapt their farming practices given adequate compensation (Sangkapitux *et al.*, 2009). While building awareness about ethnic communities' sustainable practices around ecosystem service projects could help change lowland perceptions, discrimination against upland minorities continues due to power and control over the uplands residing with lowland Thai policy-makers.

Wunder *et al.* (2008b) reviewed a sample of programmes involving payments for environmental services that included several studies from tropical South America, and found user- as opposed to government-financed programmes were better in terms of fit to targeted beneficiaries, local conditions and needs, and monitoring. China, Mexico and Costa Rica each have large programmes giving payments to landowners for changing land uses (Sanchez-Azofeifa *et al.*, 2007; Jack *et al.*, 2008).

Rewards or compensation do not have to be direct cash payments to individuals; they can be non-monetary payments to groups or guarantees of privileged or secure access (such as land tenure) to services or other resources like training (Leimona *et al.*, 2009; Neef & Thomas, 2009). Non-financial incentives may be more important to poverty alleviation than direct cash payments (Leimona *et al.*, 2009).

In the complex resource management situations typically found in upland watersheds, the introduction of new markets for ecosystem services needs to consider carefully existing access rights as well as who is excluded and who will benefit or be at a disadvantage (Corbera *et al.*, 2007; Mollinga *et al.*, 2007). Poor, marginalized and otherwise vulnerable groups are often more dependent on ecosystem services and have relatively lower opportunity costs than others (Jack *et al.*, 2008) but their capacities to engage may also be limited. Poor farmers in Vietnam's uplands with small holdings were unlikely to join reforestation schemes unless compensation was adequate to cover loss of food production (Jourdain *et al.*, 2009). Moreover, when there are many poor, small providers, transaction costs can be high and thus not competitive (Jack *et al.*, 2008). Studies of two carbon sequestration projects in Mexico showed how women and the poorest were excluded from designs and that outcomes reflected political affiliations with project managers (Corbera *et al.*, 2007). Non-participants in ecosystem services projects may also be impacted adversely, for instance when landless farmers lose access to common pool resources (Wunder, 2008). Although evidence about welfare impacts remains modest the emerging findings suggest that PES programmes on balance have had relatively small positive effects, and are unlikely to become central to poverty alleviation efforts (Wunder, 2008).

Ultimately, how rules are arrived at may matter as much as their final form. Thus, who runs a project is a crucial feature of PES schemes (Corbera *et al.*, 2007; Wunder *et al.*, 2008b). Intermediaries may be created by service buyers or sometimes a third party.

Non-governmental organizations may be helpful where farmers' groups (as providers) are not formally recognized or buyers are unfamiliar with negotiating directly with farmers (Neef & Thomas, 2009). Reliability of the organization and the ability to build trust in schemes are crucial (Koellner *et al.*, 2008; Neef & Thomas, 2009). An assessment of the management capacity of seven organizations that sell ecosystem services from tropical forests in Latin America, for example, found that marketing and client satisfaction were often neglected and that different market actors have very different criteria and preferences, making it necessary for suppliers to target offers carefully (Koellner *et al.*, 2008). The role of marketing in successful PES activities has not received adequate attention.

Monitoring of policies and projects is important: to detect incomplete or distorted implementation; to assess compliance with agreements; to evaluate actual impact; and to learn from the past to improve future interventions (Lebel & Daniel, 2009). Payments must be based on what can be monitored, usually land use, but in case of carbon sequestration projects, more precise accounting is often possible (Wunder *et al.*, 2008b). Sometimes the evidence base that links land use to delivery of particular environmental services is weak (Wunder, 2007). The typical assumption that 'forests' provide the necessary ES is a good example. Another problem is permanence: how to ensure ecosystem services continue to be protected, especially after payments from a particular programme or policy end (Wunder *et al.*, 2008b). Donors may be worried about financing long-term projects and how to handle non-compliance given their traditional role as aid providers (Wunder *et al.*, 2008a).

Despite some significant limitations, PES and related schemes are an important addition to the set of policy options and instruments to integrate conservation and development. The quality of such schemes ultimately rests on achieving a shared understanding of ecosystem services, benefits and burdens.

15.4.2 Certification

Certification is a practice intended to give a 'green' seal of approval to tree planting or 'reforestation' projects and includes chain-of-custody monitoring. One of the foremost timber certifiers is the Forest Stewardship Council, whose members comprise forestry and forestry-related corporations that are part of its 'economic chamber'. In the 1990s, successful NGO campaigns, particularly regarding unsustainable logging practices in the tropics, led to increased consumer awareness about the consumer's role in forest destruction. When consumers began to ask their suppliers for certified wood, a number of NGOs, together with businesses, decided to promote a process for enabling companies to offer and consumers to choose a 'green' product, and this resulted in the establishment of the FSC. The FSC allows certifier companies to inspect, then certifies logging and plantation companies, who can then sell timber with a FSC-certified label.

FSC's certifier companies are active in Southeast Asia and the Mekong region: In 2006, two forest areas in central Laos covering about 50,000 ha in the provinces of Khammouane and Savannakhet, were the first forests in Indochina to achieve FSC certification (WWF, 2006). But the independence and credibility of the FSC has increasingly come under question with FSC's failure to prevent the certification of non-compliant companies (Carrere, 2006; Butler & Laurance, 2008).

Instead of limiting FSC to forest management certification, organizations and businesses participating in the process decided to also include plantation management as part of its mission, lending FSC support to contentious large-scale monocultures that have resulted in severe livelihood impacts on many indigenous and local communities. In the Mekong region, FSC chain of custody has been no guarantee against illegal timber smuggling, for instance, from Laos to Vietnam.

15.5 Information

Increasing public awareness of ecosystem services can help garner wider support for their conservation at the level of policies and for improving management practices in watershed areas (Patterson & Coelho, 2009). This range from simply identifying and communicating previously unknown or under-appreciated services, through efforts to value benefits, to integrated assessment of social and ecological impacts and responses (Braumann *et al.*, 2007; Lele, 2009).

Publishing plans and information for scientific review and consultation is vital to ensure that any relevant policy options are not missed or ignored. This is especially true, for instance, with fire management where the conventional view of fire is of a destructive agent requiring immediate suppression. Fire management is crucial for some forest ecosystems to thrive and ensure continued carbon stocks and fluxes (Murdiyarso & Lebel, 2007). Fire and disease management can be used to meet land management goals under certain ecological conditions.

Forest management practices for carbon conservation and sequestration range from slowing down deforestation and assisting regeneration in the tropics to afforestation schemes and agro-forestry (Canadell & Raupach, 2008). Carbon sequestration can help with climate change protection, but there are many constraints to effective climate-forestry policies. Accidental or deliberate spread of fires into forests can result in huge emissions of CO_2 destroying carbon stocks and affecting other watershed services (Murdiyarso & Lebel, 2007). The best option for reducing carbon emissions in tropical regions is to avoid deforestation and degradation in the first place (Canadell & Raupach, 2008).

The 15th Conference of the Parties in Copenhagen under the United Nations Framework Convention on Climate Change (UNFCCC) agreed to pursue a mechanism for reduced emissions from deforestation and degradation, or REDD (UNFCCC, 2009). REDD claims to have the potential to benefit developing countries, including swidden cultivators (Mertz, 2009), but neglects underlying drivers of deforestation such as logging and tree plantations (whether for agrofuels, oil palm or pulpwood) as well as illegal logging and corruption. Moreover, questions remain about whether the funding channels and projects are designed in ways that prevent abuse (Europol, 2009) and whether the poor actually benefit – an outcome highly contingent on quality of governance within countries.

A formal meta-analysis of studies exploring the watershed services provided by tropical forests and plantations (Locatelli & Vignola, 2009) found evidence – contrary to public perceptions and official policies of several countries – for lower base flows under planted forests than non-forest land uses. It should be emphasized that this evidence comes from studies of only pine and *Eucalyptus* plantations; how plantations with other or native species affect the hydrological cycle is not known. This is consistent with earlier reviews (Calder, 2002; Bruijnzeel, 2004). At the same time some evidence was also found for lower total flow and higher base flow under natural forests than under non-forest land uses when a subset of data from small watershed studies with large differences in forest cover were analysed (Locatelli & Vignola, 2009). The authors acknowledged important constraints in a number of available studies of different forest types, and noted that when a larger dataset including larger watersheds was analysed no significant differences were found.

Valuation studies have to make many assumptions, but may help convince decision-makers on the needs for conservation of certain land covers and land uses because of the ecosystem services they provide. Most studies are hopeful about policy impact rather than demonstrating it.

As an example of a typical study, Chanhda and colleagues (2009) divided land use in Luang Namtha province of northern Laos into six classes and derived ecosystem service values for each category based on 11 biomes in a global ecosystem service valuation model (Costanza *et al.*, 1998). On this basis, the authors estimated that the forest land cover changes in Luang Namtha province resulted in a net decline of US$8.9 million in ecosystem services between 1992 and 2002 from potential forests. The decline in the value was due to soil erosion, flooding, drought and other impacts. The authors concluded that the high rate of loss of such services will have serious long-term negative ecological consequences and recommended that land reclamation projects be controlled and based on rigorous environmental impact analysis that includes assessment of impact on ecosystem services (Chanhda *et al.*, 2009). Use of such indirect estimates for services has many limitations given that values for even an individual service from a particular forest type can vary widely among places (Lele, 2009).

A much more detailed and convincing analysis was done by Pattanayak and Kramer (2001) investigating drought mitigation services of forests in Ruteng Park, Flores, Indonesia. They found evidence that the park provided a drought mitigation service in the form of base flow to farmers downstream and were able to estimate its value in terms of coffee and rice products. Using scenarios for re-establishment of forests consistent with goals of park management, they were also able to show that further increases in forest cover would result in both increases and decreases in base flow depending on local conditions and land uses in different watersheds. From a policy perspective this could help target specific watersheds with the right mixture of climatic and other features where increased protection will yield benefits of this service (Pattanayak & Kramer, 2001).

The complexity of landscape changes in many upland watersheds and diversity of services and interests they provide within and beyond the watershed precludes holding much faith in the absolute numerical findings of valuation studies. At least as important as understanding aggregate economic welfare is addressing the question of who wins and who loses under different scenarios of landscape change and why (Lele, 2009). van Beukering and colleagues (2003) noted after their valuation of multiple services from Leuser National Park on Sumatra, Indonesia, that although conservation benefits many stakeholders, the 'political power of logging and plantation industries' means that benefits continue to accrue to just the few stakeholders that favour logging and deforestation. Public awareness building, consultations and negotiations may be critical to secure support for management strategies (Pretty & Smith, 2004).

15.6 Discussion

Upland watersheds in Southeast Asia provide a diverse range of ecosystem services that are often highly specific to the particular land covers and ecosystems present, the landscape configuration and social organization. A diverse range of projects, policies and other initiatives have aimed to alter how these services are governed. In this chapter we have explored these in terms of plans, rules, incentives and information.

Planning has conventionally been led by government bureaucrats relying on neat physical and institutional separation into conservation and use and instrumentally driven definitions of classes. Forest and watershed classifications and zoning schemes have been constructed assuming and asserting strong relationships between particular classes and ecosystem services without attention to service users, alternative land uses or how ecosystem services are actually provided. This has led to unnecessary conflict between state agencies and local communities, disincentives for local conservation actions, and missed opportunities for activities that would maintain ecosystem services and also contribute to poverty alleviation.

Meaningful participation of local resource users alongside the conventional planning done by managers and, more recently, ecosystem experts, should lead to more informed and appropriate land-use plans for upland watersheds that have a chance of being implemented. Deliberative approaches to planning and assessment should help deal with competing knowledge claims about relationships of different land covers and uses with ecosystem services while recognizing that uncertainties of understanding may not be easily reducible. Local engagement is particularly crucial as the benefits of improved watershed management are often localized, and without cooperation and partnerships long-term management goals are hard to pursue (Aylward, 2005; Warner, 2006).

Regulations important for managing ecosystem services can be top-down, self-generated or more frequently a combination of local, informal rules and national, formal regulations (Lebel & Daniel, 2009). Interventions where there are multiple ecosystem services and derived benefits invariably create winners and losers. Projects and policies to improve watershed management are undertaken in the context of pre-existing institutions (Mollinga *et al.*, 2007) and as a consequence entail power relations that modify implementation and help shape outcomes. There is therefore a need to strengthen legal support and information sources for disadvantaged groups – like ethnic minorities – in dealing with formal legal processes. We suggest that it is important for future land and forest policies to leave some flexibility for local government and communities to adjust property rights systems to local cultural and ecological contexts of upland watersheds. This can take place within a wider framework, for example, that strongly restricts commercial exploitation of certain forest products, but not their subsistence uses.

Incentives can encourage provision of desired ecosystem services and protection of the ecosystems that underlie provision of those functions. Experience with payments or rewards for ecosystem services is growing in the region, suggesting that when used appropriately they will be helpful additions to the set of policy options and instruments to integrate conservation and development in particular places. Outstanding issues include ensuring equitable access, legitimacy of process, and that changes in watershed management actually contribute to the well-being of those in most need (Chan *et al.*, 2007; Corbera *et al.*, 2007; Wunder, 2008). These challenges are not restricted to instruments like payment or rewards for ecosystem services. A review of 103 ecosystem service projects – from 37 countries – implemented by The Nature Conservancy and the World Wildlife Fund (WWF), looked at projects using traditional conservation tactics such as land purchase and restoration, but also adopting new approaches such as targeting working landscapes, using new financial tools, and involving corporate funding and partners (Tallis *et al.* 2009). The review showed that the many kinds of projects often did not meet priority socio-economic needs.

Voluntary approaches are often more flexible than regulations and plans but only work if incentives are adequate or messages and social norms are persuasive enough. Building awareness about ecosystem services is invariably an important part of any intervention, locally or externally led. At the same time many projects and policies have been pursued in the absence of detailed understanding of ecosystem functions and services (Carpenter *et al.*, 2009; Daily *et al.*, 2009). Unvalidated assumptions about the relationship between land covers and hydrological services from watersheds, in particular, abound (Bruijnzeel, 2004; Aylward, 2005). Much more research on ecosystem services is needed across Southeast

Asia, ranging from understanding of ecosystem processes through to benefits and impacts on people. For many watersheds where decisions are being made, significant knowledge uncertainties will remain. Critically, drawing on diverse knowledge sources with an expectation that there is a need to negotiate, learn and adapt is the best strategy.

Projects and policies that hope to successfully improve the management of ecosystem services should seek, and expect, to learn from past interventions. Actual management practices in operation for timber, water and other ecosystem services from upland watersheds frequently do not match the plans on paper; they do not follow agency rules or fit government or other expectations based on simple incentives. Positive changes to ecological sustainability and human well-being are rarely demonstrated directly. In our review, we found monitoring to be the least well developed area of governance: independent and timely post-evaluations of projects and policies are necessary but rare.

There are many reasons, including uncertainties in how ecosystems and people will respond to interventions (Berkes, 2009), as well as more insidious ones, like systemic abuse, deception and corruption. Institutions need to be flexible enough to update rules to fit new knowledge or emerging conditions rather than assuming the solution is identifying the best practice, land use or allocation (Lebel *et al.*, 2004). The spread of invasive species, climate change and other larger-scale environmental changes with potential impacts on ecosystem services (Chopra *et al.*, 2005) provided by upland watersheds implies that adaptive responses will be imperative.

15.7 Conclusions

The upland watersheds of Southeast Asia provide a range of ecosystem services and derived benefits important to the well-being of people within them and beyond. The specific services provided and how they are valued vary hugely from place to place, underlying the importance of both ecological and social contexts. Communities, governments and firms have taken different approaches to negotiating and sharing these benefits, as well as trying to deal with trade-offs between them. In this chapter the governance of services has been explored through the lenses of plans, rules, incentives and information.

Four broad conclusions emerge. First, multi-stakeholder planning improves the assessment of under-appreciated services and users, but does not eliminate the importance of power relations or contestation of knowledge claims by stakeholders with divergent interests. Second, efforts to regulate the management of specific or an ambiguous set of ecosystem services with externally imposed rules invariably create winners and losers with outcomes that often depend on pre-existing institutions, political contexts and dominant beliefs about relationships between land covers and ecosystem services. Third, incentives to conserve ecosystem services are closely related to perceived benefits of doing so regardless of whether direct monetary payments are involved or rewards are in other forms. Fourth, shared understanding of the evidence for, and uncertainties around, particular ecosystem process, service and benefit relationships is crucial to progress, underlying the importance of monitoring and more adaptive approaches to integrating ecological and social understanding.

Taken together these findings underline the need to pay greater attention to issues of governance in the design and implementation of policies and projects to manage ecosystem services from upland watersheds. Improving governance of services is both a technical challenge of improving understanding of ecosystem processes and how they relate to benefits, and a social challenge of ensuring that interventions allocate benefits and burdens fairly while also improving the well-being of vulnerable and marginalized peoples.

Acknowledgements

This review was supported by grant ARCP2008-18NMY-Braimoh from the Asia Pacific Network for Global Change Research and grant PN67 from the Challenge Program on Water and Food. Thanks also to members of the M-POWER watershed governance working group and other team members of the ECOSMAG project.

References

Armitage, D., Plummer, R., Berkes, F., *et al.* (2009) Adaptive co-management for social-ecological complexity. *Frontiers in Ecology and Environment* **7**: 95–102.

Aylward, B. (2005) Towards watershed science that matters. *Hydrological Processes* **19**: 2643–7.

Barr, C. (2001) The political-economy of fiber, finances, and debt in Indonesia's pulp and paper industries. *Indonesian Quarterly* **29**: 173–88.

Bennagen, M.E., Indab A., Amponin A., *et al.* (2006) Designing payments for watershed protection services of Philippine upland dwellers Poverty Reduction and Environmental Management (PREM) programme. Quezon City, Philippines: Resources, Environment and Economics Center for Studies. Available at: http://www.premonline.nl/archive/5/doc/PWS%20Philippines%20final%20report.pdf

Bennett, M. (2008) China's sloping land conversion program: Institutional innovation or business as usual? *Ecological Economics* **65**: 699–711.

Berkes, F. (2009) Evolution of co-management: Role of knowledge generation, bridging organizations and social learning. *Journal of Environmental Management* **90**: 1692–702.

Bhagwat, S., Willis, K., Birks, H.J.B., Whittaker, R. (2008) Agroforestry: A refuge for tropical biodiversity? *Trends in Ecology and Evolution* **23**: 261–7.

Billon, P.L. (2000) The political ecology of transition in Cambodia 1989–1999: War, peace and forest exploitation. *Development and Change* **31**: 785–805.

Blaikie, P.M., Muldavin, J.S.S. (2004) Upstream, downstream, China, India: The politics of environment in the Himalayan region. *Annals of the Association of American Geographers* **94**: 520–48.

Braumann, K., Daily, G., Duarte, T., Mooney, H. (2007) The nature and value of ecosystem services: An overview highlighting hydrological services. *Annual Review of Environment and Resources* **32**: 67–98.

Bruijnzeel, L.A. (2004) Hydrological functions of tropical forests: Not seeing the soil for the trees. *Agriculture, Ecosystems and Environment* **104**: 185–228.

Bryant, R.L., Rigg, J., Stott, P. (1993) The political ecology of southeast Asian Forests: transdisciplinary discourses. *Global Ecology and Biogeography Letters* (special issue) **3**: 101–296.

Butler, R., Laurance, W. (2008) New strategies for conserving tropical forests. *Trends in Ecology and Evolution* **23**: 469–72.

Cairns, M. (ed.) (2007) *Voices From the Forest: Integrating Indigenous Knowledge into Sustainable Upland Farming*. Washington, DC: Resources for the Future.

Calder, I. (2002) Forests and hydrological services: Reconciling public and science perceptions. *Land Use and Water Resources Research* **2**: 1–12.

Canadell, J.G., Raupach, M.R. (2008) Managing forests for climate change mitigation. *Science* **320**: 1456–7.

Carpenter, S.R., Mooney, H., Agard, J., *et al.* (2009) Science for managing ecosystem services: Beyond the Millennium Ecosystem Assessment. *Proceedings of the National Academy of Sciences of the USA* **106**: 1305–12.

Carrere, R. (2006) Greenwash: Critical analysis of FSC certification of industrial tree monocultures in Uruguay. URL: http://wrm.org.uy/wp-content/uploads/2013/02/text.pdf (accessed 27 May 2014). World Rainforest Movement.

Chan, K., Pringle, R., Ranganathan, J., *et al.* (2007) When agendas collide: Human welfare and biological conservation. *Conservation Biology* **21**: 59–68.

Chanhda, H., Ci-fang, W., Ayumi, Y. (2009) Changes of forest land use and ecosystem service values along Lao-Chinese border: A case study of Luang Namtha Province, Lao PDR. *Forestry Studies in China* **11**: 85–92.

Chankaew, K. (1996) Kaan Kamnod Chan Kunnapaap Lumnam Korng Prathet Thai (Watershed Classification in Thailand). Kasetsart University, Bangkok. Available at: http://www.rdi.ku.ac.th/Ku-research60/ku60/watershed.html

Chopra, K., Leemans, R., Kumar, P., Simons, H. (eds) (2005) *Ecosystems and Human Well-being: Policy Responses*, vol. 3. Washington, DC: Island Press.

Contreras, A. (2003) *The Kingdom and the Republic: Forest Governance and Political Transformation in Thailand and the Philippines*. Quezon City: Ateneo de Manila University Press.

Cooper, D.E., Palmer, J.E. (eds) (1992) *The Environment in Question: Ethics and Global Issues*. London: Routledge.

Corbera, E., Brown, K., Adger, N.W. (2007) The equity and legitimacy of markets for ecosystem services. *Development and Change* **38**: 587–613.

Costanza, R., Andrade, F., Antunes, P., *et al.* (1998) Principles for sustainable governance of the oceans. *Science* **281**: 198–9.

Daily, G., Polasky, S., Goldstein, J., *et al.* (2009) Ecosystem services in decision making: time to deliver. *Frontiers in Ecology and Environment* **7**: 21–8.

Daniel, R.N. (ed.) (2005) *After the Logging Ban*. Bangkok: Foundation for Ecological Recovery (FER).

Daniel, R., Lebel, L. (2006) Land policy, tenure and use: Institutional interplay at the rural-forest interface in Thailand. USER Working Paper WP-2006-1. Chiang Mai: Unit for Social and Environmental Research, Chiang Mai University.

Daniel, R., Ratanawilailak, S. (2011) Local institutions and the politics of watershed management in the uplands of northern Thailand. In: Lazarus, K., Badendoch, N., Resurreccion, B., Dao, N. (eds), *Water Rights and Social Justice in the Mekong Region*. London: Earthscan, pp. 91–113.

Dauvergne, P. (2001) *Loggers and Degradation in the Asia-Pacific: Corporations and Environmental Management*. Cambridge: Cambridge University Press.

Dick, J. (1991) *Forest Land Use, Forest Use Zonation, and Deforestation in Indonesia: A Summary and Interpretation of Existing Information*. Environmental Management Development in Indonesia, EMDI Publications.

EarthTrends (2003) *Biodiversity and Protected Areas – Indonesia*. Washington, DC: World Resources Institute.

Engel, S., Pagiola, S., Wunder, S. (2008) Designing payments for environmental services in theory and practice: An overview of the issues. *Ecological Economics* **65**: 663–74.

Europol (European Law Enforcement Agency) (2009) Carbon Credit fraud causes more than 5 billion euros damage for European Taxpayer. Europol, 9 December 2009.

FAO (2002) An overview of forest products statistics in South and Southeast Asia. Bangkok: United Nations Food and Agriculture Organisation.

Fisher, B., Christopher, T. (2007) Poverty and biodiversity: Measuring the overlap of human poverty and the biodiversity hotspots. *Ecological Economics* **62**: 93–101.

Fisher, B., Turner, K., Zylstra, M., *et al.* (2008) Ecosystem services and economic theory: Integration for policy-relevant research. *Ecological Applications* **18**: 2050–67.

Flaherty, M., Jengjalern, A. (1995) Differences in assessments of forest adequacy among women in northern Thailand. *Journal of Developing Areas* **29**: 237–54.

Forestry Act (1941) *Praraachabanyat Paamai Po. So. 2484* [Forestry Act of 1941]. *Matraa 4(1)* [article 4, Section 1].

Forsyth, T. (1996) Science, myth and knowledge: Testing Himalayan environmental degradation in Thailand. *Geoforum* **27**: 275–92.

Forsyth, T. (1998) Mountain myths revisited: Integrating natural and social environmental science. *Mountain Research and Development* **18**: 126–39.

Forsyth, T., Walker, A. (2008) *Forest Guardian, Forest Destroyers: The Politics of Environmental Knowledge in Northern Thailand*. University of Washington Press.

Garcia-Fernandez, C., Casado, M.A. (2005) Forest recovery in managed agroforestry systems: The case of benzoin and rattan gardens in Indonesia. *Forest Ecology and Management* **214**: 158–69.

George, A., Pierret, A., Boonsaner, A., Valentin, C., Orange, D., Planchon, O. (2009) Potential and limitations of payments for environmental services (PES) as a means to manage watershed services in mainland Southeast Asia. *International Journal of the Commons* **3**: 16–40.

Heinimann, A., Breua, T., Kohler, T. (2005) Watershed classification in the Lower Mekong Basin. *Mountain Research and Development* **25**: 181–2.

Hirsch, P., Lohmann, L. (1989) Contemporary politics of environment in Thailand. *Asian Survey* **29**: 439–51.

ICEM (2003a) Lao PDR National Report on Protected Areas and Development. Review of Protected Areas and Development in the Lower Mekong River Region. Indooroopilly, Queensland: International Centre for Environmental Management.

ICEM (2003b) Protected areas and development: Lessons from Cambodia. In: *Review of Protected Areas and their Role in Socio-economic Development in the Four Countries of the Lower Mekong River Region*. Indooroopilly, Queensland: International Centre for Environmental Management, pp. 15–29.

ICEM (2003c) Thailand National Report on Protected Areas and Development. Review of Protected Areas and Development in the Lower Mekong River Region. Indooroopilly, Queensland: International Centre for Environmental Management.

IUCN (2007) World Database on Protected Areas. UNEP-WCMC in partnership with the IUCN World Commission on Protected Areas (WCPA).

Jack, B., Kousky, C., Sims, K. (2008) Designing payments for ecosystem services: Lessons from previous experience with incentive-based mechanisms. *Proceedings of the National Academy of Sciences of the USA* **105**: 9465–70.

Jenerrete, G., Marussich, W., Newell, J. (2006) Linking ecological footprints with ecosystem valuation in the provisioning of urban freshwater. *Ecological Economics* **59**: 38–47.

Jourdain, D., Pandey, S., Tai, D.A., Quang, D.D. (2009) Payments for environmental services in upper-catchments of Vietnam: Will it help the poorest? *Internaional Journal of the Commons* **3**: 64–81.

Kashio, M. (1995a) *Forestry in Thailand*. Association for the International Cooperation of Agriculture and Forestry.

Kashio, M. (1995b) Sustainable forest management in Asia and the Pacific. In: *Proceedings of a Regional Expert Consortium on Implementing Sustainable Forest Management in Asia and the Pacific*. Bangkok: Food and Agriculture Organisation (FAO), pp. 17–31.

Kerr, J. (2007) Watershed management: Lessons from common property theory. *International Journal of the Commons* **1**: 89–109.

Knupfer, J. (2002) Survival strategies of ethnic minorities in the highlands of northern Thailand – case studies from Chiang Rai Province. International Symposium on Sustaining Food Security and Managing Natural Resources in Southeast Asia – Challenges for the 21st Century, January 8–11, 2002, Chiang Mai, Thailand. University of Hohenheim, Germany. Available at: https://www.uni-hohenheim.de/fileadmin/einrichtungen/sfb564/events/uplands2002/Full-Pap-S2-2_Knuepfer.pdf

Koellner, T., Sell, J., Gahwiler, M., Scholz, R. (2008) Assessment of the management of organization supplying ecosystem services from tropical forests. *Global Environmental Change* **18**: 746–57.

Lang, C. (2000) The national hydropower plan study: Planning and damming in Vietnam. *Watershed: People's Forum on Ecology* **5**: 48–51.

Lang, C. (2002) *The Pulp Invasion: The International Pulp and Paper Industry in the Mekong Region*. Montevideo: World Rainforest Movement.

Lasco, R., Pulhin, J.M. (2006) Environmental impacts of community-based forest management in the Philippines. *International Journal of Environment and Sustainable Development* **5**: 46–56.

Laungaramsri, P. (2000) The ambiguity of "watershed": The politics of people and conservation in northern Thailand. *Sojourn* **15**: 52–75.

Laungaramsri, P. (2002a) On the politics of nature conservation in Thailand. *Kyoto Review* URL: http://kyotoreview.cseas.kyoto-u.ac.jp/issue/issue1/article_168.html

Laungaramsri, P. (2002b) *Redefining Nature: Karen Ecological Knowledge and the Challenge to the Modern Conservation Paradigm*. Chennai: Earthworm Books.

Lebel, L. (2005) Institutional dynamics and interplay: Critical processes for forest governance and sustainability in the mountain regions of northern Thailand. In: Huber, U.M., Bugmann, H.K.M., Reasoner, M.A. (eds), *Global Change and Mountain Regions: An Overview of Current Knowledge*. Berlin: Springer-Verlag, pp. 531–40.

Lebel, L., Daniel, R. (2009) The governance of ecosystem services from tropical upland watersheds. *Current Opinion in Environmental Sustainability* **1**: 61–8.

Lebel, L., Contreras, A., Pasong, S., Garden, P. (2004) Nobody knows best: Alternative perspectives on forest management and governance in Southeast Asia: Politics, Law and Economics. *International Environment Agreements* **4:** 111–27.

Lebel, L., Daniel, R., Badenoch, N., Garden, P. (2008) A multi-level perspective on conserving with communities: Experiences from upper tributary watersheds in montane mainland southeast Asia. *International Journal of the Commons* **1:** 127–54.

Leimona, B., Joshi, L., van Noordwijk, M. (2009) Can rewards for environmental services benefit the poor? Lessons from Asia. *International Journal of the Commons* **3**: 82–107.

Lele, S. (2004) Beyond state-community polarisations and bogus "joint"ness: Crafting institutional solutions for resource management. In: Spoor, M. (ed.), *Globalisation, Poverty and Conflict*. Dordrecht: Kluwer Academic Publishers, pp. 283–303.

Lele, S. (2009) Watershed services of tropical forests: from hydrology to economic valuation to integrated analysis. *Current Opinion in Environmental Sustainability* **1**: 148–55.

Locatelli, B., Vignola, R. (2009) Managing watershed services of tropical forests and plantations: Can meta-analyses help? *Forest Ecology and Management* **258**: 1864–70.

Manivong, V., Cramb, R.A. (2008) Economics of smallholder rubber expansion in Northern Laos. *Agroforest Systems* **74**: 113–25.

Mann, C.C. (2009) Addicted to rubber. *Science* **325**: 564–6.

Mertz, O. (2009) Trends in shifting cultivation and the REDD mechanism. *Current Opinion in Environmental Sustainability* **1**: 156–60.

Millennium Ecosystem Assessment (2005) *Ecosystems and Human Well-being: Synthesis.* Washington, DC: Island Press.

Mollinga, P., Meinzen-Dick, R., Merrey, D. (2007) Politics, plurality and problemsheds: A strategic approach for reform of agricultural water resources management. *Development Policy Review* **25**: 699–719.

Murdiyarso, D., Lebel, L. (2007) Local to global perspectives on forest and land fires in Southeast Asia. *Mitigation and Adaptation Strategies for Global Change* **12**: 2381–6.

Naughton-Treves, L., Holland, M., Brandon. K. (2005) The role of protected areas in conserving biodiversity and sustaining local livelihoods. *Annual Review of Environment and Resources* **30**: 219–252.

Neef, A. (2001) Land tenure and soil conservation practices – evidence from West Africa and Southeast Asia. In: Stott, D.E., Mohtar, R.H., Steinhardt, G.C. (eds), *Sustaining the Global Farm: Selected Papers from the 10th International Soil Conservation Organization Meeting held May 24–29, 1999 at Purdue University and the USDA-ARS National Soil Erosion Research Laboratory*, Purdue University, pp. 125–30. Available at: http://topsoil.nserl.purdue.edu/nserlweb-old/isco99/pdf/ISCOdisc/SustainingTheGlobalFarm/P238-Neef.pdf

Neef, A., Thomas, D. (2009) Rewarding the upland poor for saving the commons? Evidence from Southeast Asia. *International Journal of the Commons* **3**: 1–15.

Neef, A., Sangkapitux, C., Kirchmann, K. (2000) Does land tenure security enhance sustainable management? Evidence from mountainous regions of Thailand and Vietnam. Stuttgart: Institute of Agricultural Economics and Social Sciences in the Tropics and Subtropics.

Pasong, S., Lebel, L. (2000) Political transformation and the environment in Southeast Asia. *Environment* **42**: 8–19.

Pattanayak, S. K., Kramer, R.A. (2001) Worth of watersheds: A producer surplus approach for valuing drought mitigation in Eastern Indonesia. *Environment and Development Economics* **6**: 123–46.

Pattanayak, S., Wendland, K.J. (2007) Nature's care: Diarrhea, watershed protection, and biodiversity conservation in Flores, Indonesia. *Biodiversity Conservation* **16**: 2801–19.

Patterson, T., Coelho, D. (2009) Ecosystem services: Foundations, opportunities, and challenges for the forest product sector. *Forest Ecology and Management* **257**: 1637–46.

Phongpaichit, P. (1999) Corruption: Is there any hope at all? Paper for the Prajadhipok Institute Workshop on Governance, Pattaya.

Poffenberger, M. (ed.) (1999) *Communities and Forest Management in Southeast Asia.* Gland: IUCN – The World Conservation Union.

Pretty, J., Smith, D. (2004) Social capital in biodiversity conservation and management. *Conservation Biology* **18**: 631–8.

RECOFTC (2009) Community forest management in Sam Pak Nam, Thailand. Bangkok: Regional Community Forestry Training Centre for Asia and the Pacific.

Rerkasem, K., Rerkasem, B., Kaosa-ard, M., *et al.* (1994) *Assessment of Sustainable Highland Agricultural Systems*. Bangkok: Thailand Development Research Institute (TDRI).

Rerkasem, K., Lawrence, D., Padoch, C., Schmidt-Vogt, D., Ziegler, A.D., Bruun. T.B. (2009) Consequences of swidden transitions for crop and fallow biodiversity in Southeast Asia. *Human Ecology* **37**: 347–60.

Ross, W. (2003) Sustainable tourism in Thailand: Can ecotourism protect the natural and cultural environments? In: Second Meeting of the Academic Forum for Sustainable Development, International Sustainability Conference, Fremantle, Western Australia.

Roth, R. (2004a) On the colonial margins and in the global hotspot: Park-people conflicts in highland Thailand. *Asia Pacific Viewpoint* **45**: 13–32.

Roth, R. (2004b) Spatial organization of environmental knowledge: Conservation conflicts in the inhabited forest of Northern Thailand. *Ecology and Society* **9**: 5; URL: http://www.ecologyandsociety.org/vol9/iss3/art5

Sanchez-Azofeifa, G., Pfaff, A., Robalino, J., Boomhower, J. (2007) Costa Rica's payment for environmental services program: Intention, implementation, and impact. *Conservation Biology* **21**: 1165–73.

Sangkapitux, C., Neef, A., Polkongkaew, W., Pramoon, N., Nonkiti, S., Nanthasen, K. (2009) Willingness of upstream and downstream resource managers to engage in compensation schemes for environmental services. *Internationl Journal of the Commons* **3**: 41–63.

Sato, J. (2003) Public land for the people: The institutional basis of community forestry in Thailand. *Journal of Southeast Asian Studies* **34**: 329–46.

Schmidt-Vogt, D. (1998) Defining degradation: The impacts of swidden on forests in northern Thailand. *Mountain Research and Development* **18**: 135–49.

Sharma, E., Xu, J. (2007) Land-use, landscape management and environmental services in the mountain mainland Asia: Introduction. *Tropical Ecology* **48**: 129–36.

Sidle, R., Ziegler, A., Negishi, J., Nik, A.R., Siew, R., Turkelboom, F. (2006) Erosion processes in steep terrain – truth, myths and uncertainties related to forest management in Southeast Asia. *Forest Ecology and Management* **224**: 199–225.

Sowerine, J.C. (2004) Territorialisation and the politics of highland landscapes in Vietnam: negotiating property relations in policy, meaning and practice. *Conservation and Society* **2**: 97–136.

Sowerine, J., Dzung, N.H., Poffenberger, M. (1998) Ba Vi National Park and the Dzao. stewards of Vietnam's upland forests. A collaborative study by the Asia Forest Network and the Forest Inventory and Planning Institute. Research Network Report, Number 10. Asia Forest Network; Center for Southeast Asia Studies, Berkeley.

Spash, C. (2007) Deliberative monetary valuation (DMV): Issues in combining economic and political processes to value environmental change. *Ecological Economics* **63**: 690–9.

Sturgeon, J.C., Menzies, N., Lagervist, Y., *et al.* (2013). Enclosing ethnic minorities and forests in the golden economic quadrangle. *Development and Change* **44**: 53–79.

Swallow, B., Leimona, B., Yatich, T., Velarde, S.J., Puttaswamaiah, S. (2007) The conditions for effective mechanisms of compensation and rewards for environmental services. Working Paper 38. Nairobi: World Agroforestry Centre.

Tallis, H., Goldman, R., Uhl, M., Brosi, B. (2009) Integrating conservation and development in the field: implementing ecosystem service projects. *Frontiers in Ecology and Environment* **7**: 12–20.

Thomas, D.E. (2005) *Participatory Watershed Management in Upper Northern Region of Thailand*. Chiang Mai: ICRAF.

Thomas, D.E. (2006) Participatory watershed management in Ping watershed: Final report. Bangkok: Office of Natural Resources and Environmental Policy and Planning, Ministry of Natural Resources and Environment.

Thomas, D.E., Ekhasing, B., Ekhasing, M., *et al.* (2008) Comparative assessment of resource and market access of the poor in upland zones of the Greater Mekong Region. Report submitted to the Rockefeller Foundation under Grant No. 2004 SE 024. Chiang Mai: World Agroforestry Centre.

Trisurat, Y., Pattanavibool, A. (2008) Classification of Protected Areas system in Thailand for conservation priority. Paper presented at the annual meeting of the International Congress for Conservation Biology, Convention Center, Chattanooga, TN, July 13, 2008.

Turner, R., Daily, G. (2008) The ecosystem services framework and natural capital conservation. *Environmental and Resource Economics* **39**: 25–35.

UNEP-WCMC (2003) World Database on Protected Areas (WDPA) Version 6. Cambridge: World Database on Protected Areas Consortium.

UNFCCC (2009) Copenhagen Accord. Draft decision – CP.15. 18 December 2009. FCCC/CP/2009/L.7. United Nations Framework Convention on Climate Change.

van Beukering, P., Cesar, H., Janssen, M.A. (2003) Economic valuation of the Leuser National Park on Sumatra, Indonesia. *Ecological Economics* **44**: 43–62.

Vandergeest, P., Peluso, N.L. (1995) Territorialization and state power in Thailand. *Theory and Society* **24**: 385–426.

van Dijk, A., van Noordwijk, M., Calder, I. *et al.* (2009) Forest-flood relation still tenuous – comment on 'Global evidence that deforestation amplifies flood risk and severity in the developing world' by CJA Bradshaw, NS Sodi, KS-H Peh, and BW Brook. *Global Change Biology* **15**: 110–15.

Van Noordwijk, M., Leimona, B., Emerton, L., *et al.* (2007) Criteria and indicators for ecosystem service reward and compensation mechanisms: realistic, voluntary, conditional and pro-poor. Working Paper 37. Nairobi: World Agroforestry Centre.URL: http://www.worldagroforestrycentre.org/SEA/Publications/files/workingpaper/WP0082-07.ZIP

van Vliet, N., Mertz, O., Heinimann, A., *et al.* (2012) Trends, drivers and impacts of changes in swidden cultivation in tropical forest-agriculture frontiers: A global assessment. *Global Environmental Change* **22**: 418–29.

Walker, A. (2003) Agricultural transformation and the politics of hydrology in northern Thailand. *Development and Change* **34**: 941–64.

Walker, A. (2004) Seeing farmers for the trees: Community forestry and the arborealisation of agriculture in northern Thailand. *Asia Pacific Viewpoint* **45**: 311–24.

Walker, A. (2006) Community forests in the uplands: Good for forests, but what about farmers? In: Lebel, L., Xu, J., Contreras, A. (eds), *Institutional Dynamics and Stasis: How Crises Alter the Way Common Pool Resources are Perceived*. Chiang Mai: Regional Centre for Social Science and Sustainable Development (RCSD), Chiang Mai University, pp. 110–32.

Walker, A., Farrelly, N. (2008) Northern Thailand's specter of eviction. *Critical Asian Studies* **40**: 373–97.

Wanishpradist, A. (2005) Dynamic of local knowledge and alternative management of resources on the highlands: A case study of Hmong Mae Sa Mai community in Mae Rim district, Chiang Mai province (in Thai). Chiang Mai: Chiang Mai University.

Warner, J.F. (2006) More sustainable participation? Multi-stakeholder platforms for integrated catchment management. *Water Resources Development* **22**: 15–35.

Wataru, F. (2003) Dealing with contradictions: Examining national forest reserves in Thailand. *Southeast Asian Studies* **41**: 206–38.

Wittayapak, C. (1996) Political ecology of the expansion of protected areas in northern Thailand. In: 6th International Conference on Thai Studies, 14–17 October, Chiang Mai, Thailand.

Wittayapak, C., Dearden, P. (1999) Decision-making arrangements in community-based watershed management in Northern Thailand. *Society and Natural Resources* **12**: 673–91.

Wunder, S. (2007) The efficiency of payments for environmental services in tropical conservation. *Conservation Biology* **21**: 48–58.

Wunder, S. (2008) Payments for environmental services and the poor: Concepts and preliminary evidence. *Environment and Development Economics* **13**: 279–97.

Wunder, S., Campbell, B., Frost, P., Sayer, J., Iwan, R., Wollenberg, L. (2008a) When donors get cold feet: The community conservation concession in Setulang (Kalimantan, Indonesia) that never happened. *Ecology and Society* **13**: 12 [online] URL: http://www.ecologyandsociety.org/vol13/iss1/art12/

Wunder, S., Engel, S., Pagiola, S. (2008b) Taking stock: A comparative analysis of payments for environmental services programs in developed and developing countries. *Ecological Economics* **65**: 834–52.

WWF (2006) Forests in Indochina receive FSC certification. Posted 26 Jan 2006. URL: http://www.panda.org/what_we_do/where_we_work/greatermekong/news/?57420/Forests-in-Indochina-receive-FSC-certification (accessed 13 May 2014).

Xu, J. (2006) The political, social, and ecological transformation of a landscape: The case of rubber in Xishuangbanna, China. *Mountain Research and Development* **26**: 254–62.

Xu, J., Melick, D. (2007) Rethinking the effectiveness of public protected areas in Southwestern China. *Conservation Biology* **21**: 318–28.

Xu, J., Yang, Y., Fox, J., Yang, X. (2007) Forest transition, its causes and environmental consequences: Empirical evidence from Yunnan of Southwest China. *Tropical Ecology* **48**: 1–14.

Xu, J., Lebel, L., Sturgeon, J.C. (2009) Functional links between biodiversity, livelihoods and culture in a Hani swidden landscape in Southwest China. *Ecology and Society* **14**: 20 [online] URL: http://www.ecologyandsociety.org/vol14/iss22/art20/.

Ziegler, A.D., Fox, J.M., Xu, J.C. (2009) The rubber juggernaut. *Science* **324**: 1024–5.

16 Socio-Economic Impacts of a Wetland Restoration Program in China's Poyang Lake Region

Fen Li[1], Lin Zhen[2], He Qing Huang[2], Yunjie Wei[2], Li Yang[2] and Sandra Uthes[3]

[1]*Shenzhen Institute of Building Research Co., Ltd, Shenzhen, China*
[2]*Institute of Geographic Science and Natural Resources Research, Chinese Academy of Sciences, Beijing, China*
[3]*Leibniz-Centre for Agricultural Landscape Research (ZALF), Müncheberg, Germany*

16.1 Introduction

Many countries faced with the pressures of balancing economic development with environmental protection have recognized that uncontrolled development may lead to an irreversible loss of ecologically vulnerable regions (EVRs) (Ndubisi *et al.*, 1995). EVRs are characterized by fragile ecosystems that are sensitive to outside disturbances such as economic activities and land use policies. Human consumption of resources in such regions can easily assume a non-sustainable direction. EVRs are often located in transition zones such as urban-rural, agro-pasture and water-land areas (X. Li, 1996).

Since the 1990s, the consequences of land use changes in EVRs have increasingly influenced decision-making processes at local, regional and global scales, and they have therefore also become an important topic in the scientific community. A number of studies have analyzed the effects of land use change and land use functional change, for example, Tanrivermis *et al.* (2003). These authors analyzed agricultural land use change and the sustainable use of land resources in the Mediterranean region of Turkey. Musacchio and Grant (2002) developed an integrated systems model of the coastal prairie ecosystem to simulate

Vulnerability of Land Systems in Asia, First Edition. Edited by Ademola K. Braimoh and He Qing Huang.

the effect of alternative resources policies on land use decisions. Ademola Braimoh (2009) integrated remote-sensing data with social surveys to identify the role of demographic, technological and market-based variables in cropland expansion during economic reforms in Ghana. The results of such studies may contribute to a better understanding of how land use cover change (LUCC) and GIS methods can provide better and more targeted information that may, in turn, help to improve conservation policies and land use planning strategies (Alejandro *et al.*, 2003). In China, Zhen *et al.* (2005) analyzed land use changes, their driving forces and related policy implications upstream from the Jinghe watershed. Spatial and temporal GIS data taken from remote sensing images were used to investigate the dynamics of land use change. A structured questionnaire survey was distributed to investigate factors affecting land use changes. Y. Li (2006) utilized mathematical statistics to quantify the driving factors behind land use change in Xi'an city and predicted future land use changes. Yang and Liu (2008) evaluated the development stage of sustainable land use in hilly areas and suggested corresponding countermeasures according to the evaluating index system of sustainable land use.

Most studies of land use changes have focused on the differences between land use change in different regions, the driving factors of land use change, the affected land use types and the causes of land desertification. The above research has focused primarily on the natural and ecological dimensions of land use change, with only a few studies of its social and economic dimensions. At present, land use changes drive the process of change in ecological systems to a large extent and significantly impact their ability to provide ecosystem services (Working Group of the Millennium Ecosystem Assessment, 2003). For instance, land degradation on arable lands has a negative impact on crop productivity. A decrease in forest and grassland areas has a negative impact on vegetation and regulation functions such as the water and carbon cycles. Furthermore, land use changes also have an impact on the livelihoods of local residents. The question of how to manage the trade-offs between environmental protection and maintaining the livelihoods of the rural population is therefore an important research field.

China is one of the nations with the most EVRs in the world. EVRs cover approximately 60% of the entire land area in China (according to one spokesperson from the Ministry of Environmental Protection of the People's Republic of China). Meanwhile, most of these regions are also poverty-stricken areas in which the sensitivity of the ecosystems and the need for economic development clash. Poyang Lake is the largest freshwater lake in China. Its wetland ecosystem is of global importance and, therefore, is receiving increasing attention from the Chinese government and the international community. The occurrence of severe flood events and certain political decisions have had severe negative effects on rural livelihoods (Jiang *et al.*, 2008). Between 1948 and 1998, the land use policy for the lake area in the downstream portion of the Yangtze River changed from 'Reclaiming parts of lakes for farmland' to 'Converting farmland to lakes (CFTL)'. Whereas in earlier times wetlands (water and unused land) were converted to farmland to increase agricultural area, the policy at this time calls for the re-conversion of agricultural land to wetlands. The CFTL policy is therefore an important driver of land use change in the downstream lake area of the Yangtze River, and its implementation has had significant effects on regional land use patterns. Since the 1990s, the Poyang Lake region has also experienced considerable economic growth, which poses a number of challenges for its long-term sustainable development.

The aim of the study is to analyze both the land use changes resulting from wetland restoration as well as their impact on the socio-economic situation of the Poyang Lake region. In addition, regional farmers' willingness to accept eco-compensation and the eco-compensation burden of the various local governments are calculated. The results of these analyses are used to derive recommendations regarding the future implementation of eco-compensation.

16.2 Study area

16.2.1 Background

Poyang Lake is an important wetland area with many ecosystem functions. As the largest lake connected to the Yangtze River, Poyang Lake plays an important role in the regulation of floods and guards against floods in the middle and downstream areas. The lake is located in the north of Jiangxi Province, on the south bank of the Yangtze River in the red-soil hilly area (Figure 16.1). Its surface area varies between 3000 and 4000 km^2 depending on the season. Relevant land categories in the area around the lake include farmland, grassland, orchard/plant nurseries, water supply and aquaculture ponds, wetlands, settlements, and industrial land. The total wetland area covers 2787 km^2. The area for this study includes 10 counties and two districts administered by the three cities of Nanchang, Jiujiang and Shangrao (cf. Table 16.3). The case study area is 21,000 km^2, with a total population of 9.07 million inhabitants (calculated in 2004) and a population density of 432 inhabitants per km^2.

The CFTL policy was launched both to protect the environment and reduce the risk of floods, particularly after the floods in 1998. According to the statistics, the region was home to 9.21 million inhabitants by the end of 2005. The arable land area per capita was 0.045 ha, which was only 42.4% of China's average and was lower than the FAO figure for the lowest acceptable amount of arable land per capita (= 0.053 ha).

Three key challenges for sustainable development exist:

1. High population density, the limited availability of agricultural land and surplus labor. As a result of the wetland restoration, the arable land per capita and total agricultural employment have decreased. Non-agricultural industries that could provide new employment have not yet been developed. Rising surplus labor rates have therefore

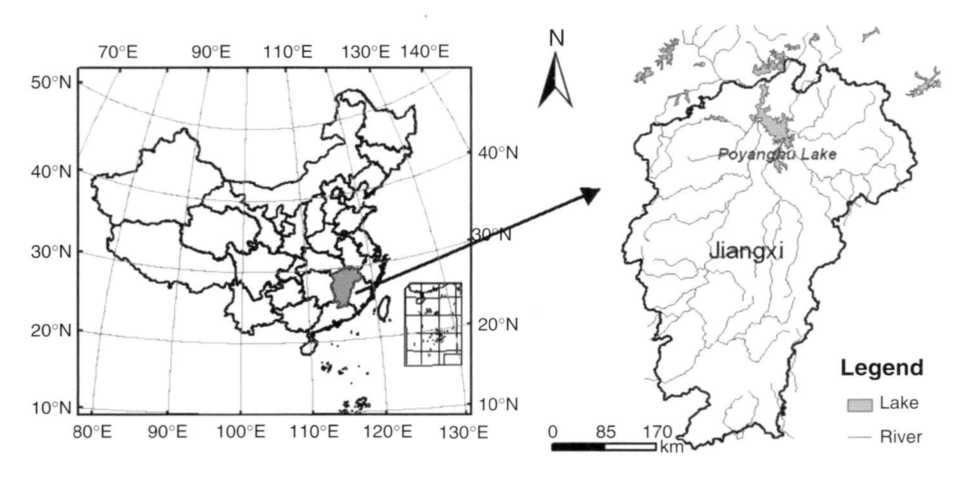

Figure 16.1 Location of Poyang Lake in China and in Jiangxi Province. *Source*: Data Center for Resources and Environmental Sciences, Chinese Academy of Sciences, 2004

become a severe problem. The rate increased from 17.1% in 1997 to 40% in 2005 (based on regional experts' estimates).

2. A lack of reasonable eco-compensation mechanisms (F. Li *et al.*, 2009). The Poyang Lake region covers two districts and 10 counties, and there are management conflicts among the different governmental organizations. Unless responsibilities are not clearly defined, eco-compensation mechanisms cannot be implemented efficiently.
3. An aging rural population and non-sustainable traditional farming. Based on field surveys, it is known that most of the people who remain in the villages are elderly people and children. For example, the average age of the elderly people who remain is about 61, accounting for 23% of the whole population in the study area. As for young people, their average age is only 10.3, accounting for 12.7%. More and more laborers are leaving the rural areas for off-farm employment in the hope of improving their income.

16.3 Methods

16.3.1 Analysis of land use and economic data

We used existing statistical data from the Yearbook of Jiangxi Province from 1997 to 2006 and GIS spatial data (between 1997 and 2005) to analyze both the land use changes in the region before and after the implementation of the wetland restoration policy as well as the importance of the different economic sectors in the region. The available data were separated into data for two phases: the first phase, which occurred before policy implementation (from 1997 to 2000), and the second phase, which occurred after policy implementation (from 2000 to 2005).

16.3.2 Stakeholder analysis

Stakeholder analysis is an approach to understanding a system and its changing processes by identifying key actors or stakeholders and assessing their respective interests in that system (Grimble & Chan, 1995). The stakeholder analysis for this study involved semi-structured interviews with relevant experts. The experts interviewed ($n = 35$) included government officials, representatives of public administration and environmental economists from the Policy Research Center of the Ministry of Environmental Protection, the State Forestry Administration, the Academy of Macroeconomic Research of the Reform Committee, the Institute of Geographic Science and Natural Resources Research of the Chinese Academy of Science, the Academy of Environmental Sciences of Anhui Province, and other government departments and institutions of environmental management. The interviews were completed in September 2008; 30 questionnaires were usable. We used the SPSS 13.0 software to analyze the questionnaires. The statistical methods utilized included descriptive statistics, mean comparison, paired samples t-tests and single factor analysis of variance. Nine different types of stakeholders were defined: government agencies, farmers, environmental NGOs, the media, the public, community, research institutions, investment institutions and firms.

In the interviews, the experts were asked to rank the following issues on a scale from 1 and 10:

1. Positive attitude towards eco-compensation: some stakeholders volunteer to participate in the eco-compensation program and their ideas or behavior may positively impact the eco-compensation mechanism, while other people will have a negative

attitude to the implementation of the eco-compensation mechanism and passively accept its impact. Thus, our aim was to ask the experts to score the level of positive attitude towards eco-compensation on the part of the nine stakeholders.
2. Decision-making power with regard to the implementation of eco-compensation: some stakeholders have more power with regard to eco-compensation implementation, and policy-makers would like to take their influence into consideration. They contribute to many aspects of such programs, including providing funding, management, and program control.
3. Importance of stakeholder interests: the interests and needs of the stakeholders are different and not equally important to the eco-compensation mechanism.

After the three-dimensional analysis of these stakeholders, we grouped them into three types: core stakeholders, secondary stakeholders and marginal stakeholders.

16.3.3 Household surveys

The next step included household surveys in sub-parts of the region in which the land use and economic structure had obviously changed as a result of the program. We selected four towns in Jiujiang and Hukou counties of Duchang city and two towns in Poyang County of Shangrao City. From March 25 to April 11 2008, we received 270 usable questionnaires. The household surveys had two main purposes: (i) to understand the relationships and functions of the rural population in the environmental protection and compensation process, and (ii) to estimate the farmers' willingness to accept ecological compensation for program participation.

The questionnaire addressed the following topics:

- The basic situation of the respondents and their family members, including information such as gender, age, occupation, income and the income spent on food;
- Land use, including information such as land use types, soil quality, and distance from the lake.
- Economic conditions and changes.
- Eco-compensation, including information such as awareness and attitudes regarding Poyang Lake protection and compensation, stakeholders' relationship, participation, and compensation use.

16.3.4 Farmers' willingness to accept eco-compensation (WTA)

Economists have traditionally addressed environmental goods valuation by adopting methodologies that rely on survey responses, as is the case with the Contingent Valuation Method (CVM) (Mitchell & Carson, 1989). The CVM typically uses survey techniques to determine individuals' willingness to pay (WTP) for the hypothetical provision of an environmental good or their willingness to accept (WTA) compensation for the economic development related to the hypothetical provision of a public good. Thus, these values are considered to represent the economic benefits of the proposed change and can be aggregated using a cost-benefit framework to determine the social benefits of public policies that usually improve social welfare (Salvador *et al.*, 2009). The willingness to accept eco-compensation (with the household as the unit) was estimated based on the following equation:

$$E(WTA) = \sum_{i=1}^{5} p_i\, b_i \tag{16.1}$$

where p_i is the percentage of the totality on each bid point, b_i is the amount of willingness for each bid point, and *WTA* is the willingness to accept eco-compensation.

The farmers were asked whether they support the continued implementation of the CFTL policy. Farmers who were in favor of continuation were asked how much annual compensation per hectare they would need to accept the policy. According to the surveyed results, about 35% of the respondents did not support continuation of the program because of the low level of compensation and their worries about losing the land. We used five bid points between 12,000 and 15,000 yuan per ha. The bid scale was based on the estimates of regional experts.

16.3.5 Estimation of the eco-compensation burden of the local governments

16.3.5.1 Land-based benefit coefficient The availability of farmland is important to the basic living conditions of the rural population in the Poyang Lake region, notably in that it provides food. We used the total available farmland area in the 12 counties of Poyang Lake as a proxy for direct benefits from the land. Counties with more agricultural land obtain more direct benefits from the land than do counties with less agricultural land.

16.3.5.2 Engel coefficient The Engel coefficient expresses the proportion of total family income spent on food. This parameter may be obtained directly from household surveys. We asked the farmers how much of their total income they spend on food and then calculated the average.

16.3.5.3 Payment ability coefficient We used economic development level and willingness to pay (WTP) to estimate the public payment ability for ecosystem services. The WTP is simulated by the R. Pearl growth curve model and the Engel coefficient. At different stages of economic development, people's awareness of the value of the environment and their needs are different. Once the basic need for items such as food and clothing has been satisfied, along with economic development, other non-physiological needs become increasingly important, such as the valuation of environmental services, which also leads to an increased WTP for environmental services. The WTP process for ecosystem services is similar to the process described by the R. Pearl growth curve. Therefore, we use this model to simulate the public WTP for ecosystem services (J. Li *et al.*, 1999). The R. Pearl growth model expression is as follows:

$$y = \frac{k}{1 + a\,\mathrm{e}^{-bt}} \tag{16.2}$$

where y is society's WTP for the ecological benefit, k is the maximum of y, t is the development level of society as indicated by the reciprocal of the Engel coefficient, a, b are constants, and e is the base of natural logarithms. To simplify the problem, the parameter k is set as the maximum value of y, which is 1; a and b are also 1. In the above model, we considered the development level of society to determine WTP for the ecological benefit. We used the reciprocal of the Engel coefficient to represent the time axis, t (abscissa), and did the necessary conversion ($T = t + 3$); in this way, we can determine WTP:

$$y_i = \frac{1}{1 + e^{-\left(\frac{1}{E_{ni}} - 3\right)}} \qquad (16.3)$$

16.3.5.4 Calculation of the eco-compensation burden The total compensation burden to be paid by each county is determined by multiplying its benefit coefficient by its payment ability coefficient.

16.4 Results

16.4.1 Land use changes

The most significant changes were a decrease in farmland area and an increase in water area. The farmland area decreased by 19,416 ha between 1997 and 2005 (accounting for 4.68% of total arable land in 1997) and increased by 5956 ha from 2000 to 2005 (1.43%). In total, the farmland area decreased by 13,460 ha (3.25% of total arable land in 1997). At the same time, the total area of water increased by 14,661 ha (+3.86% compared to 1997). Over the entire 8 years, the forest area decreased by 1066 ha (−0.21% compared to 1997). The grassland area decreased by 1340 ha (−1.87%) in the first phase. In the second phase, the decrease was three times as high as in the first phase. The built-up land continued to increase, changing from 59,148 ha in 1997 to 62,957 ha in 2005 (+6.44%), and the unused land increased slightly, changing from 89,376 ha in 1997 to 89,617 ha in 2005 (+0.42%) (Figure 16.2).

In general, the period between 1997 and 2005 was characterized by an increase in human disturbance due to population growth, an increase in residential areas, the presence of government policies, and a growing fishing industry. As a result, the average farmland area per

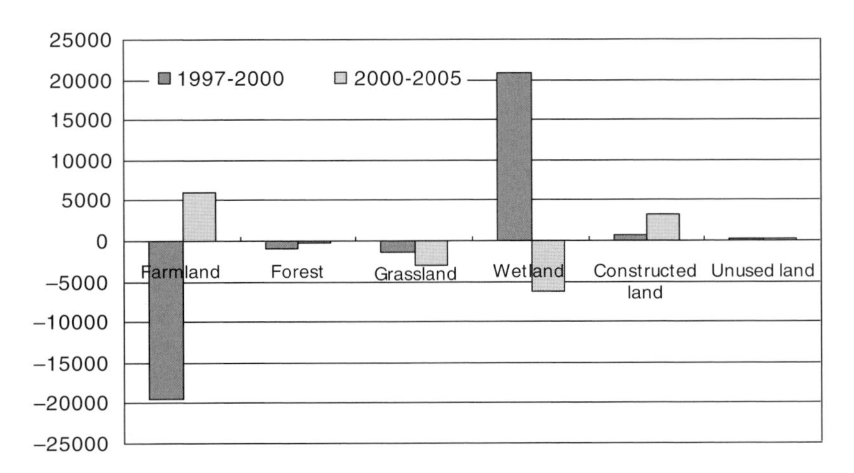

Figure 16.2 Changes in land use structure from 1997 to 2000 and 2000 to 2005 in the Poyang Lake region. *Source*: Data Center for Resources and Environmental Sciences, Chinese Academy of Sciences, 2000–2005.

capita decreased, the landscape became more simplified and fragmented, and the stability of the regional ecosystems decreased. These land use changes also caused a change in land use functions, which are intimately connected to the social and economic status of the local residents.

Along with economic development and population growth, the share of built-up land expanded, leading to a change in the rural structure. The farmland area per household in 2004 decreased by 63.21% compared to 1999, including a 76.92% decrease in paddy fields and a 46.48% decrease in agricultural dry land. Along with the improvement in quality of life, the annual per capita expenditure on food grains decreased from 303.61 yuan to 265.00 yuan. As a result, the agricultural production function became less important. The decrease in forest and grassland area contributed to a decrease in the net productivity of vegetation and reduced the water supply so that the importance of the supply functions decreased gradually. In contrast, the water area increased by 14,661 ha between 1997 and 2005. Theoretically, the functions of water-saving and flood regulation should have increased due to the increase in water area, and the function of climate regulation should also have increased due to the intense evaporation from water surfaces and vegetation. The landscape of Poyang Lake is special, and is an important habitat for a number of wild animals because of its diversity of natural resources and favorable climate. For example, the region is the habitat for 159 water bird species that together account for 53% of the total number in China. Based on the existing research results (L. Li *et al.*, 2008), it seems that the main functions of the Poyang Lake ecosystem are supporting and regulating functions.

16.4.2 Changes in the economic structure

Between 1997 and 2005, the relevance of the primary sector decreased from 19.3% to 9.6%, while the secondary and tertiary sectors became more important (from 45.5 to 51.1% and 35.3 to 39.3%, respectively).

The income structure of farmers in the region also changed considerably (Figure 16.3). In 1997, before the implementation of the policy, agricultural and fishing activities accounted

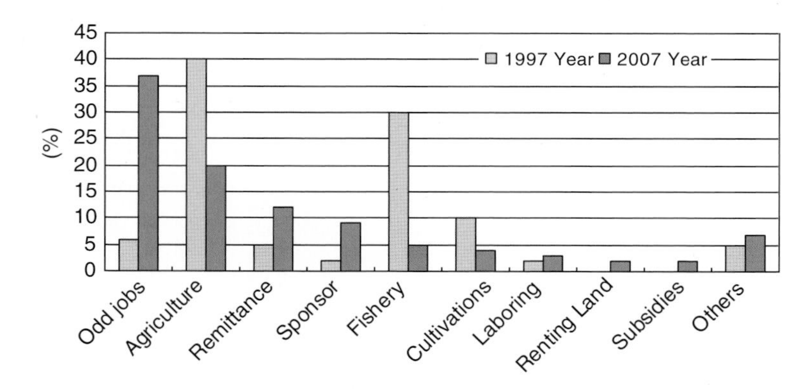

Figure 16.3 Farmers' income structure in 1997 and 2007 in the Poyang Lake region. Data were obtained from household surveys in 2008. The study included a total of nine villages in six counties, and we surveyed their residents between 25 March and 11 April 2008. We used simple random sampling to select the households, and included a total of 270 householders

for about 80% of the total income of the local residents, while off-land employment was only of minor importance. By 2007, the share of income from off-land work had increased to 37%, while cultivation and remittances from sons, daughters or younger relatives became the second-most important sources of income. Through the implementation of the policy, more than two-thirds of the younger population had left the villages for off-land employment, while elderly people of 50 years and older remained in the villages and continued land-based production. This age group was found to display a generally lower educational level and to lack the skills necessary for off-land employment.

Based on an analysis of the per capita GDP, one can see that the economic development level of the eight counties (cities) on the west bank of Poyang Lake was significantly higher than that of the four counties in the eastern section. The average per capita GDP of the western portion was almost three times higher than that of the eastern portion (7934 yuan and 2704 yuan, respectively; data source: *Jiangxi Statistical Yearbook 2006*). The highest per capita GDP was found in Nanchang County (9613 yuan) (excluding Nanchang and Jiujiang City); this figure is almost four times higher than the lowest value, which was associated with Poyang County (2137 yuan).

16.4.3 Stakeholder groups

The stakeholder analysis had the purpose of determining potential benefits and conflicts of core and less relevant stakeholders involved in the planning and implementation of the ecological compensation mechanism. The scoring results from the expert interviews are shown in Table 16.1.

We can identify three types of stakeholders:

1. Primary stakeholders, those with scores of 6 or more in the three dimensions, including central and local governments, farmers, research institutions and local enterprises. These stakeholders are an indispensable group for the eco-compensation mechanism.

Table 16.1 Results of the stakeholder analysis

	Scoring		
Dimension	1 to 4	4 to 6	6 to 10
Positive attitude	Firms Public	Investment institutes Media Community	Government Households Scientific institutes Environmental NGOs
Decision-making power	Media Public	Households Investment institutes Environmental NGOs Community	Government Enterprises Scientific institutes
Importance of interests	Public Media	Investment institutes Environmental NGOs Scientific institutes	Government Households Enterprises Community

Source: Expert questionnaires and interview, 2008.

They either are directly affected by the compensation mechanism or have at least a close relationship to it.

2. Secondary stakeholders, those with scores of 4–6 points; these were investment institutions, environmental NGOs and community-based organizations. They are relatively close to the ecological compensation mechanism but not immediately affected by it. However, if their needs can at some point no longer be satisfied, they will also be affected by eco-compensation and thus will play a greater role in the implementation process.
3. Marginal stakeholders, those with a score of less than 4 points, including the public and the media. These actors are passively exposed to the impact of eco-compensation, and they are less important in the implementation process.

To simplify the research problem, we selected only the core stakeholders (the government and the farmers) for further analysis.

16.4.4 Farmers' willingness to accept eco-compensation

The average WTA of the farmers was 13,912 yuan/ha per household and 15,525 yuan/ha on average (Figure 16.4). The total WTA for the entire Poyang Lake region ranged between 5.76 and 6.45 billion yuan (there are 1.84 million households and 0.225 ha farmland per household in the Poyang Lake region).

An above-average WTA was more often found in close-to-lake households, in low-education households with a generally lower income level and fewer skills useful for non-agricultural employment, and in households with more family members than average. Educational level, income, family size and the distance to the lake all had a statistically significant impact on the WTA of the farmers, while age, gender and other factors were less important (Table 16.2).

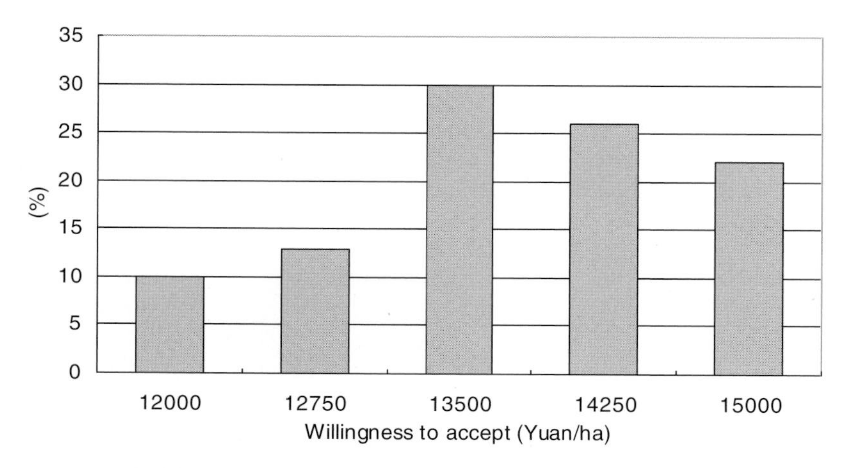

Figure 16.4 Probability of willingness to accept (WTA) in the Poyang Lake region. Data were obtained by household surveys in 2008. To gain the necessary data for estimation of WTA we used a single bound dichotomous Contingent Valuation Method (CVM). The bid values were given to those who agreed to accept eco-compensation amounts of 12,000, 12,750, 13,500, 14,250, 15,000 CNY (Chinese yuan) per hectare

Table 16.2 Relationship between socio-economic information and the willingness to pay (WTA) of the farmers in the Poyang Lake region

	Gender		Age				Education			
	Male	Female	<35	36–50	51–65	>66	None	Primary	Middle	High
Numbers[a]	83	22	12	35	48	10	15	40	34	16
Mean WTA[b]	15,345	16,485	15,435	14,865	16,815	12,360	16,800	15,930	14,925	14,985
P	0.594		0.406				0.001			

	Family scale				Income level (yuan)[c]				Distance to lake (m)			
	1–2	3–4	5–6	>7	1	2	3	4	<500	500–1500	1500–2500	>2500
Numbers[a]	8	36	43	18	16	59	18	6	44	21	113	16
Mean WTA[b]	16,410	15,420	16,545	13,260	19,080	15,105	15,000	12,480	16,170	15,195	12,960	18,645
P	0.001				0.000				0.005			

[a]The number of the whole sample for the WTA is 105.
[b]The mean WTA overall is 15,525 yuan/ha.
[c]Income level (yuan): 1 = <5000; 2 = 5001–20,000; 3 = 20,001–40,000; 4 = >40,001.
Source: Household survey, 2008. The study included a total of nine villages in six counties, and we surveyed their residents between 25 March and 11 April 2008. We used simple random sampling to select the households, and included a total of 270 householders.

16.4.5 Eco-compensation burden of the local governments

The environmental services provided by the CFTL have the character of a public good. It is therefore suggested that the local governments might bear the compensation costs as representatives of the public.

The calculated Engel coefficients of Nanchang, Jiujiang and Shangrao were 0.455, 0.464 and 0.474, respectively. We simulated the WTP coefficients of the urban population of these cities in considering how to pay for the ecological value of wetland restoration based on Equation 16.3. The WTP coefficients were 0.310, 0.300 and 0.291, respectively. Finally, we used per capita county GDP to weigh the coefficients. The total compensation burden figures for Nanchang, Jiujiang and Shangrao were 0.715, 0.174 and 0.111, respectively. This means that Nanchang should pay 71.5%, Jiujiang 17.4% and Shangrao 11.1% of the total eco-compensation costs in the case study area (Table 16.3).

16.5 Discussion

The main reason for the sizable difference in per capita GDP between the western and eastern regions of Poyang Lake is the existence of different local conditions. For instance, the six western counties are influenced by the industrial development zone around Nanchang and Jiujiang and are therefore considered a key development region for Jiangxi Province. The four counties in the east of the region are less economically advanced than those in the western part because of low investment rates and insufficient infrastructure. In combining this information with the model results, we can conclude that the payment ability of Nanchang and Jiujiang is higher than that of Shangrao. After the field study, we found that the water infrastructure in the western area was better than that in the east because of location and economic conditions. Therefore, the capacity for resistance to natural disasters is greater in Nanchang and Jiujiang than in Shangrao. For instance, the economic losses resulting from floods in Nanchang and Jiujiang cities were only one-quarter of those of the counties in the eastern area, namely the Poyang and Yugan counties of Shangrao City. Thus, the local governments of these counties have to spend a large part of their financial resources to mitigate the impacts of natural disasters instead of using their funding to develop the regional economy. We also compared the total arable land area in each county and used this figure as a proxy for potential wetland restoration benefits. The result showed that the benefits for Nanchang and Jiujiang are greater than those for Shangrao. Nanchang and Jiujiang cities should therefore bear a higher share of the total eco-compensation costs than Shangrao City.

The farmers in the region received compensation for the farmland used for wetland restoration, but due to budgetary constraints of the local governments, the compensation level was less than the loss of income. The farmers are, therefore, struggling to maintain their living standards. Their dependence on the compensation is relatively high. Farm income in the region is very low due to the decreasing arable land per capita and poor soil fertility. A large number of migrant workers are therefore seeking jobs in cities, while the on-farm labor force is aging, and this phenomenon may be causing many social and economic problems. Farmers have become used to the payments and rely on continued compensation instead of looking for alternative sources of income. The cost-effectiveness of the existing compensation mechanism is therefore not very high. Another difficulty is the existing institutional and stakeholder conflicts in the region, which need to be resolved. The

Table 16.3 Benefit of wetland restoration, payment ability and total compensation burden of the counties in the Poyang Lake region

Cities	Counties	Arable land (ha)	Direct beneficiary	GDP per capita (yuan)	Indirect beneficiary	Compensation burden	Total[a]
Nanchang	Nanchang	9204	0.0219	17,238	0.2359	0.0695	0.715
	Nangchang	74,666	0.1778	9613	0.1315	0.3144	
	Xinjian	52,370	0.1247	7738	0.1059	0.1775	
	Jinxian	55,117	0.1313	6376	0.0872	0.1539	
Jiujiang	Jiujiang	5824	0.0139	7728	0.1023	0.0191	0.174
	Yongxiu	24,499	0.0584	6036	0.0799	0.0626	
	De'an	7614	0.0181	5079	0.0672	0.0164	
	Xingzi	9129	0.0217	3667	0.0486	0.0142	
	Duchang	39,727	0.0946	2220	0.0294	0.0374	
	Hukou	13,854	0.0330	4254	0.0563	0.0250	
Shangrao	Yugan	47,901	0.1141	2209	0.0283	0.0434	0.111
	Poyang	79,924	0.1904	2137	0.0274	0.0700	

[a]Total compensation burden of the counties Nanchang, Jiujiang, and Shangrao.

Source: Bureau of Statistics of Jiangxi (2008) *Jiangxi Statistical Yearbook*. Beijing: China Statistics Press [in Chinese].

responsibilities of the 12 counties within the eco-compensation process must be clarified to avoid further conflicts and allow for effective policy implementation.

Based on our analyses, we present the following suggestions regarding the future implementation of the eco-compensation mechanism in the Poyang lake region.

- Transform the current 'blood transfusion' mechanism into a 'blood generating' mechanism. Successful use of the eco-compensation mechanism requires that the current compensation-oriented mechanism be combined with additional measures that encourage awareness and personal engagement among the local farmers and thus improve their self-development ability.
- The local governments should bear the burden of compensation for two reasons. First, the CFTL policy was implemented at central and local levels. Second, the ecosystem services have the character of a public good.
- Promote cooperation and dialogue among stakeholders in the region to solve existing conflicts.
- Develop non-agricultural industries. Land use changes in the region have caused a labor surplus of almost 40%. Absorbing the surplus labor force in other sectors is, therefore, key to the sustainable development of the Poyang Lake region. Thus, increasing non-agricultural employment should be the main eco-compensation strategy. We suggest the following countermeasures. First, it is necessary to develop small and medium-sized enterprises to absorb the rural labor surplus. The fishing resources in the Poyang Lake region provide a basis for the development of water bird production and fish processing. However, the current enterprise landscape is under-developed in comparison with existing resources. Investment aids for such industries should therefore be promoted. Second, natural resources and regional cultural heritage provide a good basis for the development of regional tourism. A regional development plan should be enacted for a travel route around the perimeter of Poyang Lake. This plan could provide more employment opportunities and promote local economic development.

16.6 Conclusions

Wetland restoration is an important practice for reducing the occurrence of floods. However, these environmental benefits come at a cost. Farmers must give up their land for the benefit of this restoration process and thus lose a proportion of their agricultural income. This is particularly the case in poorer areas where the agricultural sector is of high importance and there are few alternative income sources, as in the Poyang Lake region. The implementation of a fair eco-compensation mechanism is therefore a challenging task.

We have analyzed farmers' willingness to accept eco-compensation and the eco-compensation burden of local governments. The results show that the policy of converting farmland to lakes has caused significant land use changes. Such changes have an impact on the industrial structure at the macro-level and the income structure of farmers at the micro-level. The primary industrial sector (agriculture) has gradually decreased in importance, while the second and tertiary sectors (industry and services) have significantly improved in this respect. The farmers' income structure has changed in recent years. Income from agriculture has steadily fallen, while off-land income has become more important. Given the severe socio-economic problems of the region, including low education levels, limited land resources, surplus labor, an aging rural population and a non-sustainable traditional agriculture, eco-compensation is an important stabilizing factor that can mitigate severe structural changes.

However, we have also identified a number of crucial aspects that will require further attention in the future if we are to improve the implementation of the eco-compensation mechanism in Poyang Lake. These include resolving existing stakeholder conflicts, distributing the eco-compensation burden based on both potential benefits and the payment ability of the different counties around Poyang Lake and developing accompanying measures to absorb the rural labor surplus, such as vocational training, the development of the fisheries industry, and regional plans for tourism and industry development.

Acknowledgements

This study was funded by the EU 6th Framework Programme for Research, Technological Development and Demonstration, Priority 1.1.6.3 Global Change and Ecosystems (European Commission, DG Research, contract 003874 (GPCE)), the National Basic Research Program of China (No. 2009CB421106) and the National Key Project of Science and Technical Supporting Programs of China (No. 2006BAC08B06). The authors thank the staff at the Mountain-River-Lake Development Office of Jiangxi Province and other government authorities in the study areas for sharing their time and expertise.

References

Alejandro, V., Elvira, D., Isabel, R., *et al.* (2003) Land use-cover change process in highly bio-diverse areas: the case of Oaxaca, Mexico. *Global Environmental Change* **13**: 175–84.

Braimoh, A.K. (2009) Agricultural land-use change during economic reforms in Ghana. *Land Use Policy* **26**: 763–71.

Bureau of Statistics of Jiangxi (2007) *Jiangxi Statistical Yearbook 2006*. Beijing: China Statistics Press.

Grimble, R., Chan, M.K. (1995) Stakeholder analysis for natural resource management in developing countries. *Natural Resources Forum* **19**:113–24.

Jiang, L., Bergen, K.M., Brown, D.G., Zhao, T., Tian, Q., Qi, S. (2008) Land-cover change and vulnerability to flooding near Poyang Lake, Jiangxi Province, China. *Photogrammetric Engineering and Remote Sensing* **74**:775–86.

Li, F., Zhen, L., Huang, H., *et al.* (2009) Impacts of land use functional change on WTA and economic compensation for core stakeholders: a case study in Poyang Lake. *Resources Science* **31**: 580–9 [in Chinese].

Li, J., Jiang, W., Jin, L. (1999) *Ecological Value*. Chongqing: Chongqing University Press [in Chinese].

Li, L., Deng, X., Zhan, J., Li, Y. (2008) The valuation of the importance function of ecosystem services in Poyang lake – based on the AHP approach. *Journal of Anhui Agricultural Sciences* **36**: 8786–7 [in Chinese].

Li, X. (1996) A review of the international researches on land use/land cover change. *Acta Geographica Sinica* **51**: 553–7 [in Chinese].

Li, Y. (2006) *Studies on land use dynamic changes of Xi'an City*. Xi'an: Chang' An University [in Chinese].

Mitchell, C.R., Carson, T.R. (1989) *Using Surveys to Value Public Goods: The Contingent Valuation Method*. Washington DC: RFF/The Johns Hopkins University Press.

Musacchio, R.L., Grant, E.W. (2002) Agricultural production and wetland habitat quality in a coastal prairie ecosystem: simulated effects of alternative resource policies on land-use decisions. *Ecological Modeling* **150**: 23–43.

Ndubisi F., DeMeo, T., Ditto, D. (1995) Environmentally sensitive areas: a template for developing greenway corridors. *Landscape and Urban Planning* **33**: 159–77.

Salvador, D.S., Francese, H., Sala-Garrido, R. (2009) The social benefit of restoring water quality in the context of the Water Framework Directive: A comparison of willingness to pay and willingness to accept. *Science of the Total Environment* **407**: 4574–83.

Tanrivermis, H., Sertac, G., Bülbül, M. (2003) Land use change analysis and sustainable use of land resources in Turkey. *Journal of Arid Environment* **54**: 553–65.

Working Group of the Millennium Ecosystem Assessment (2003) *Ecosystems and Human Well-Being: A Framework for Assessment.* London: Island Press.

Yang, Z., Liu, Y. (2008) Research on land use change and sustainable land use in hilly areas. *Journal of Anhui Agricultural Sciences* **36**: 8198–201 [in Chinese].

Zhen, L., Xie, G., Yang, L., Cheng, S., Guo, G. (2005) Land-use change dynamics, driving forces and policy implications in Jinghe watershed of Western China. *Resources Science* **27**: 33–7 [in Chinese].

17 China's Sloping Land Conversion Program: Are the Farmers Paid Enough?

Shubhechchha Thapa[1], Xing Lu[2] and Ademola K. Braimoh[3]

[1] *Texas A&M University, Texas, USA*
[2] *Yunnan University, Kunming, China*
[3] *The World Bank, Washington, DC, USA*

17.1 Introduction

With a total budget of RMB 337 billion (over US$48 billion), the Sloping Land Conversion Program (SLCP) of China is one of the biggest payment-for-ecosystem services (PES) schemes in the world (Bennett, 2009; Liu *et al.*, 2008; Mullan & Kontoleon, 2009). Also known as the Grain for Green Program, the SLCP was initiated in 1999 with the aim of increasing the vegetation cover by 32 million ha by converting around 14.67 million hectares (4.4 million hectares estimated with slopes of 25° or above) of cropland into forest and grassland, and the remainder by afforesting barren land, by the end of 2010. The program aimed to combat environmental problems (flooding, soil erosion, desertification, etc.) as well as alleviate poverty by giving communities the opportunities of alternative economic activities (Liu *et al.*, 2008; Xu *et al.*, 2004; Yao *et al.*, 2010). At the outset of this program, the government compensated farmers with an annual grain subsidy of 2250 and 1500 kg/ha and a cash subsidy of 300 yuan for converted cropland in the upper reach of the Yangtze River Basin, and in the upper and middle reaches of the Yellow River Basin, respectively, plus a one-off subsidy of 750 yuan/ha for seedlings (Liu *et al.*, 2008; Xu *et al.*, 2004). Later, the compensation was given in cash equivalent to the market value of grain (1.40 yuan per kg), administered by the central government and subject to decisions of each provincial government. Both the grain and cash subsidies are for 8, 5 and 2 years

Vulnerability of Land Systems in Asia, First Edition. Edited by Ademola K. Braimoh and He Qing Huang.

if the cropland is converted into, respectively, ecological forests(timber-producing), economic forests (orchards, or plantations of trees with medicinal values) and grassland (Xu *et al.* 2004).

The SLCP is centered in the western part of China because of its environmental fragility (>360 million ha affected with soil erosion, 174 million ha with desertification, 600 million ha of cropland with slope >25°) and severe poverty (60% population under the poverty line) (Liu *et al.*, 2008). The pilot phase (1999–2001) was implemented in Sichuan, Shanxi and Gansu provinces, and was then extended to 20 provinces, 400 counties and 27,000 villages in 2001, during which period the transition occurred on an average of 408,000 ha of cropland per year (Bennett, 2008; Xu *et al.* 2004). By the end of 2006, 9 million ha of cropland had been converted, and 11.7 million ha of barren land afforested, directly benefitting 120 million farmers (Liu *et al.*, 2008). By the end of 2008, 151.36 billion yuan had been invested in the program, which involved around 26 million households in 25 provinces (Liu & Wu, 2010).

Natural vegetation provides ecosystem services like climate regulation, carbon sequestration, soil formation, etc., which are crucial for human wellbeing. Computing the value of such ecosystem services is often a challenging task, particularly for those not freely traded in the market and rarely considered to be economically productive (Brauman & Daily, 2008; Tallis & Polasky, 2009). Recently, the value of terrestrial-based carbon has been recognized in the marketplace by the Kyoto Protocol, which supports carbon emission trading projects for which land use, land-use change, and forestry (LULUCF) activities that sequester carbon or avoid carbon emissions are considered to be a solid basis. The sloping land conversion program of China is similar to the activities defined under Article 3, paragraph 3 of the LULUCF sector that covers direct, human-induced, afforestation, reforestation and deforestation activities (UNFCCC, 2008). So, the SLCP provides the opportunity for forest carbon payments, which can be either for carbon sequestration or for protecting existing carbon stocks.

The main objectives of this chapter are to evaluate carbon sequestration under the SLCP policy framework and compare the economic value of net carbon sequestration with the payments received by farmers in southwest China under the SLCP. Insights from the analysis provide strategies for ensuring the sustainability of SLCP and similar PES schemes for ecosystem services management.

17.2 The study area

The study was conducted in a 2005 -km^2 area within Yunnan Province in southwest China. It includes the Supa River watershed area, which drains about 667 km^2 and accounts for 23.8% of the whole county. The area falls in the upper reach of the Yangtze River Basin, and is one of the priority areas for the SLCP, where agriculture on sloping land was very intense before the implementation of the program. The Supa watershed has a population density of 96 persons per km^2. Livelihood activities of the local population are primarily farming, livestock and forestry, with an average farmland holding of about 0.133 ha per person, generating a per capita income of $185/year (Lu & Li, 2009). The Supa River valley is an integration of agricultural development, nature conservation and hydropower exploitation, whose ecosystem service functions maintain the local agricultural population and biological diversity as well as providing resources for hydropower development (Lu & Li, 2009). Soil erosion intensity is not very high, but due to the fragile ecology in some areas, there exists the potential for ecological disruption. Areas of slight erosion account for about half of the drainage area. Although the proportion of intensively and extremely intensively eroded areas is relatively low, they are more localized in certain areas, which results in serious soil erosion in some parts. The property rights mechanism in the Supa watershed is ambiguous. There is state control over forests and water rights, and community control over collective forests/agricultural land, but it is difficult to identify individuals who provide watershed services (Lu & Li, 2009). This ambiguity calls for proper coordination of watershed management activities, including the distribution of services due to various ecosystem services providers.

17.3 Data sources and analysis

Land-cover maps of the study area for 1989 and 2001 were prepared from Landsat TM images. Supervised classification was used with training areas limited to areas that did not experience any land-cover change between the periods covered by the images. The map was classified into five land cover classes: dense forest, sparse forest, water, grassland and farmland (Figure 17.1). The overall change from one land type to another between 1989 and 2001 was determined by pixel to pixel comparison resulting in a land-cover change matrix in which each element in the matrix is the proportion of the total land area of its class in 1989 that changed to another class in 2001 (Table 17.1).

The sequestered and lost carbon in the landscape between 1989 and 2001 was estimated using the Integrated Valuation of Ecosystem Services and Tradeoffs (InVEST) software (Nelson *et al.*, 2009). The InVEST tool consists of a suite of models that use land-use and land-cover maps for quantifying and mapping the levels and economic values of ecosystem

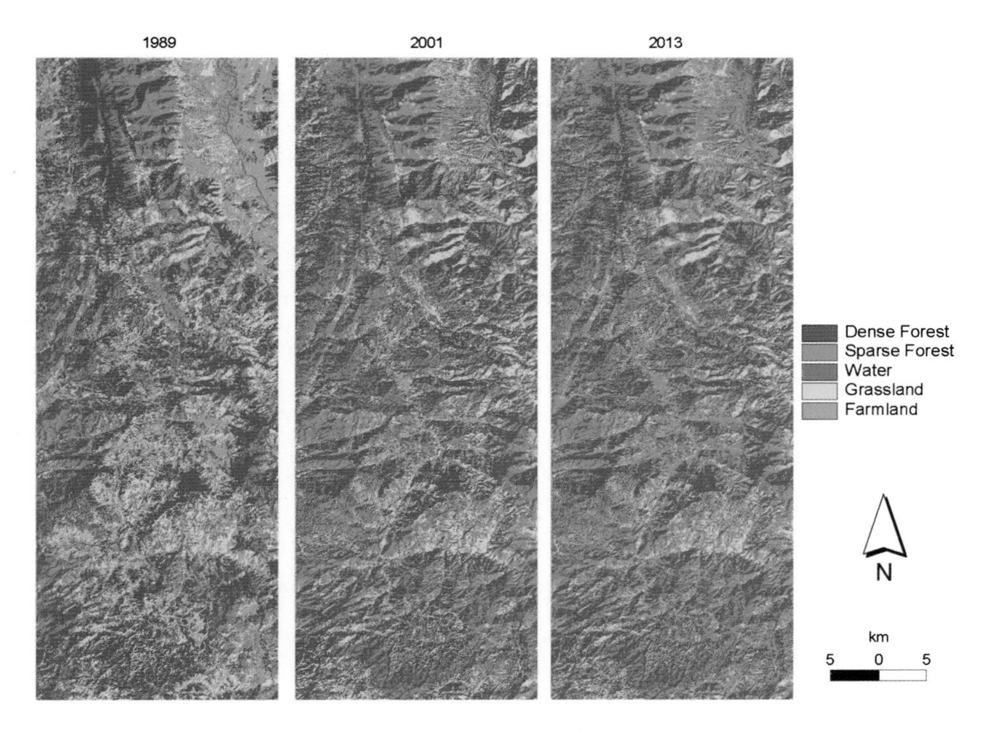

Figure 17.1 Land cover map of the study area

Table 17.1 Land cover transition matrix 1989–2001 (%)

	2001						
1989	Dense forest	Sparse forest	Water	Grassland	Farmland	Total 1989	Loss
Dense forest	37.10	5.14	0.00	1.69	1.10	45.03	7.93
Sparse forest	2.20	6.84	0.00	2.48	0.40	11.92	5.08
Water	0.00	0.00	1.92	0.00	0.00	1.92	0.00
Grassland	1.76	2.97	0.00	10.45	0.90	16.09	5.64
Farmland	13.19	1.84	0.00	4.59	5.43	25.04	19.61
Total 2001	54.25	16.79	1.92	19.21	7.83	100.00	38.26
Gain	17.16	9.95	0.00	8.76	2.40	38.26	

Table 17.2 Carbon pool estimates (Mg/ha)

	Above ground	Below ground	Soil	Dead organic matter
Dense forest	180	90	60	40
Sparse forest	75	25	25	20
Water	0	0	0	0
Grassland	15	35	15	4
Farmland	3	2	2	0

services (e.g. carbon storage) provided by the landscape (Tallis *et al.*, 2010). InVEST is largely data-driven with results presentable in both biophysical and monetary terms. To model C storage, we estimated four fundamental carbon pools: above ground biomass, below ground biomass, soil and dead organic matter; then we aggregated the C in each of the pools to derive the estimated C storage in each grid cell. Carbon pool estimates for the land classes (Table 17.2) were established following guidance from the literature (Chen *et al.*, 2009; Shanmughavel, 2001; Tallis *et al.*, 2010). Carbon sequestration/loss over time was computed by subtracting the estimated C storage in the entire landscape in 1989 from that in 2001. To enable a comparison of PES under the SLCP with the social value of sequestered carbon[1], we computed C storage between 2001 and 2013 by generating a land-cover map for 2013. The land-cover map in 2013 assumes that all farmland areas in 2001 will be converted to sparse forest by 2013 in line with the SLCP policy. In reality the land-cover distribution by 2013 may be more complex due to other intervening driving factors.

17.4 Results and discussion

17.4.1 Quantitative data on land-cover change

The land covered by natural vegetation increased between 1989 and 2001 (Table 17.1). The most dramatic change was the increase in the dense forest cover and the decrease in the farmland cover. Dense forest increased from 45% of total land area in 1989 to 54% in

[1] The social value of sequestered C refers to the social damage avoided by not releasing on extra ton of C into the atmosphere (Stern, 2007). Tol (2005) estimated the value to be $43 per ton of carbon.

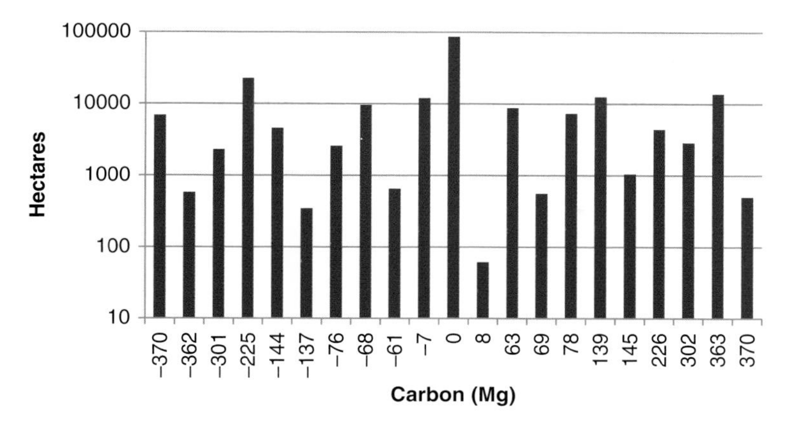

Figure 17.2 Total carbon sequestered between 1989 and 2001. Columns to left of zero signify losses (negative values) and columns to right of zero signify gains (positive values)

2001, whereas farmland decreased from 25% to 8% during the same period. Farmland was converted to dense forest and grassland over 13% and 5% of the land area, respectively. Forest cover and grassland experienced almost twice as much gain as loss, whereas farmland experienced almost nine times more loss than gain. There was a gain of forest cover (dense and sparse forest) over 27% of the landscape compared to a loss of cropland over 20% of the landscape. At the same time, there was a loss in 8% of the dense forest and a gain in 2% of the farmland. Apart from the forest conservation policies of the Chinese government, the communal initiatives by the people of Supa watershed to manage the forest area may be responsible for the remarkable conversion of cropland to forests and grassland. Gain in biomass helps to reduce soil erosion as well as improve other watershed ecosystem services including water provision and landslide prevention.

17.4.2 Carbon dynamics in the landscape

Between 1989 and 2001, around 60,000 ha of landscape lost carbon at a rate varying from 7 to 370 Mg/ha whereas there was carbon sequestration at rates of 8 to 370 Mg/ha in 50,000 ha of land (Figure 17.2). There was a net loss of 7692 Mg/ha/year of carbon during that period as a result of land-cover changes from higher to lower biomass, i.e. from forest to grassland and cropland (Table 17.1). Based on the projected land-cover changes, a net gain of 898,028 Mg/ha/year of carbon was estimated for the period between 2001 and 2013. The social value of carbon based on the unit cost of $43 and a market discount rate of 7% is depicted in Figure 17.3. The figure shows that by 2013, around 70,000 ha of the landscape would have sequestered carbon worth from $1912 to $11,268.

17.4.3 Landscape value

At a rate of US$1 = 7RMB, the compensation given to the farmers is estimated at $567/ha/year (Table 17.3). This is based on the market price of rice (2010RMB/ton; www.cereal.com.cn, 27 May 2010), wheat (2030RMB/ton; www.cereal.com.cn, 27 May 2010) and maize (1990RMB/ton; www.pigol.cn, 1 June 2010). Between 2001 and 2013,

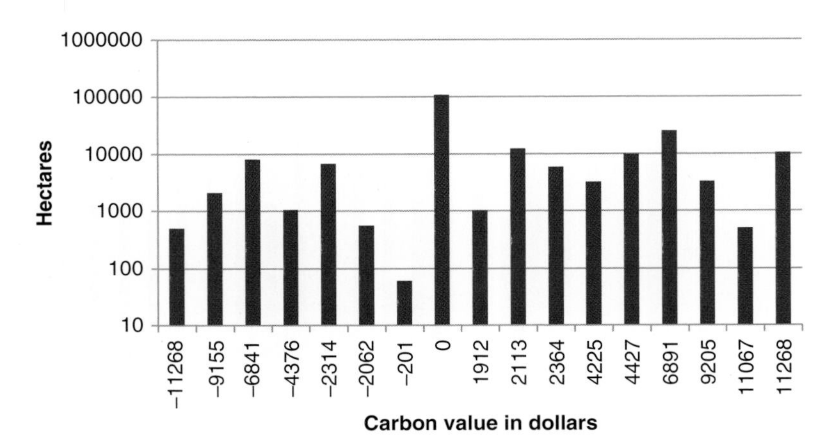

Figure 17.3 Social value of carbon (in dollars) sequestered between 2001 and 2013. Columns to left of zero signify losses (negative values) and columns to right of zero signify gains (positive values)

Table 17.3 Compensation to farmers per year under the Sloping Land Conversion Program (Yunnan)

Subsidy	Yuan/ha	Dollar/ha
Grain	3600	514
Cash	300	43
Seedlings	68	10
Total	4968	567

15,697 ha of cropland was projected to be converted into forest, i.e. 1308 ha/year. Thus, the estimated compensation to farmers was $0.9 million/year. However, the social value of sequestered carbon is estimated at $3.5 million/year.

Thus, even though government compensation under the SLCP is higher than farmers' per capita income of $185 (Lu & Li, 2009), the payments are far lower than the actual value of services provided by the landscape. Carbon sequestration is the only ecosystem service considered in this chapter. If the monetary value of other ecosystem services was taken into account, the value would be much higher, overshadowing the compensation the farmers received.

17.5 Conclusion

While payments under the SLCP cover the opportunity cost of converting cropland to forest, they are insignificant compared to the ecosystem services value. Payment for ecosystem services programs can suffer from a lack of consistency (permanence) in implementation, which may cause a reversal of the carbon gained. Because the compensation received by farmers is not perpetual, there is the possibility of converting forest back to cropland after the subsidy ends (after 5 or 8 years). This problem is most likely to occur amongst farmers with limited alternative choices for other economic activities. This may lead to accelerated erosion and other environmental degradation. Also in some regions the strategy of providing subsidies varies with the market price of grain. For instance, when the price of grain was high, farmers were paid cash; otherwise they were paid in grain (Xiaping *et al.*, 2009), and in some cases stale grain was supplied (Xu *et al.*, 2004). Xu *et al.* (2004) and Bennett (2008) reported some records of shortfall in the delivery of the total stipulated compensation. These limitations markedly influence the sustainability and permanence of the payment for ecosystem services program. In addition to this, the economic analysis of carbon reveals that the social value of carbon sequestered as a result of converting the farmland into forest is very high. Under this scenario, the PES scheme of SLCP seems theoretically to undervalue the services provided by the farmers. So, the actual payment of SLCP is just a tiny fraction of the maximum payments that farmers should get for regulating various ecosystem services. However, there was no existing PES market except for the government program.

Nevertheless, the SLCP has led to remarkable vegetation recovery in most parts of the program areas, and has improved carbon storage and other ecosystem services. To increase benefits to the service providers and ensure the sustainability of the carbon benefits, alternative mechanisms that better reward farmers are required. Given the fact that compensation is not the only solution for ecosystem service conservation, there is the need to complement it with other policy instruments such as regulation, fiscal incentives and information provision to influence farmers' behavior that affects the ability of ecosystems to provide ecosystem services.

References

Bennett, M.T. (2008) China's sloping land conversion program: institutional innovation or business as usual? *Ecological Economics* **65**: 699–711.

Bennett, M.T. (2009) *Markets for Ecosystem Services in China: an Exploration of China's "Eco-Compensation" and other Market-based Environmental Policies*. A Report from Phase I Work on an Inventory of Initiatives for Payment and Markets for Ecosystem Services in China. *Forest Trends*, pp. 34–36.

Brauman, K.A., Daily, G.C. (2008) Ecosystem services. In: Jorgensen, S.E., Fath, B.D. (eds), *Human Ecology*. Oxford: Elsevier, pp. 1148–54.

Chen, X., Lupi, F., He, G., Ouyang, Z., Liu, J. (2009) Factors affecting land reconversion plans following a payment for ecosystem service program. *Biological Conservation* **142**: 1740–7.

Houghton, R.A. (1990) The global effects of tropical deforestation. *Environmental Sciences and Technology* **24**: 414–22.

Liu, C., Wu, B. (2010) Grain for Green Programme in China: Policy Making and Implementation? Briefing series-Issue 60. University of Nottingham, China Policy Institute.

Liu, J., Li, S., Ouyang, Z., Tam, C., Chen, X. (2008) Ecological and socioeconomic effects of China's policies for ecosystem services. *Proceedings of the National Academy of Sciences of the USA* **105**: 9477–82.

Lu, X., Li, H. (2009) *Study on Ecological Compensation of Supa River Watershed Payment for Environment Services: China's Experiences of Rewarding Upland Poor*. Yunnan University Press, pp 188–213.

Mullan, K., Kontoleon, A. (2009) Participation in Payments for Ecosystem Services programmes in developing countries: The Chinese Sloping Land Conversion Programme.

Environmental Economy and Policy Research Working Papers 42. University of Cambridge, Department of Land Economics.

Nelson, E., Mendoza, G., Regetz, J., *et al.* (2009) Modeling multiple ecosystem services, biodiversity conservation, commodity production, and tradeoffs at landscape scales. *Frontiers in Ecology and the Environment* **7**: 4–11.

Shanmughavel, P., Zheng, Z., Liqing, S., Min, C. (2001) Floristic structure and biomass distribution of a tropical seasonal rain forest in Xishuangbanna, southwest China. *Biomass and Bioenergy* **21**: 165–75.

Stern, N.H. (2007) *The Economics of Climate Change: the Stern Review*. Cambridge University Press.

Tallis, H., Polasky, S. (2009) Mapping and valuing ecosystem services as an approach for conservation and natural-resource management. *Annals of the New York Academy of Sciences* **1162**: 265–83.

Tallis, H.T., Ricketts, T., Nelson, E., *et al.* (2010) InVEST 1.004 beta User's Guide. The Natural Capital Project, Stanford.

Tol, R.S.J. (2005) The marginal damage costs of carbon dioxide emissions: an assessment of the uncertainties. *Energy Policy* **33**: 2064–74.

UNFCCC (2008) Kyoto Protocol Reference Manual on Accounting of Emissions and Assigned Amount.

Xiaping, C., Xiaoqing, Z., Jingcong, S. (2009) *Study on Ecological Benefits of Upland Conversion. Payment for Environment Services: China's Experiences of Rewarding Upland Poor.* Yunnan University Press, pp. 99–120.

Xu, Z., Bennett, M., Tao, R., Xu, J. (2004) China's sloping land conversion program four years on: current situation, pending issues. *The International Forestry Review. Special Issue: Forestry in China – Policy, Consumption and Production in Forestry's Newest Superpower* **6**(3-4): 317–26.

Yao, S., Guo, Y., Huo, X. (2010) An Empirical analysis of the effects of China's land conversion program on farmers' income growth and labor transfer. *Environmental Management* **45**: 502–12.

18 Community-Based Peatland Management for Greenhouse Gas Reduction Based on Fire-Free Land Preparation

Bambang Hero Saharjo
Forest Fire Laboratory, Forest Protection Division, Department of Silviculture, Faculty of Forestry, Bogor Agricultural University, West Java, Indonesia

18.1 Introduction

Transboundary haze pollution arising from burning during land preparation has become a perennial problem in Indonesia during the dry season, especially in the last 10 years (Saharjo 2007). Smoke pollution reached a record high in 1997 and 1998, during which more carbon dioxide was released than from the cars and power stations of Western Europe in one year (WWF, 1997). In total, the fires contributed 22% of the world's carbon dioxide released in that year, or more than 700 million tonnes, which made Indonesia one of the largest contributors of carbon emissions in the world (UNCHS, 1999). Most of the smoke originated from illegal burning for land preparation on oil palm and industrial forest plantations as well as from shifting cultivation systems, often associated with burning (Saharjo, 2005). Goldammer (1993) stated that fire, being a cheap and easy practice, has been used for land preparation in shifting cultivation for thousands of years, without causing environmental problems because of the low human population pressure on forest resources. The forest provided a sustainable subsistence for indigenous forest inhabitants, and their impact had little effect on overall forest ecosystem stability (Nye & Greenland, 1960).

Vulnerability of Land Systems in Asia, First Edition. Edited by Ademola K. Braimoh and He Qing Huang.

The practice of burning releases minerals from ash, which is rich in organic carbon, phosphorus, magnesium, potassium, and sodium. The amount of nutrients in the soil temporarily increases after burning, but declines rapidly by leaching during rainy periods (Saharjo, 1995). Although burning increases the amount of nutrients in the soil, which temporarily enhances growth performance, it also has negative effects. Fire is a significant source of gases and particulates for the atmosphere. Environmentally important gases produced by fire include carbon dioxide, carbon monoxide, methane, non-methane hydrocarbons and oxides of nitrogen. Fire also produces large amounts of small, solid particles, or 'particulates', which absorb and scatter incoming solar radiation, as well as provoking a variety of human health problems (Levine, 1996). Fire can therefore be considered as one of the main agents of multiple relationships between humans and the environment. Changes in fire patterns can be taken as an indicator of change in land-use patterns and overall environmental conditions (Malingreau & Gregorie, 1996).

Until the year 2006, forest and land fires burning in Indonesia were quite significant as they had negative impacts. The situation changed in 2007 and 2008, when forest and land fires reduced by about 70%, based on a Forestry Department official report. In the same period, the ASEAN Specialized Meteorological Centre (ASMC) reported that burning reduced by 50% in Singapore. In 2007, of all the hotspots detected in Indonesia, about 10,280 (64.1%) originated in the community; 2644 (16.5%) were from estate crops; 1691 (10.54 %) were in forest concessions; and 1430 hotspots (8.91%) were in forest plantation areas. From previous experience, banning the use of fire on community land has resulted in more fires and damage. Therefore, it becomes imperative to find a clear solution to this problem whilst ensuring that the livelihood needs of the communities that depend on the land are still satisfied.

One promising solution is zero burning. To realize a zero burning policy as a tool for greenhouse gas reduction through community involvement, and to produce charcoal and organic carbon as end products, a 3-ha experimental plot was established in the high-fire-risk area of Mangsang Village, Musi Banyuasin District, South Sumatra Province, Indonesia.

18.2 Greenhouse gas emissions

18.2.1 Southeast Asian greenhouse gases emissions

In 2000, Southeast Asia contributed 12% of global greenhouse gas (GHG) emissions, amounting to 5187.2 Mt CO_2-eq, including emissions from the land-use change and forestry (LUCF) sector (ADB, 2009). The region's total emissions increased by 27% during 1990–2000, faster than the global average. On a per capita basis, the region's emissions are considerably higher than the global average, but are still relatively low when compared to developed countries. The LUCF sector has been the largest source of GHG emissions in the region, contributing 75% of the total in 2000. The region in 2000 accounted for about half of global LUCF GHG emissions. Sources included the decrease in biomass stocks of forestland through deforestation, logging and fuel wood collection; and the conversion of forestland to other uses such as cropland, grassland or pasture, and settlements (ADB, 2009). The energy sector is also a key source of emissions in the region (15%). For agriculture, accounting for 8% of emissions, the major sources are livestock production, rice cultivation, use of nitrogenous fertilizer, and burning of agricultural residues.

Southeast Asia's GHG emissions from the energy sector increased by 83% during 1990–2000, the highest among the major emission sources (ADB, 2009). Greenhouse gas emissions from the energy sector have increased significantly since 1990, and are expected to continue to increase rapidly as demand for energy grows and as the region seeks to maintain high economic growth. Agriculture-related emissions increased by a more modest 21% during 1990–2000, whereas total emissions from the LUCF sector increased by 19%. About 59% of Southeast Asia's GHG emissions in 2000 came from Indonesia, mainly due to LUCF emissions (ADB, 2009). Covering almost 42% of the region's land area and containing 40% of its population, Indonesia is the biggest contributor of GHG emissions and is therefore a key player in the struggle against the adverse impacts of climate change.

18.2.2 Indonesian greenhouse gas emissions

The amount of carbon stored in tropical areas is about 83.3 Gt, with 44.5 Gt (53.1%) found in Indonesia. In Indonesia, most of the carbon is found on the three main islands: Sumatra, which stored 18.3 Gt (41.1%); Kalimantan, 15.1 Gt (33.8%); and West Papua, 10.3 Gt (23%). From 1990 to 2002, about 3.5 Gt carbon from peatlands was lost on Sumatra alone. In the 1997 peat and vegetation fires, about 0.81–2.56 Gt, representing 13–40% of estimated mean annual global carbon emissions from fossil fuels, were released to the atmosphere.

The total amount of carbon in peatlands in Southeast Asia is at least 42,000 Mt (depending on assumptions of peat thickness and carbon content), which is equal to at least 155,000 Mt in potential CO_2 emissions (Hoojier *et al.*, 2006). At present, likely CO_2 emissions (fires excluded) from drained peatlands are calculated to be between 355 and 874 Mt/year, with a likely average value of 632 Mt/year. If current rates and practices of peatland development and degradation continue, this may increase to 823 Mt/year (most likely value) in 10 to 30 years. Current emissions from Indonesia alone are 516 Mt/year, representing 82% of peatland emissions in Southeast Asia (fires excluded), 58% of global peatland emissions (fires excluded), and almost twice the emissions from fossil fuel burning in Indonesia (Hoojier *et al.*, 2006) (Figure 18.1).

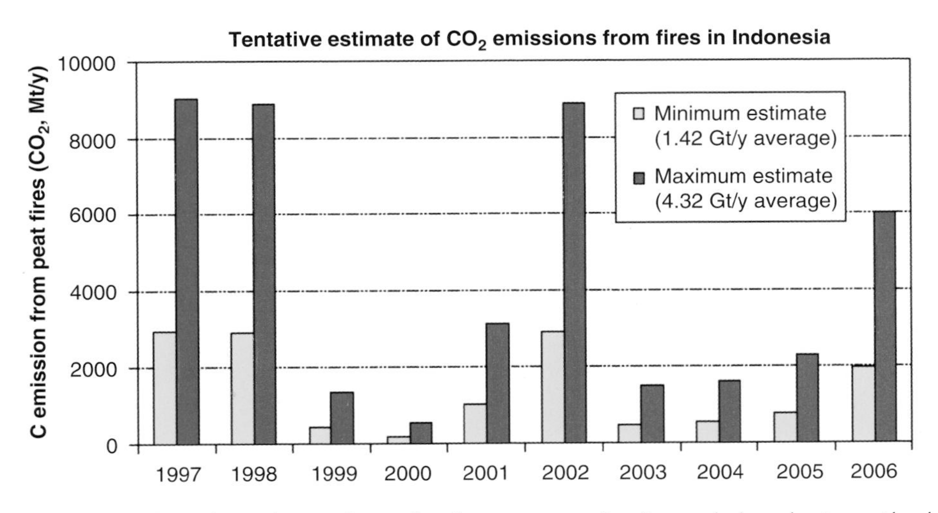

Figure 18.1 Tentative estimate of annual and average annual carbon emissions due to peatland fires, determined on the basis of hotspot counts for Borneo and the carbon emissions collected by Page *et al.* (2002) for 1997 (based on Hoojier *et al.*, 2006, reproduced with permission of Springer)

If emissions from peatland fires (which are also caused by deforestation and drainage) are included, the total CO_2 emissions are significantly higher. Over 1997–2006, CO_2 emissions from peatland fires in Indonesia were several times those due to peat decomposition in drained peatland areas: 1400 Mt/year to possibly as much as 4300 Mt/year (Hoojier *et al.*, 2006). The lower (and more likely) figure, added to current likely emissions from peat decomposition, yields a total CO_2 emission figure for Southeast Asian peatlands of 2000 Mt/year (over 90% of which are from Indonesia), equivalent to almost 8% of global emissions from fossil fuel burning. As the emissions come from only 0.2% of the global land area, these are probably the most concentrated land-use-related CO_2 emissions in the world. If emissions from peatland drainage and degradation (including fires) are included, Indonesia takes third place in global CO_2 emissions, behind the USA and China (Hoojier *et al.*, 2006). Without peatland emissions, Indonesia takes 21st place. Interestingly, the annual CO_2 emission of 2000 Mt/year calculated for 2005 is supported by an independent study conducted by Wetlands International (Hoojier *et al.*, 2006). They estimated an average annual emission of 1480 Mt/year between 1990 and 2002, based on mapping of lost peat areas and measurement of reductions in peat thickness in remaining peatlands. They found an area of 3.7 million hectares of historically mapped peatland had been totally lost by 2002, i.e. all the peat had been removed and the soil should now be classified as 'mineral' (Figure 18.2).

It should be noted that, while peat fire emissions currently exceed those from slower peat decomposition, this does not mean that the problem can be solved by fire fighting (Hoojier *et al.*, 2006):

- First and foremost, peatland fires are promoted by deforestation and forest degradation, and by drying of peat linked to peat land drainage; they can be stopped in the longer term only if these root causes are dealt with.

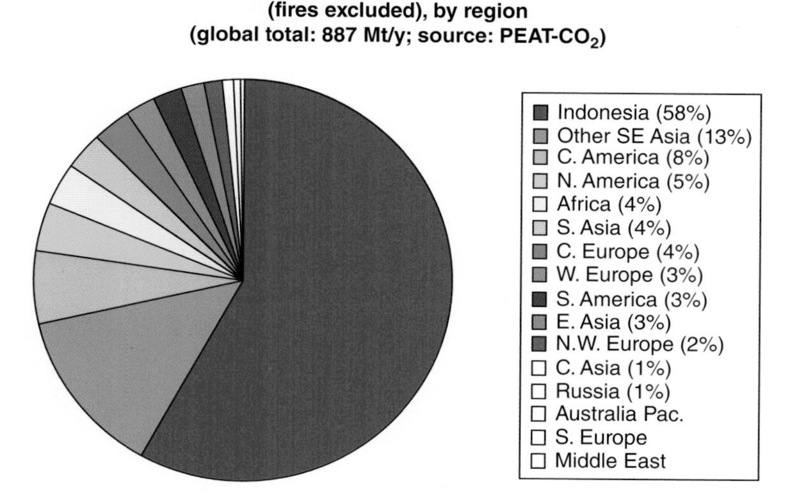

Figure 18.2 CO_2 emissions from peatlands in Indonesia and the rest of Southeast Asia as compared with emissions for other peatland regions of the world (based on Hoojier *et al.*, 2006, reproduced with permission of Springer)

Table 18.1 Number of detected hotspots during 2005 and 2006 as monitored by the Department of Forestry

		Number of hotspots detected		
No.	Province	2005	2006	% change
1	North Sumatra	3380	3581	−6.50
2	Riau	22,630	35,426	56.54
3	Jambi	1208	6948	475.17
4	South Sumatra	1182	21,734	1738.75
5	West Kalimantan	3022	29,266	864.43
6	Central Kalimantan	3147	40,897	1199.56
7	South Kalimantan	758	6469	753.43
8	South Sulawesi	133	1201	803.01

- Secondly, stopping only the fires but not the drainage simply means it will take longer for the carbon resources to be released to the atmosphere. Climate scientists look at total emissions over long time intervals, e.g. 100 years, and may consider the timing of peatland emissions (with or without fires) less relevant.

18.3 The current Indonesian forest fire situation

Data from the Department of Forestry (2007) showed that between 2005 and 2006 the increase in hotspots was significant, affecting eight fire risk provinces in Indonesia. Only North Sumatra Province produced fewer hotspots (−6.50%), whereas other provinces produced more hotspots, varying from 56.5% more for Riau, to 1738.8% more for South Sumatra (Table 18.1).

However, according to the Department of Forestry data, the number of hotspots fell significantly between 2006 and 2007, by an average of 71.39% (Table 18.2) (KNLH, 2008), while according to data from ASMC Singapore a 50% decrease in hotspots was detected (Table 18.3).

Table 18.2 Number of hotspots detected during 2006 and 2007 by the Department of Forestry (KNLH, 2008)

		Number of hotspots detected		
No.	Province	2006	2007	% change
1	North Sumatra	3581	936	−73.86
2	Riau	11,526	4169	−63.83
3	Jambi	6948	3120	−55.09
4	South Sumatra	21,734	5182	−76.16
5	West Kalimantan	29,266	7561	−74.16
6	Central Kalimantan	40,897	4800	−88.26
7	South Kalimantan	6469	928	−85.65
8	South Sulawesi	1201	551	−54.21
	Average			−71.39

Table 18.3 Number of hotspots detected during 2006 and 2007 taken by ASMC Singapore (KNLH, 2008)

		Number of hotspots detected		
No.	Province	2006	2007	% change
A: Sumatra Island				
1	Bangka Belitung	953	477	−49.95
2	Bengkulu	233	118	−49.36
3	Nanggroe Aceh Darussalam	336	172	−48.81
4	Jambi	2617	1310	−49.94
5	Riau Island	67	34	−49.25
6	Lampung	947	474	−49.95
7	Riau	4654	2361	−49.27
8	West Sumatra	361	181	−49.86
9	South Sumatra	5057	2532	−49.93
10	North Sumatra	1015	512	−49.56
	Total Sumatra	16,240	8171	−49.69
B: Kalimantan				
1	West Kalimantan	6197	3103	−49.93
2	South Kalimantan	1079	540	−49.95
3	Central Kalimantan	5580	2801	−49.80
4	East Kalimantan	2842	1430	−49.68
	Total Kalimantan	15,698	7874	−49.76
	Total A+B	31,938	16,045	

Most of the hotspots detected between January and December 2007 were in the community, whereas forest plantations had the least hotspots (Table 18.4)

Hotspots detected in peatland in 2007 numbered 3127, for both Sumatra (2305) and Kalimantan (912), which represented about 20.05% of the entire 16,045 hotspots detected (KNLH, 2008). Hotspots detected in mineral soil were 12,828 – in Sumatra (5866) and Kalimantan (6962) – or 79.95% of all hotspots detected (Table 18.5).

18.4 Greenhouse gas emissions reduction

18.4.1 Smoke management

Smoke management practices include reducing fuel loads, improving combustion efficiency, igniting fires effectively, and mopping-up efficiently. The specific practices used in a locale or region depend largely on the fire management objectives to be satisfied and the types of fuel to be burned (Debano *et al.*, 1998). Emissions reduction involves the use of fuel management practices to reduce smoke when fire occurs. Burning less fuel reduces smoke and particulate emissions. Nearly 50% of the emissions from prescribed burning can result from logging residues. Something as simple as lowering the wood-utilization standards and as a result, harvesting more residues, can significantly reduce smoke emissions (Debano *et al.*, 1998).

Isolating large pieces of wood residue (such as stumps) that are prone to smolder for long periods can also reduce the level of emissions. Burning smaller parcels of land and keeping

Table 18.4 Number of hotspots detected during January to December 2007 according to land use (KNLH, 2008)

No.	Province	Forest concession	Forest plantation	Estate crop	Community	Total
A: Sumatra Island						
1	Bangka Belitung	0	0	0	477	477
2	Bengkulu	2	0	13	103	118
3	Nanggroe Aceh Darussalam	7	15	23	127	172
4	Jambi	72	172	119	947	1310
5	Riau Island	0	0	2	32	34
6	Lampung	0	99	35	340	474
7	Riau	323	422	698	918	2361
8	West Sumatra	8	1	32	140	181
9	South Sumatra	12	172	136	2212	2512
10	North Sumatra	6	23	34	439	512
	Total Sumatra	430	904	1092	5735	8161
B: Kalimantan Island						
1	West Kalimantan	350	271	675	1807	3103
2	South Kalimantan	74	51	36	379	540
3	Central Kalimantan	598	91	430	1682	2801
4	East Kalimantan	229	113	411	677	1430
	Total Kalimantan	1251	526	1552	4545	7874
	Total A+B	1691	1430	2644	10,280	16,045
	%	10.54	8.91	16.48	64.07	100

soil and forest floor material out of slash piles can also reduce emissions. Prescribed fire can produce a wide range of particulate emissions, with the magnitudes of these depending on whether the fire is dominated by flaming combustion or smoldering.

Smoldering fires produce the greatest emissions for nearly all combustion products (Debano *et al.*, 1998). Avoiding potential smoke problems is achieved by scheduling burning during the season when fuel moisture and meteorological conditions are likely to result in reduced smoke emissions. Dilution measures involve controlling emission rates by fire management techniques or scheduling prescribed fires when atmospheric processes favor unstable conditions and therefore maximum mixing of the lower atmosphere A key to good smoke dilution and dispersal is accurate weather forecasting, particularly about wind speeds and direction, atmospheric stability and mixing heights (Debano *et al.*, 1998).

18.4.2 Greenhouse gas emission reduction through land preparation without fire: an example from the community

18.4.2.1 Research site In order to solve the problem of high greenhouse gas emissions from fire-based land preparation, a 3-ha agricultural plot belonging to the local community located in peatland was established. The research site was located at Mangsang village, Bayung Lencir subdistrict, Musi Banyuasin district, South Sumatra Province, Indonesia. Bayung Lencir subdistrict stretches between 103°26′4″ to 104°41′35″E and 1°45′5″ to 2°28′55″S. It covers 6699.12 km^2 in area, equal to about 26% of Musi Banyuasin district.

Table 18.5 Number of hotspots detected during January to December 2007 over peatland areas compared with non-peatland areas (KNLH, 2008)

No.	Province	Peatland	Mineral soil	Number	%
A: Sumatra Island					
1	Bangka Belitung	31	446	477	5.84
2	Bengkulu	3	115	118	1.44
3	Nanggroe Aceh Darussalam	44	128	172	2.11
4	Jambi	109	1201	1310	16.03
5	Riau Island	6	28	34	0.42
6	Lampung	73	401	474	5.80
7	Riau	1242	1119	2361	28.89
8	West Sumatra	41	140	181	2.22
9	South Sumatra	580	1952	2532	30.99
10	North Sumatra	176	336	512	6.27
	Total Sumatra	2305	5866	8171	
B: Kalimantan Island					
1	West Kalimantan	420	2683	3103	39.41
2	South Kalimantan	75	465	540	6.86
3	Central Kalimantan	293	2008	2801	35.57
4	East Kalimantan	124	1306	1430	18.16
	Total Kalimantan	912	6962	7874	
	Total A+B	3127	12,828	16,045	
	%	20.05	79.95	100	

The area is lowland with an average altitude less than 15 m above sea level. The soil type is predominantly organosol in the low-lying areas, whereas more elevated parts are dominated by red-yellow podsolic soil. The land of Bayung Lencir is subject to fire as a result of human activity, which has markedly degraded the native forest. The degraded forest has been replaced by shrubs and alang-alang grassland (*Imperata cylindrica*), which increases the risk of fire especially during the dry seasons. Historical data show that in 1997, about 5000 ha of the land in Bayung Lencir was burned.

Most of the fires that occurred in Bayung Lencir, especially during dry seasons, resulted from burning of cleared and dried shrubs and grassland. Community members believe that ash from burning will increase soil nutrients and crop yields. Unfortunately the effects of the ash are not long-lasting, although the negative impact of smoke in the environment persists. This situation is common for areas that often use fire for land preparation. Actually burnt materials can generally be processed into charcoal, charcoal briquette and organic fertilizer. This processing minimizes smoke and reduces greenhouse gases, and farmers can benefit economically from the products.

There are two oil palm companies located near Mangsang village that have a positive impact on the economic development of the village. The community has learned much from the estate on how to establish oil palm, rubber, and horticultural crops. Oil palm and rubber plantations occupy 20,248 ha of Bayung Lencir land; 10,000 ha is degraded land, being ex-forest concession areas; and another 100 ha is swampy and not cultivated yet.

There were five Dusun (sub-villages) encompassed by Mangsang village with a total population of about 11,900 people, of which 60% worked for the estate company, 25% worked as farmers and the remaining 15% worked as teachers, traders or were self-employed.

There are three main ethnic groups in Mangsang: Jawa, OKI (Ogan Komering Ilir) and Palembang. Apart from its road links, Mangsang village is located along the Lalan River, which is used for access to other villages and sub-districts.

18.4.2.2 Research methodology

18.4.2.2.1 Soil sampling, biomass and greenhouse gas analyses Five soil samples taken from the 3-ha sample plot were used for chemical and physical analysis. Fuel load was estimated from the amount of plant material, both living and dead, found in 10 subplots of 1 × 1 m, spread randomly in the 3-ha plot. The samples were collected, separated and weighed. Gaseous emissions from forest fires were calculated from consumed biomass. First, the consumed biomass was estimated using the following equation (Seiler & Crutzen, 1980):

$$M = A \times B \times E$$

where M is the total mass of forest consumed by burning (tonnes), A is the burned area (ha), B is the biomasss loading (t/ha) and E is the burning efficiency. Assuming that carbon accounts for 45% of the mass of the forests, and 90% of the carbon is released in the form of CO_2, then the mass of CO_2 released to the atmosphere is:

$$M(CO_2) = 0.9 \times 0.45 \times M$$

Once the mass of CO_2 produced by burning is known, the mass $M(x_i)$of any other species x_i can be calculated using the CO_2-normalized species emission ratio, $ER(x_i)$. The emission ratio is the ratio of the production of species x_i to the production of CO_2 in the fire, given by:

$$ER(x_i) = \frac{M(x_i)}{M(CO_2)}$$

where x_i = CO, CH_4, NO_x, or NH_3.

18.4.2.2.2 Charcoal and briquette making The fuel load found on agricultural land belonging to the community was dominated by grass, shrubs, litter, branches, leaves and stumps. All these materials were collected and separated into woody and non-woody materials. Woody materials were used for making charcoal for briquettes, and non-woody materials were converted to organic fertilizers. Using simple equipment, the woody materials were put into a simple drum and burnt with low levels of oxygen for about 8 hours. This process produces charcoal, which can then be used to make charcoal briquettes by mixing with tapioca using a high-pressure machine.

18.4.2.2.3 Organic fertilizer making Fuel materials such as shrubs, litter and grass were chipped using simple machinery. The entire mixture was then mixed with EM-4 (Effective Microorganism-4), as an agent for decomposition processes, to produce the organic fertilizer after 3 weeks.

18.4.2.3 Results

18.4.2.3.1 Soil analysis Based on soil chemical analysis it was found that pH varied from 4.00 to 4.20; organic C varied from 11.42 to 17.71%; total N from 0.19 to 0.36%; P from 1.2 to 1.5 ;ppm and cation exchange capacity (CEC) from 29.16 to 36.22 cmol/kg

Table 18.6 Chemical properties of peat

Plot	pH H_2O	Organic C (%)	N total (%)	P (ppm)	Ca (cmol/kg)	Mg (cmol/kg)	K (cmol/kg)	Na (cmol/kg)	CEC (cmol/kg)	BS (%)
1	4.00	17.71	0.36	1.5	1.08	1.53	0.38	0.56	36.22	9.8
2	4.20	11.42	0.19	1.4	0.92	1.12	0.40	0.53	29.16	10.2
3	4.10	12.96	0.23	1.2	0.80	0.98	0.26	0.44	32.35	7.7

CEC, cation exchange capacity; BS, Base Saturation.

(Table 18.6). Analysis of the physical properties of the peat showed that bulk density varied from 0.48 to 0.55 g/cm^3, porosity varied from 79.25 to 81.88%, water content from 21.67 to 27.34%, available water from 14.14 to 18.98% and permeability from 29.08 to 48.80 cm/h (Table 18.7). These data show that the quality of the soil is low and something should be done to make the land more productive. One of the most effective measures adopted by the local people is to use fire for land preparation in order to obtain the mineral-rich ash. In this research the role of ash was replaced by organic fertilizer made from the fuel load derived from the land.

18.4.2.3.2 Fuel load To reach the objective, research was conducted in the field and in the laboratory. Ten sub-plots of 1 m^2 (1 × 1 m) each were established within the 3-ha plot in the field.

Vegetation found in each sub-plot, such as grass, shrubs and dead material such as litter, branches, leaves and stumps (woody and non-woody material), was taken and weighed to obtain its fresh weight. For each of the different materials 200 g of fresh samples were taken for analysis in the laboratory to determine their moisture content. Finally, the dry weight of the material/biomass in the plot calculated – it was about 44 t/ha.

18.4.2.3.3 Greenhouse gas emissions By using the Seiler and Crutzen formula (1980), the calculation revealed that if 44 t/ha fuel load is burnt, it will produce 3.465 t CO_2, 0.036 t CH_4, 0.0014 t NO_x, 0.044 ton NH_3, 0.0367 t O_3 and 0.641 t CO.

18.4.2.3.4 Charcoal and briquettes The fuel load of grass, shrubs, litter, branches, leaves and stumps, with an average fresh weight of about 44 ton/ha, which is usually burnt to get nutrient-rich ash, was instead converted to charcoal and briquettes (from woody materials) and organic fertilizers (from non-woody materials). While the charcoal and briquettes became daily energy sources replacing fossil fuels, the organic fertilizer was

Table 18.7 Physical properties of peat

Plot	Bulk density (g/cm^3)	Porosity (%)	Water content pF 4.2	Available water (%)	Permeability (cm/h)
1	0.50	81.09	25.84	16.40	39.87
2	0.48	81.88	21.67	18.98	48.80
3	0.55	79.25	27.34	14.14	29.08

used to enrich the soil, replacing the chemical fertilizers that often produce high emissions of greenhouse gases.

The results from the 3-ha demonstration plot showed the importance of the community in minimizing greenhouse gas emissions through participatory peatland management without burning. Using such an approach the community can better manage their land, prepare the land without burning, and obtain organic fertilizer, thereby making the land more useful, productive and efficient. The fuel load materials such as grass, shrubs, litter, branches, leaves and stumps are freely available, making the processing convenient for farmers.

18.4.3 Peatland management and restoration of organic soils

These mitigation practices have the potential to sequester carbon at a rate of 7.33–139.33 t CO_2/ha/year and reduce N_2O emissions by 0.05–0.28 t CO_2-eq/ha/year (ADB, 2009). The sequestration of carbon can be achieved by avoiding drainage of organic or peaty soils that are known to contain high densities of carbon, or by re-establishing a high water table in the area (Freibauer *et al.*, 2004). Furthermore, emissions of GHGs from drained organic soils can be reduced by avoiding the planting of row crops and tubers, avoiding deep ploughing, and maintaining a shallower water table (IPCC, 2007).

Restoring peatland areas or organic soils can reduce the runoff from agricultural fields and settlements, which causes eutrophication, algal blooms, and hypoxic dead zones in lakes, estuaries, bays and seas. It can also reduce flood damage, stabilize shorelines and river deltas, retard saltwater seepage, recharge aquifers, and improve wildlife, waterfowl and fish habitats. Restoration of organic soils can also improve soil quality and aesthetic and amenity values, promote biodiversity and wildlife habitats, and support energy conservation.

18.5 Conclusion

Based on the official data, it has been found that communities are among the most important players in producing high emissions of greenhouse gases through burning of peatland during land preparation. To overcome this problem, land preparation without fire is one of the best solutions. Results of research in the agricultural peatland belonging to one such community show that the fuel load composed of grass, shrubs, litter, branches, leaves and stumps, with an average fresh weight of around 44 t/ha, can be used to produce charcoal, briquettes and organic fertilizers, therefore preventing release of greenhouse gases during burning. It was found that 3.465 t CO_2, 0.036 t CH_4, 0.0014 t NO_x, 0.044 t NH_3, 0.0367 t O_3 and 0.641 t CO would be released if the same amount of fuel load is burnt. This demonstrates that the community could play an important role in reducing greenhouse gas emissions through land preparation without burning especially on peatland. To empower the community in this role, and persuade them not to use fire in land preparation, there should be incentives or real action from the government to initiate community-based fire management.

Acknowledgements

This research was supported by a grant provided by the Directorate General of High Education, Ministry of National Education of the Republic of Indonesia.

References

ADB (2009) *The Economics of Climate Change in Southeast Asia: Regional Review*. Asian Development Bank.

Debano, L., Neary, D.G., Ffolliott, P.F. (1998) *Fire's Effects on Ecosystems*. John Wiley & Sons, Inc.

Department of Forestry (2007) *Statistical Development of the Year 2006*. Jakarta: Directorate General of Land Rehabilitation and Social Forestry.

Freibauer, A., Rounsevell, M., Smith, P., Verhagen, A. (2004) Carbon sequestration in the agricultural soils of Europe. *Geoderma* **122**: 1–23.

Goldammer, J.G. (1993) Fire management. In: Pancel, L. (ed.), *Tropical Forestry Handbook*, Vol. 2. Berlin: Springer Verlag, pp. 1221–67.

Hoojier, A., Silvius, M., Wosten, H., Page, S. (2006) *PEAT-CO2, Assessments of CO2 emissions from drained peatland in SEA*. Delft Hydraulics report Q3943.

IPCC (2007) *Climate Change 2007: Mitigation of Climate Change. Contribution of Working Group III to the Fourth Assessment Report of the Intergovernmental Panel on Climate Change*. Cambridge, UK: Cambridge University Press.

KNLH (2008) *Status Lingkungan Hidup Indonesia (SLHI)*. Kementerian Negara Lingkungan Hidup.

Levine, J.S. (1996) Introduction. In: Levine, J.S. (ed.), *Biomass Burning and Global Change*, Vol. 1. The MIT Press, pp. 1–2.

Malingreau, J.P., Gregorie, J.M. (1996) Developing a global vegetation fire monitoring system for global change studies: A framework. In: Levine, J.S. (ed.), *Biomass Burning and Global Change*, Vol. 1. The MIT Press, pp. 14–24.

Nye, P.H., Greenland, D.J. (1960) *The Soil Under Shifting Cultivation*. Tech. Comm. 51. Commonwealth Bureau of Soils, UK.

Page, S.E., Siegert, F., Rieley, J.O., Boehmn, H.D.V., Jaya, A., Limin, S. (2002) The amount of carbon released from peat and forest fires in Indonesia during 1997. *Nature* **420**: 61–5.

Saharjo, B.H. (1995) The changes in soil chemical properties following burning in a shifting cultivation area in South Sumatra. *Wallaceana* **75**: 23–6.

Saharjo, B.H. (2005) Indonesia forest fire. Paper presented at the International Symposium on Forest Fire, Tokyo, Japan.

Saharjo, B.H. (2007) Shifting cultivation in peatlands. *Mitigation and Adaptation Strategies for Global Change*. **12**: 135–46.

Seiler, W., Crutzen, P.J. (1980) Estimates of gross and net fluxes of carbon between the biosphere and the atmosphere from biomass burning. *Climate Change* **2**: 407–27.

UNCHS (1999) Inter-agency report on Indonesian forest and land fires and proposal for risk for reduction in human settlements. UNCHS (Habitat), UNDP, UNHD and ADPC.

WWF (1997) The year the world caught fire. WWF International Discussion paper by Nigel Dudley. WWF International.

19 Structuring Climate Finance for Adaptation Measures in Vulnerable Ecosystems: Lessons from India

A. Damodaran
Indian Institute of Management Bangalore, Bangalore, India

19.1 Introduction

Adaptation is the main way in which ecosystems render themselves less vulnerable to stresses and strains imposed by natural or human-induced factors. In the ecological sense of the term, adaptation refers to the manner by which an organism or species adjusts to its bio-physical environment (Lawrence, 1995; Abercrombie *et al.*, 1997). From a social science perspective, adaptation is the process by which individuals adjust to changes in the collective behavior of socioeconomic systems (Denevan, 1983; Hardesty, 1983). Smit *et al.* (2000) consider adaptation as involving adjustment of natural or human systems to experienced or future climatic conditions or their effects or impacts, which may be beneficial or adverse.

In its operational sense, adaptation involves both the *process* and *condition* of getting adapted (IPCC, 2007). Technologies, community structures, institutional arrangements, and public policies all contribute to the process and condition by which human communities get adapted (Downing *et al.*, 1997; UNEP, 1998).

Adaptation measures can be placed into two categories, namely 'natural' (or 'autonomous') and 'planned'. While the former involves a natural adjustment process to interim

Vulnerability of Land Systems in Asia, First Edition. Edited by Ademola K. Braimoh and He Qing Huang.

variability in climate factors, the latter involves proactive intervention on a larger scale to address 'secular' changes in climate.

The real challenge of adaptation activities arises from the separation of the adaptation process (which entails economic costs) from its end results or benefits (adjustment to global warming) over time. Planned intervention measures call for large investments and entail large-scale impacts, and are required to be undertaken in the short run for anticipated 'inter-generational', 'long-term' benefits. However, the long-term nature of benefits may be perceived by the present sacrificing generation as of no consequence to them. Indeed inter-generational 'bias' against investments that have only long-term benefits is normal for communities that suffer from high time preference (and hence high discount rates on future benefits) when it comes to investments that have only long-term benefits. Climate financing systems that aim at providing 'new' and 'additional' resources for planned adaptation measures should ideally aim to overcome the temporal bias of the kind mentioned. This does not appear to be the case with climate financing systems in different parts of the world, including the cases from India that are described here. With reference to the tropical coastal zones of north Kerala and the semi-arid ecosystems of rural Bangalore in Karnataka, this chapter demonstrates that adaptation suffers from temporal bias on account of its scale of complexity and the possible loss of economic welfare it entails for farming and local community households.

19.2 Approach

The approach to the study is based on testing the co-benefit proposition in actual field conditions. At the core of the chapter is the proposition that mitigation yields the co-benefit of reducing the adaptation burden amongst local communities in developing countries. The greater the mitigation commitments, the less will be the probability of local communities carrying out costly planned adaptation measures. This argument draws sustenance from the findings of the field surveys conducted in the states of Kerala and Karnataka in south India.

The first finding emerging from field studies is that the transition to a full and effective adaptation process by local communities is not a smooth process and entails a few discrete steps. The higher the 'adaptation step' climbed, the greater is the technological complexity of the adaptation process and the greater the economic costs incurred. This finding serves to highlight the importance of the Tol and Fankhauser (1997) thesis about the need to incorporate transition costs in the assessment of equilibrium adaptation costs.

The second finding is based on the point that adaptation needs to be viewed in its 'bio-physical' and 'socio-economic' senses. This proposition is in the spirit of what Hurd *et al.* (1997) highlight to be the importance of market and non-market adaptation measures in the assessment of impact costs. Indeed the core of the study is that non-market adaptation measures may be achieved by vulnerable ecosystems at the cost of 'market mal-adaptation'.

19.3 Methodology of field studies

Field studies were carried out in the coastal area of Dharmadam in North Kerala during 2002–03 and in semi-arid ecosystems in the rural hinterlands of Bangalore District (2007). In the case of the former nearly 540 fisherman households and three field staff members from public works and the local fisheries departments were chosen to explore resource utilization patterns in relation to the local ecosystem as well as assessing costs and benefits of various adaptation works and activities in the coastal area, both at the household and ecosystem levels. As far as the Bangalore District studies were concerned, the field surveys were carried out in three villages. Initially the focus was on a survey of the technological interventions and resources management and utilization patterns for forests, water regimes, agriculture systems and degraded lands. Subsequently a detailed study was carried out of three types of farming systems: traditional–subsistence, part-commercial and purely commercial farming systems. A total of 70 households were chosen for the study based on purposive, stratified sampling in 2007. Farmers who practiced natural adaptation systems, had a keen sense of future adaptation technology possibilities and were inclined to participate in the survey, were chosen from a larger population of 165 originally identified. The final sample comprised 10 marginal farmers with less than 1 acre of dryland, 25 small farmers (holding between 1 and 2.5 acres of land), 25 medium farmers (with holdings of 2.5 to 5 acres of land) and 10 farmer households owning more than 5 acres of groundwater-irrigated land holdings that were cultivated with market-oriented horticultural and vegetable crops.

Questionnaires to the surveyed households were designed to assess actual costs and benefits of different types of adaptation possibilities trialled on earlier occasions (both natural and planned) or proposed for the future. Where farmers were clueless about new 'planned adaptation technologies', questions on possible costs and benefits were posed on hypothetical terms, after a clear description was provided to farmers about the nature of the types of adaptation technologies that could be tried out.

Table 19.1 sums up the properties of the ecosystems chosen for study in terms of habitats/ecological regimes, dominant crop/plant species and vulnerability factors. As is evident from Table 19.1, while a combination of natural (weather variability) and human-induced factors (land use change, institutional weaknesses and biotic pressures) account for vulnerability of the semi-arid agricultural and forest ecosystems, in the case of the coastal ecosystem, human-induced change has been the dominant factor accounting for change.

19.4 Co-benefits approach to adaptation financing and equity

Agricultural systems in the developing world are already threatened by climate variability, which increases their vulnerability to food insecurity (Bates *et al.*, 2008). Hydrological systems in the developing countries of Asia and Africa are threatened by climate change, resulting in serious imbalances in water budgets. Coastal areas in developing countries are threatened by sea level rise. While a few natural adaptation measures are undertaken by agricultural and coastal communities in countries like India, the desired extent of adaptation measures is not attained on account of financial constraints.

Fund flows for adaptation programs are primarily driven by national resources and grants. India spends 2% of its GDP on adaptation and is expected to spend more as the worst impacts of climate change intensify (Anonymous, 2009a). For undertaking the full range of adaptation measures this scale of funding may still be inadequate. The much hoped for flow of international financing for adaptation projects has not materialized. Thus there are clear limits to the extent to which adaptation can be pursued in developing countries.

Article 2 of the United Nations Framework Convention on Climate Change (UNFCCC) states that the 'ultimate objective of this Convention and any related legal instruments that the Conference of the Parties may adopt is to achieve, in accordance with the relevant provisions of the Convention, stabilization of greenhouse gas concentrations in the atmosphere at a level that would prevent dangerous anthropogenic interference with the climate system. Such a level should be achieved within a time frame sufficient to allow ecosystems to

Table 19.1 Features of semi-arid and coastal ecosystems in the study area

Ecosystem type	Regimes	Dominant species	Vulnerability factors
Semi-arid agro ecosystems (rural Bangalore, Kolar)	Wet arable land Tanks/ponds Dry arable land Pastures Dry waste lands Forests/tree groves Plantations	*Eleusine coracana* (finger millets) dry crop legumes *Cynodon dactylon* *Alsinoides* spp. *Morus alba* *Tamarindus indica* *Eucalyptus* spp. *Grevillea robusta*	Rainfall variability Breakdown of grazing cycle Land-use change Extraction of groundwater Construction of dwellings and infrastructure Encroachment Unauthorized biomass removal
Forests (Bangalore (rural)	Thorny scrub Dry tropical Mixed deciduous	*Albizzia amara* *Acacia pinnata* *Acacia auriculiformis* *Pongamia pinnata*	Livestock grazing Development projects Encroachment of land-use change Biomass removal for industrial use Illicit felling
Coastal areas (north Kerala)		*Cocos nucifera* *Casuarina equisetifolia* Mangroves	• Land-use change • Discharge of industrial pollution on shorelines and in shallow seas • Mechanized trawlers • Construction of new dwellings and tourism complexes

Modified from Damodaran (2001).

adapt naturally to climate change, to ensure that food production is not threatened and to enable economic development to proceed in a sustainable manner.'

What comes out clearly from this proviso is that ecosystems in developing countries need to be provided with the opportunity to adapt naturally to climate change. For this to be achieved there is a need to stabilize greenhouse gas (GHG) concentration in the atmosphere. This can be done only if mitigation measures are undertaken.

Therefore the key paradigm advanced here is that apart from its direct GHG reduction impact, mitigation affords the co-benefit of less burdensome adaptation by local communities. Low levels of mitigation efforts by developed countries, besides being against the spirit of obligations enshrined in the Kyoto Protocol, would only serve to further increase the adaptation costs for developing countries (Lewellyn & Chaix, 2007). The Third Assessment Report of the IPCC states that until 2050, global mitigation efforts that are designed to cap GHGs at 550 ppm would benefit developing countries significantly, particularly when combined with enhanced adaptation (IPCC, 2007).

Indeed, para 1 of the Copenhagen Accord of December 2009 while recognizing the scientific view that 'the increase in global temperature should be below 2 degrees Celsius', also underlines the fact that this should be achieved 'on the basis of equity and in the context of sustainable development' through 'long term co-operative action to combat climate change' (Anonymous, 2009b). In reality, the two provisos provide ample legal and moral justification for pursuing the co-benefit approach to mitigation in future climate governance architectures.

The problem is that the range of adaptation options is large with the economic burden of high-gradient adaptation options being unbearable to farming households. Given the fact that local communities in the agriculture sectors subsist on low incomes, have high marginal utilities for current income and hence have high time preference, the shadow cost of marginal adaptation will be far higher than what is reckoned in nominal, undiscounted monetary terms.

The second factor is that at the household level, undertaking complex adaptation measures could be at the cost of current income. This is particularly true of agricultural systems that specialize in commercial cropping. The concept of equity does not weaken even if the commercial segments of developing country agriculture are focused upon. Though traditional farming systems involve low costs of production and have higher 'cost-effective' adaptation capacity compared with commercial agriculture, the case of the latter cannot be ignored on equity considerations. In reality, the social costs of low returns for commercial farms can be higher on account of the large segment of landless labor households dependent on them by way of wage labor. Thus the adaptation burden needs to be reduced for all segments of the agricultural communities in developing countries in the interests of livelihood protection.

This would imply that the co-benefits of mitigation on adaptation form an important and more universally acceptable principle for the future progress of the Climate Change Convention. The following findings from field surveys in India amplify the importance of tapping the co-benefit principle.

19.5 Adaptation gradients

Field surveys conducted in the study areas revealed that adaptation is not a standard one-fits-all fix that can risk-proof agricultural communities. There are different gradations of adaptation depending on the experienced or anticipated climate stress. Table 19.2 outlines the different gradients of adaptation actions that are possible for coastal and semi-arid ecosystems. The adaptation measures indicated in Table 19.2 are those that the local communities and public agencies have carried out or plan to carry out to tackle climate variability of various intensities. Modal values of costs and benefits were worked out at current prices, based on information provided by the surveyed farmers, public works department staff (for costs on sea wall construction) and related public agencies. The relative loss of benefits and incremental costs are indicated in columns 5 and 6. To work out costs, amortized values for capital and operational costs were worked out while benefits were estimated in terms of damages avoided.

As can be seen from Table 19.2, high-gradient adaptation measures entail reduced benefits in the agricultural sector or cause higher costs to be incurred by local communities when it comes to coastal zones, hydrological regimes and land resources. In the case of agriculture, the extreme event of a shift from conventional cropping to a new crop that is short duration and highly tolerant to water stress can induce high incremental costs by way of reduced yield and higher costs of plant materials and seeds. Similarly for the coastal zones, construction of sea walls, which is resorted to in areas that are prone to cyclones, entails high capital expenditure besides reducing access to beaches, land close to the sea, and fishing buoys (Damodaran, 2006, 2009, 2012). Land degradation can be handled by simple re-clothing of degraded lands with species that are endemic to the area. However, in the extreme event of secular change in climate, it would become necessary to develop

Table 19.2 Relative benefits or costs from different gradations of adaptation projects

Serial no.	Sector	Adaptation activity	Mitigation level	Modal values at current prices in INR	
				Benefit[b]	Cost[c]
1	Agriculture	Drought-resistant short-duration plant varieties	Low	0.30	–
		Just-in-time sustainable irrigation	Medium	0.60	–
		Business-as-usual[a]	High Status quo (irrigated crops)	1.00	–
2	Coastal zone	Sea walls	Low	–	1.00
		Bio-shields (mangrove plantations)	Medium	–	0.17
		Business-as-usual[a] (routine protection measures)	High	–	0.003
3	Degraded land	Low-transpiration crops	Low	–	1.00
		Medium-transpiration crops	Medium	–	0.38
		Business-as-usual (high-transpiration crops)	High	–	0.25
4	Water (groundwater)	Artificial recharge	Low	–	1.00
		Natural recharge	Medium	–	0.09
		Business-as-usual[a]	High	–	0.01

Based on field surveys in 2002–03 and 2007. *Source*: Damodaran (2009).
[a]Business-as-usual: current level of utilization of the natural resource without any adaptation activity being undertaken.
[b]Benefit: optimal yield of finger millets in Karnataka (probability) in dryland and irrigated conditions.
[c]Cost of undertaking the adaptation activity.

low-transpiration varieties that can adjust to the edaphic attributes of the area. Given the fact that breeding new varieties of trees, vines and arboreal crops is time consuming and capital intensive, advanced adaptation measures (like the introduction of low-transpiration crops, construction of sea walls, and artificial groundwater recharge) will involve either high costs or significant loss of benefits.

19.6 Adaptation possibility trends for agro and coastal ecosystems: preliminary assessment

The second round of field surveys was conducted to look more at the specifics of adaptation possibilities for different farming systems in the semi-arid tracts of the rural fringes of

Bangalore. It was seen that, except for marginal farmers who practiced traditional cropping systems based on finger millets and leguminous inter-crops, sections of small and medium farmers practiced a combination of traditional and commercial cropping aided by irrigation. The big farmers were predominantly reliant on commercial horticultural crops, which were all sold through large organized markets. These farmers were entirely dependent on ground water for irrigation purposes and frequently encountered bore-well failures due to sinking ground water levels. Despite these signals, farmers in the small, medium and big categories did not aim to conserve water. Any adaptation measure that was based on low water usage was seen as threatening their yield and hence net income. Rather, faced with water scarcities, farmers practicing commercial cropping attempted to tap groundwater from deeper aquifers.

Table 19.3 sums up the adaptation gradients for farming systems in the semi-arid ecosystems. It summarizes the principal crops for different systems of cultivation, namely traditional, part-commercial and commercial, and the corresponding gradients of adaptation measures (Damodaran, 2009 and 2012). As can be seen from Table 19.3, 'planned adaptation measures' that are technically advanced are likely to be economically impracticable. It is also noteworthy that the degree of planned adaptation required for traditional farmers is less, even for medium-level adaptation measures, as these farming systems are more tuned to handle climate variability in the form of droughts. Paradoxically it is the traditional farmers who are in a position to secure public financing for droughtproofing and related measures in view of the large number of national rural development programs that are targeted at them.

The survey yielded the statistically significant finding that there is a trade-off between the different gradients of adaptation measures and net income of households. For a typical farmer, adaptive action undertaken is a function of availability of technologies and income. Preliminary data on possible income losses reveal that the trade-off between adaptation gradients and income was least for traditional cropping and part-commercial farms on account of their inherently high natural adaptation levels. In the case of commercial crops, the two variables pulled in different directions even when it involved a minor increase in adaptation activities. This has been partly due to the market-dependent income generation system characteristic of commercial farms. Unlike traditional and part-commercial farmers, commercial farmers were indifferent about moving over to the highest level of adaptation once they had crossed the medium adaptation level.

19.7 Financing systems for adaptation to climate change

The benchmark of an efficient and just system of financing climate change redress centers on the following parameters: 'new and additional, predictable, adequate and equitable' (Damodaran, 2009). The concept of 'appropriateness' is also considered to be a parameter by experts, though in practice this is closely linked to equity (Damodaran, 2009). Despite the importance accorded by the world community to financing activities that address climate change issues, nothing much has been achieved in tangible terms (Damodaran, 2009). The concept of 'adequacy' of environmental financing can be judged by the supply of funds relative to needs. Alternatively, it can be taken to mean the flow of funds that is sufficient to cover relevant costs (Muller, 2008, as cited in Damodaran, 2009). The flow of funds for development and global environmental causes has not been encouraging, even in years when the global economy was booming. In the year 2007, Official Development Assistance (ODA) flows through bilateral and multilateral channels was only of the order of US$103.7 billion (UNFCCC, 2008, p. 91). This represented an average of 0.23% of the

Table 19.3 Gradients of adaptation techniques for semi-arid agro-ecosystems in Karnataka (India)

Agro-ecosystem regime	Dominant crop/species	Gradients of adaptation	Adaptation measures	Nature of adaptation measure (natural/planned)
Subsistence farming systems	Finger millets and leguminous inter-crops	Low	Emergency sowing and planting of short duration when pre-monsoon rains fail	Natural
		Medium	Water conservation through storage in ponds and tanks; reduced sowing of inter-crops	Natural and planned
		High	Artificial water recharge; river basin transfer; planting of low-transpiration crops	Planned
Part-commercial agriculture	Millets	Low	Emergency sowing; reducing irrigation rounds	Natural and planned
	Horticulture		Crop thinning	Planned
	Crops and vegetables	Medium	Water conservation through storage in ponds and tanks	
		High	Cropping change towards low water-using crops Artificial water recharge	Planned
Commercial agriculture	Annual horticultural crops, citrus crops	Low	Plugging irrigation leakage Reducing rounds of irrigation	Planned
		Medium	Water conservation	Planned
		High	Cropping shift to low water-using food crops	Planned

Source: modified from Damodaran (2009 and 2012).

GDP of developed countries as against the Monterry ODA target of 0.7% of gross national income (Damodaran, 2009).

While funds that are available for adaptation and mitigation are inadequate, in the case of adaptation the situation is worse as private funds do not have a propensity to invest in adaptation projects compared with mitigation projects. This is evident from Table 19.4.

Nearly 60% of total global investment in environmental projects has been mobilized from domestic sources (UNFCCC, 2008). While foreign direct investment (FDI) and debts account for 20% in the European Union (EU), they account for 90% in Africa and the Middle East (UNFCCC, 2008). The annual availability of adaptation funds has been less

Table 19.4 Financing systems for different gradients of adaptation measures for semi-arid agriculture

Farm type	Gradient of adaptation measures[a]	National public financing	International public funding			Private financing	
			New	Additional	Predictable	National	International
Traditional subsistence farming	L	√	–	–	–	√	–
	M	√	–	–	–	–	–
	H	–	√	√	–	–	
Part-commercial	L	√	–	–	–	–	–
	M	√	–	–	–	–	–
	H	–	√	√	–	–	
Commercial farming	L	√	–	–	–	√	–
	M	–	–	–	–	–	√
	H	–	–	–	–	–	–

[a]L, low; M, medium; H, high.

than $500 million/year (UNFCCC, 2008, p. 92). By comparison, funds annually available for mitigation is pegged close to $1 billion/year. These figures pale into insignificance when reckoned against the figures of revenue generation of $9 billion from CDM and JI projects reported for the year 2007 (UNFCCC, 2008).

Estimates vary about the scale of funding required for adaptation. While the World Bank guesstimates current need to be of the order of $9 to $41 billion, UNDP estimates are higher at $86 billion. The UNFCCC places adaptation funding requirements to be in the range of $28–67 billion (Muller, 2008). The World Bank's estimates are based on anticipated flows from ODA, FDI and domestic investment for adaptation activities. It is reckoned by the Bank that 10–20% of the ODA and concessional flows amounting to $4–8 billion will flow for adaptation activities. The Bank likewise estimates that 10–20% of FDI flows amounting to $2–4 billion will find its way for funding adaptation activities. Gross domestic investment is expected to contribute $3–30 billion, which represents a 2–10% share of aggregate domestic investment (Muller, 2008).

Article 11 of the UNFCCC provides for a financial mechanism that functions under the aegis of the COP of the Convention. This proviso forms the basic principle underlying a financial mechanism that is set up to fund adaptation activities as envisaged by the Convention. Delivery of financial resources for adaptation will depend upon a combination of factors. The first factor involves the adoption of resource allocation criteria that are based on a mix of efficiency and equity principles. The second factor refers to the ease by which needy countries can access the financial mechanism.

Currently adaptation financing focuses on global and national environmental benefits. The Strategic Priority on Adaptation (SPA) funds under the Global Environment Facility (GEF) Trust fund requires generation of global environmental benefits. The LDCF (Least Developed Countries Fund) and the SCCF (Special Climate Change Fund) under the Global Environment Facility (GEF) apply only when climate change affects core sectors of development such as agriculture, water, health or infrastructure. It does not carry the requirement of generating global environmental benefits (UNFCCC, 2008). GEF considers costs of adaptation as costs imposed on vulnerable countries to meet their immediate adaptation needs, which in turn are understood to be the additional costs imposed by climate change to render development climate-resilient (UNFCCC, 2008). Since ex ante calculation of additional costs of adaptation is complex, the GEF has developed a sliding scale for LDCF and SCCF funding that serves as a proxy for estimating additional costs. In terms of the sliding scale, smaller projects receive proportionately more GEF funding than bigger ones, since they are assumed to have a higher adaptation component (GEF, 2006). The scale provides an indication of the possible maximum amount of GEF funding for any given project size and its application is optional (UNFCCC, 2008).

The Adaptation Trust Fund, set up under the Kyoto Protocol in 2007–08, attempts to fund 'full adaptation costs of projects and programs' that are run 'to address the adverse effects of climate change' (Anonymous, 2009c). The Fund describes the full cost of adaptation as 'costs associated with implementing concrete adaptation activities that address the adverse effects of climate change' (Anonymous, 2009c). It is noteworthy that one of the important objectives of the Fund is to maximize multi-sectoral or cross-sectoral benefits. This objective accords well with the emphasis of the Fund on funding programmatic approaches through a series of small-sized projects.

19.8 Evidence from the study area

The method employed to assess the state of climate financing in the study area was to catalog ongoing and anticipated natural resource development and droughtproofing projects that have implications for adaptation. The list of projects fell into two categories – one funded by national sources and the others funded through international sources. The list of anticipated projects include GEF-funded projects like the India Sustainable Land and Ecosystem Management (SLEM), which aims at soil and water conservation, integrated crop and livestock husbandry, agro-forestry and the introduction of alternative cash crops. This project serves as a model for other countries, and envisages the introduction of new land conservation techniques to enable communities to adapt to local challenges. The other project of the GEF that has a strong adaptation content is China's Hai River Basin project, which relies on a combination of new water-saving techniques including reduction of evapotranspiration from irrigated croplands of the river basin (GEF, 2006). It is anticipated that similar projects may find their way into the study area. Also listed were nationally and locally funded household initiatives for adaptation-relevant projects such as water conservation, which was assisted by local co-operative banks.

The main findings on the existing financing systems in the study area are that low- and medium-level adaptation measures undertaken by traditional and part-commercial farming systems are well supported by national public funding through dryland and national watershed development projects. However, there is no formal financing system for high-end adaptation projects for traditional agriculture. The advent of GEF-assisted projects that are based on 'new and additional financial resources' drawn from international public financing systems can change the scenario. Indeed, the introduction of the GEF-assisted SLEM project and river basin projects (such as the one for the Hai River in China) in the study area will contribute to the flow of international assistance for high-end adaptation activities to be carried out by traditional and part-commercial farming systems. Such projects would facilitate cropping shifts and effect structural adjustments to farming systems through changes in evapotranspiration regimes.

However, GEF-assisted projects run the risk of being unpredictable, since they may not endure beyond a given project life. In the case of commercial farming, availability of public financing, both national and international, is a major constraint (Table 19.4). Private financing is a strong possibility through national and international financing institutions for low- and medium-level adaptation activities (Table 19.4). There is a clear dearth of financing for high-end adaptation activities. The existing schemes of assistance for export crops focus almost exclusively on production, and transportation rather than on adaptation activities. Thus climate financing with its restrictive scope in terms of range and extent of activities covered, cannot cover even modest levels of adaptation activities by farmers to tide them over moderate climate variability, let alone covering the different gradients of adaptation options. The concept of compensating loss of income is totally alien to the notion of climate financing. This acts to further depress possibilities of autonomous adaptation actions by market-dependent farming systems. Given these facts, there is a greater need to interweave regular development assistance with dedicated adaptation funding as promised under the Climate Change Convention and the Kyoto Protocol.

The other method for augmenting fund flows for effective adaptation activities in vulnerable zones is to tap the second co-benefit of mitigation, namely of augmenting the adaptation fund kitty. This can be achieved by raising revenues from mandatory allowance-based carbon markets based on auctioning of a certain percentage of allowances, instead of assigning them free of cost as at present. The proceeds of the same could be devoted to helping adaptation processes in developing countries. Indeed the Norwegian proposal submitted to the UNFCCC envisages a similar mechanism to finance adaptation activities in developing countries (UNFCCC, 2008).

19.9 Lessons and implications: summing up

The field studies in the study areas are continuing. The results reflect the position emerging from the round of surveys carried out in 2002–03 (for the coastal ecosystem) and in 2007 for the semi-arid ecosystem mentioned above. The lessons emerging from the field surveys are interesting. The first major lesson is that national strategies and action plans for addressing climate change need to be premised on the realization that adaptation is not a modular strategy that can be carried out across space and time without regard to ecosystem specificities and the socio-economic dynamics of agriculture, forests and coastal zone ecosystems. There are obvious costs involved in undertaking different levels of adaptation activities and the cost-benefit calculus works to the detriment of agencies that carry out these activities. The burden of adaptation is high for local communities in rural India, particularly for those activities that are undertaken in anticipation of climate change in the distant future. The study indicates that the adaptation burden is high across different classes of farmers and different farming systems. Indeed the burden of undertaking even a modicum of adaptive adjustment can be severe for commercial cropping farm systems since crop yields of these systems are highly susceptible to diminution. The resultant fall in income or welfare needs to be compensated in case these farming systems have to contribute to anticipatory adaptation.

Thus the study indicates that adaptation cannot be a substitute for targeted mitigation. While it is possible for medium-level adaptation measures to be undertaken by farming communities in the study area, the task of undertaking the most technically advanced adaptation measures is difficult. This is precisely why mitigation action should be undertaken to address the problem of climate change.

Currently, in terms of structure, volume and the scale of international and national funding, climate financing mechanisms that are geared to adaptation do not meet the requirements for covering the entire range of adaptive actions. The issue of compensating agents for loss in income is nowhere emphasized in climate financing schemes. The Adaptation Trust Fund seeks to realize its finances by apportioning a 2% share in the proceeds from the Kyoto Protocol's Clean Development Mechanism (CDM) project activities. This is clearly inadequate to support the full range of adaptation activities listed in Tables 19.2 and 19.3. The larger challenge will be to tap the allowance-based carbon markets for financing adaptation. Getting allowance markets to contribute to adaptation will raise the stakes for solving the inefficiencies arising from over-compensation of firms in these markets (Stavins, 2007). Additionally, adaptation action can be funded by a range of development and dedicated funds. In the context of India, measures could range from building in adaptation financing components in 'market facilitation' and 'rural employment generation' funds. A slew of adaptation-sensitive trade-related measures can further incentivize the process. Commercial farms that face crop reduction on account of adaptation measures undertaken need to be provided with preferential market access and premium prices for their produce. The scheme can be based on benchmarked standards of adaptation activities undertaken, on the lines of analogous schemes designed for organic crops and food. Unfortunately proposals currently mooted to incorporate low-carbon concerns in global trade seek to incentivize mitigation action by erecting non-trade barriers on normally produced goods. There is no place for adaptation in the scheme of things. This needs to be re-examined.

References

Abercrombie, M., Hickman, C.J., Johnson, M.L. (1997) *A Dictionary of Biology*. Harmondsworth: Penguin Books.

Anonymous (2009a) *Government of India Submission to UNFCCC on Nationally Appropriate Mitigation Actions (NAMAs)*.

Anonymous (2009b) *Copenhagen Accord* Decision -/CP.15; available at: http://unfccc.int/files/meetings/cop_15/application/pdf/cop15_cph_auv.pdf

Anonymous (2009c) *Report of the Adaptation Fund Board*. Note by the Chair of the Adaptation Fund Board, Conference of Parties Serving as the meeting of the Parties to the Kyoto Protocol, Item 9 (a) of the provisional agenda, Fifth session Copenhagen, 7–18 December 2009. FCCC/KP/CMP/2009/14,19 November 2009.

Bates, B., Kundzewics, Z.W., Wu, S., Palutikof, J. (eds) (2008) Climate Change and Water. Technical Paper of the Intergovernmental Panel on Climate Change. Geneva: IPCC Secretariat.

Damodaran, A. (2001) *Towards an Agro Ecosystem Policy for India – Lessons from Two Case Studies*. New Delhi: Centre for Environment Education, Tata McGraw Hill.

Damodaran, A. (2006) Coastal resource complexes of south India: options for sustainable management. *Journal of Environmental Management* **79**: 64–73.

Damodaran, A. (2009) Climate financing approaches and systems: an emerging country perspective. Working Paper # 8(E)-2009. Graduate School of Management, St Petersburg State University.

Damodaran (2012) The Economics of Coping strategies and Financing adaptation action in India's semi-arid ecosystems. *International Journal of Climate Change Strategies and Management*, **4**(4): 386–403.

Denevan, W.M. (1983) Adaptation, variation, and cultural geography. *Professional Geographer* **35**: 399–407.

Downing, T.E., Ringius, L., Hulme, M., Waughray, D. (1997) Adapting to climate change in Africa. *Mitigation and Adaptation Strategies for Global Change* **2**: 19–44.

GEF (2004) *From Ridge to Reef – Water, Environment and Community Security – GEF Action on Transboundary Water Resources*. Washington, DC: Global Environmental Facility. Available at: http://www.thegef.org/gef/node/1544

Hardesty, D.L. (1983) Rethinking cultural adaptation. *Professional Geographer* **35**: 399–406.

Hurd, B., Callaway, J., Kirshen, P., Smith, J. (1997) Economic effects of climate change on U.S. water resources. In: Mendelsohn, R., Newmann, J. (eds), *The Impacts of Climate Change on the U.S. Economy*. Cambridge, UK, and New York: Cambridge University Press.

IPCC (2007) Climate Change 2007: Impacts, Adaptation and Vulnerability. *Contribution of Working Group II to the Fourth Assessment Report of the Intergovernmental Panel on Climate Change* (eds Parry, M.L., Canziani, O.F., Palutikof, J.P., van der Linden, P.J., Hanson, C.E.). Cambridge: Cambridge University Press.

Lawrence, E. (1995) *Henderson's Dictionary of Biological Terms*. Harlow: Longman, Scientific and Technical.

Llewellyn, J., Chaix, C. (2007) *The Business of Climate Change II*. Lehman Brothers.

Muller, B. (2008) *International Adaptation Finance: The Need for an Innovative and Strategic Approach*. Oxford Institute for Energy Studies, EV 42, June.

Reilly, J. (1999) What does climate change mean for agriculture in developing countries? A comment on Mendelsohn and Dinar. *World Bank Research Observer* **14**: 295–305.

Reilly, J., Schimmelpfennig, D. (1999) Agricultural impact assessment, vulnerability and the scope for adaptation. *Climatic Change* **43**: 745–88.

Risbey, J., Kandlikar, M., Dowlatabadi, H., Graet, D. (1999) Scale, context, and decision making in agricultural adaptation to climate variability and change. *Mitigation and Adaptation Strategies for Global Change* **4**: 137–65.

Smit, B., Burton, I., Klein, R.J.T., Wandel, J. (2000) An anatomy of adaptation to climate change and variability. *Climatic Change* **45**: 223–51.

Stavins, R. (2007) *A U.S. Cap-and-Trade System to Address Climate Change*. The Hamilton Project, The Brookings Institution.

Tol, R.S.J., Fankhauser, S. (1997) On the representation of impact in integrated assessment models of climate change. *Environmental Modelling and Assessment* **3**: 63–74.

UNEP (1998) *Handbook on Methods for Climate Impact Assessment and Adaptation Strategies, 2* (eds Feenstra, J., Burton, I., Smith, J., Tol, R.). Amsterdam: United Nations Environment Program, Institute for Environmental Studies.

UNFCCC (2008) Investment and Financial Flows to address Climate Change: An Update. Technical Paper_FCCC/TP/2008/7, November.

20 Scientific Uncertainty and Policy Making: How can Communications Contribute to a Better Marriage in the Global Change Arena?

Gabriela Litre

Center for Sustainable Development, University of Brasilia, Brasilia, Brazil. Brazilian Research Network on Global Climate Change, National Institute for Space Research (Rede CLIMA-INPE)

> It is not the voice that directs the story: it is the ears.
>
> Italo Calvino

Scholars and practitioners do not always have a happy marriage. On the one hand, policy makers often complain about a lack of policy-relevant research results; on the other, scientists frequently mention the ignorance of policy makers concerning their policy-relevant research results.

Concurrently, donors want to know precisely what results are being obtained with their money, while the general public demands both clarity and transparency on the impact scientific discoveries may have on their everyday lives. This is true in virtually every scientific arena, but particularly in the challenging global environmental change field, which requires moving urgently from scientific diagnosis to adaptation strategies (Vogel *et al.*,

Vulnerability of Land Systems in Asia, First Edition. Edited by Ademola K. Braimoh and He Qing Huang.

2007). As the International Human Dimensions Programme on Global Environmental Change (IHDP, 2009) states:

> The increased understanding of the challenges we are currently facing has shifted the focus in yet another way, from understanding the dynamics of global environmental change to using that understanding to devise ways to meet the challenges that we see emerge. This has pushed the scientific community to pay more attention to the relationship between science and policy, to include more use-inspired and policy-relevant research, and to improve communication with government, business, NGOs and the civil society at large.

If we all agree on the need to actively bridge understanding and action, why is an effective science-policy linkage still so difficult to achieve? Stone *et al.* (2001) mention a cocktail of reasons, including the inadequate supply of policy-relevant research, and the lack of access to research and data analysis. They further detail the poor policy comprehension of researchers towards policy processes; the societal disconnection of both researchers and decision makers; the ignorance of politicians concerning the existence of policy-relevant research; dismissive or unresponsive policy makers and leaders; power relations; and last but not least, the validity of research itself, as hindrances to establishing effective science-policy linkage.

As in any marriage at stake, there is a key thread underlying each of the above-mentioned factors for conflict, i.e., *lack of communication*. Derived from the Latin word *communicatio* (to share, divide out; communicate, impart, inform; join, unite, participate in; lit. 'to make common'), communication is vital in fostering effective science-policy dialogue by presenting complex issues in an accessible way to both decision makers and the general public. However, to achieve this, scientists must learn to use new communications tools both to share their results with society and manage the tight deadlines of the media and decision makers. It is not only about communication receipts, but also about creating avenues of mutual reliability between worlds that often speak different languages. In this chapter, we will analyze how those avenues of trust can be constructed through a more fluent, two-way communication between researchers and policy makers.

20.1 A case study: the establishment of marine reserves off the Californian coast

When talking about a more fluent and effective interaction between researchers and practitioners, a set of communication tools and even 'communication survival kits' are frequently offered to scholars. They usually include hints on how to deal with media deadlines, how to create engaging press releases and how to catch the attention of an already information-saturated public. Science communicators or public information officers are also mentioned as helpful in assisting scholars in learning how to both present their research more effectively and overcome boundaries and/or language/cultural barriers.

There is abundant literature about communications strategies for effective science-policy linakge, designed by scientists, information officers, journalists, communication scholars and even decision makers (Burns *et al.*, 2003; Christensen, 2007; Clark & Meidinger, 1998; Collier & Toomey, 1997; Court & Young, 2003; European Commission, 2006; Moser, 2008; Nisbet & Scheufele, 2009; Rogers, 2000; Warren *et al.*, 2007; Weber & Word, 2001).

Most of these action plans follow the pattern presented by Grorud-Colvert *et al.* (2010). Interested in fostering the policy impact of their interdisciplinary research, the authors summarized the main steps of a successful communication strategy implemented for the establishment of marine reserves off the Californian coast, and suggested that the framework used could and should be adopted more widely by the scientific community. Scientists who see communication as a top-down transmission of information run the risk of alienating key audiences – particularly those who are knowledgeable about the issue. Even within the decision-making field, the audiences in question are usually extremely diverse, with different levels of technical knowledge, values, opinions and cultural backgrounds. Hence no single form of communication will be most effective at reaching all of them.

The authors continue to outline a classical four-step strategy used in the case study, involving;

1. Getting to **'know the audience'** – identifying the needs, level of knowledge and background of different groups, and using this to tailor communication efforts.
2. Identifying **'main messages'** – including the original problem, why this should matter to the audiences, the actions required, and what benefits the audiences would derive from those actions.
3. **Choosing communication tactics** – a diverse range of communication approaches were used, including printed materials, web content and presentations to small groups.
4. **Measuring the success of communications.**

The strategy used in the case study described by Grorud-Colvert *et al.* (2010) involved collaboration between scientists, graphic designers, communication professionals, and policy experts, and the authors assert that this framework should become more prevalent. In addition, they suggest that a communication strategy should be an inherent component of research grants, and that scientists should make the most of the growing body of available resources.

Vogel *et al.* (2007) highlight the benefits of involving practitioners from the outset in the research design, problem definition or framing, choice of approaches, negotiation of credible and legitimate knowledge systems (including 'expert knowledge' and 'local knowledge'), and communication and involvement with relevant stakeholders.

When scholars present their research and its limitations, they are fostering collaborative, data-based discussions about different management options and how to implement them. This process also can benefit scientists as they demonstrate to donors the potential for their data; many foundations and agencies are interested in funding research that has a political or social connection. But, as we will see, the problem is larger than just finding the appropriate way to share a message: It is also about who produces the message, when, why, with which purposes, with what impacts, to whom the message is addressed, and with what underlying values (European Commission, 2006; Mahlman, 1998).

20.2 A matter of trust

The 'climate gate' affair of 2009 – the publication of e-mails and documents hacked or leaked from one of the world's leading climate research institutions – showed that a more concerted effort must be urgently made to explain and engage the public in understanding the processes and practices of science and scientists.

Catching the public's attention and obtaining financial support for research sometimes tempt scientific institutions to force headlines through high-impact, simplified messages.

But the other side of any aggressive communications strategy, especially in the complex global change field, is the risk of losing people's trust.

Reliability is easy to lose, while regaining it proves rather difficult. But even when respecting the factual truth, scientists are frequently unprepared to clarify how the results were obtained, how reliable they are, whether they agree with other studies, what credit is due to other scientists, and if others disagree, why this is the case. 'Moreover, results should not be emphasized more than is rightful: a public that has been disappointed once, will be skeptical forever' (European Commission, 2006, p. 34). This particularly holds true when referring to the anthropogenic causes of global change.

A more informed and demanding public has certainly moved science away from its 'ivory tower' (Baron, 2010) and has obligated scholars to learn a new language: That of communicating their work in an engaging, yet honest and accurate manner. While still in its privileged position of neutrality and objectivity, science was seen as providing hard facts and figures. However, it is often forgotten that those facts generally refer to specific temporally and geographically bounded situations. At the same time, by having scientists select and interpret policy-relevant information, their underlying commitments and values are implicitly transferred to the presented information. All people hold biases toward particular viewpoints (Klotz, 1970). Each person's particular set of biases is a result of personal life experiences, relationships, parents, schools, peers, teachers, personal practices, and the pressures of life. It is difficult for any person to deal objectively with evidence potentially destructive to one's own cherished beliefs or pride (Lewin, 1987) – or detrimental to perceived personal security, in whatever form.

Besides the difficulty in achieving objectivity, another intrinsic component of scientific enquiry – that of uncertainty – also puts effective scientific communication at stake. This is especially evident when addressing global change, where both researchers and policy makers face various sources of uncertainty. The term 'uncertainty' means different things to different audiences. Typically, scientists view uncertainty as an intrinsic component of scientific inquiry. In science, nearly all conclusions contain an element of uncertainty, because declaring something a 'certainty' requires a degree of knowledge seldom attainable in the real world (Briscoe, 2004).

For the general public, however, 'scientific' is almost synonymous with 'certain'. This perception, which is probably learnt at school, is virtually true of much old and well-established scientific knowledge. However, in many of the areas of current concern, from global change to genetic disorders, it lands very wide of the mark.

When science and society cross swords, it is often over the question of risk. Risk, as it is widely understood, has at least two dimensions: The chance of something happening and the seriousness of the consequences if it does. As is often the case with new phenomena or theories, scientists are uncertain about both of these, and further uncertainty about the chains of cause and effect supposedly at work remain. Yet the public, or the media purporting to speak for the public, may demand unqualified assurances, and may even perceive and present the response as being an unqualified assurance when it was not. By this means, the stage is set for confusion, cynicism, and even panic.

The questions of how to quantify and communicate uncertainty are currently the subject of intense study. The issues are of increasing concern in government, as government takes on ever more responsibility for regulating risky activities of all kinds, incurring political and sometimes legal liability when things go wrong (House of Lords, 2009).

Uncertainty, for scientists, points the way toward further action, whereas for policy makers, it tends to breed indecision. For example, policy makers usually apply funds after having received a complete set of facts and after all uncertainty has been eliminated, *vis-à-vis* the electorate's pressure, where funding results need to be guaranteed. However, this is an idealistic situation that in most cases the scientific community is unable to provide. According to Bradshaw and Borchers (1999) 'the idea that greater certainty can be obtained and allow for more "certain" conditions for decision making with better and faster science is based on the erroneous supposition that uncertainty is finite. ... whether or not they continue to be science-based, environmental policy formulation and decision making will be accomplished under conditions of uncertainty.' In short, uncertainty is a part of science, but this is not always easy to explain when sharing science stories with the general public. In that regard, Dronkers (2009) mentions a number of controversies involving scientific uncertainty in what he has called the 'paradox of science'. This paradox includes the following traits:

Controversy 1. Competing scientific explanations. Science is an evolutionary (and at times even revolutionary) process, often with competing explanations for why things are as they are. Science-based policy making may even become an illusion in cases of strongly conflicting scientific opinions and frequently changing insight and forecasts.

Controversy 2. Different time scales. Policy generally moves faster than science. While ongoing research produces new scientific evidence, policy decisions have to be taken on the basis of earlier preliminary insight and forecasts.

Controversy 3. Different perceptions. Policymakers base their judgments not only on scientific evidence but also on their own experience (tacit knowledge) or on information provided by non-scientific stakeholders (including the mass media). Disputes often originate from differing views on how a policy problem should be defined – policy makers and scientists approach problems from viewpoints that are basically different. Last, but not least, there exist language and cultural barriers (i.e., the production of virtually all cutting-edge scientific literature in English...).

The truth is that we can never eliminate uncertainty. In fact, new scientific discoveries often create new uncertainties (Briscoe, 2004). This does not mean, however, that there is an insurmountable disconnect between science and policy. Just because a scientist admits that we don't know everything does not mean that we don't know enough to act. The question is: How can uncertainty be communicated in a convincing way without manipulating or forcing facts?

20.3 Communicating scientific uncertainty

The need to obtain support to continue performing cutting-edge research has made global change scientists begin the process of addressing the difficulties of communicating uncertainty to non-scientists (Pittock *et al.*, 2001). They have also started to elaborate creative ideas about how to provide scientific information about uncertainty that may be useful in the formulation of policy responses to the challenges of global change. In general terms, this process involves two primary steps: (i) quantifying the uncertainties, and (ii) communicating the quantified uncertainties in such a way that policy makers might still find the underlying information useful (Webster, 2003).

20.3.1 Quantifying uncertainties

Quantifying uncertainties in global change science is, understandably, a difficult enterprise, as it boils down to determining how much is unknown about each particular variable within climate models. If scientists simply present a non-probabilistic range of climate change results, then for the data to be meaningful in a policy context, *someone* has to make a judgment about which part of the range is most likely. It is argued that the scientific community has a responsibility to offer judgments on likelihood, because, after all, they have studied the processes and phenomena that lead to uncertainty and are in the best position to quantify likelihoods, difficult as that can be. This does not preclude policy makers and the public from weighing this information and arriving at other judgments in their decisions but it gives them the judgment of experts from which to start (Webster, 2003).

20.3.2 Communicating the quantified uncertainties

Journalists frequently exaggerate controversies among global change scientists, especially in reports on the policy implications of climate change. 'It would be wrong to imagine that the science of climate change is somehow more fraught with uncertainty than many other scientific areas where society has pressing concerns.' When communicating with the public and policy makers, climate scientists should strive to avoid 'straying into one of the twin traps of ignoring the necessary caveats or producing statements so hedged in qualifications as to misrepresent what we really do understand' (Manning, 2003). At the same time, researchers need to learn to differentiate between 'science' and 'science for policy'. According to Moss and Schneider (2000), this 'involves being responsive to policymakers' needs for expert judgment at a particular time, given the information currently available, even if those judgments involve a considerable degree of subjectivity.' Above all, scientists have long realized that it is impossible to describe many aspects of climate science quantitatively, i.e., using exact numerical figures. The majority of the effects of global warming have yet to occur, so climate change projections are necessarily based on data (such as future greenhouse gas emissions) that may change. Therefore, researchers are better able to 'obtain semi-quantitative assessments of uncertainties' (Briscoe, 2004).

20.4 The need for a new language

In a world overflowing with information, how can scholars compete to obtain public attention and to receive the funding needed to continue advancing research that does not offer total certainties? The truth is that a completely new language is needed, since most scholars are only trained to produce scientific, peer-reviewed articles with empirical information and hypotheses that will have serious trouble in echoing throughout non-scientific audiences (Fahnestock, 2004; Weber & Word, 2001).

Scientific language is highly specialized and extremely concise and no one expects to find digressions or figures of speech in a peer-reviewed paper. Alternatively, the general public more frequently expects to receive emotionally engaging stories that reveal the extent to which scientific research will affect their everyday lives. While science works with hypotheses and empirical observations, the general public tends to believe stories that ring true (Weber & Word, 2001). In that regard, emotions play a great role. 'Communication among scientists is neutral and lacking in emotions. Thus, only the facts speak, motivate and convince, not the person presenting them, nor the hope that the theory is right, and not even the fascination they hold. In public communication, on the other hand, the quality of the discussion or data is not enough' (European Commission, 2006, p. 30).

In short, the 'power to tell stories' is vital when trying to catch the public eye in a world that is increasingly competing for people's attention. It is important to relate scientific news to what is already known by the general public.

If scientific news is related to a person's everyday life, then it will be easier to catch the public's attention, and their votes. Coming back to the example of climate change, since the general public does not experience global averages, to most people the idea of climate, and thus also the idea of climate variability, is bound up with their experience, in a particular locality, of variation over time, possibly measured in decades (Chapman, 2009).

Understanding the problems affecting a concrete community and presenting scientific alternatives to approach those problems is a good point of departure, not only for the general citizen, but also for decision makers. In that regard, journalists are closer to the public because as a non-specialized audience, they first need to 'digest' information and to ask the 'silly questions' to better understand the scientific message (Webster, 2002).

Journalists, as non-experts, are aided by their own experience and easily recognize these difficulties. A skilled communicator does not take anything for granted and saves readers, viewers, or listeners the problems (s)he has already had her/himself, and perhaps even conveys the same, fresh enthusiasm for what (s)he has just learned. 'If scientists want to make themselves understood, then they must make a greater effort to become an observer of their own topic from the outside (...). Given the large asymmetry that exists between scientists and their audience, they need to carefully watch their level, time and ways of explaining' (European Commission, 2006, p. 33).

Last, but not least, timing is also vital in bridging science and practice: since decision-making processes are cyclical, iterative, and ongoing, scientific input can occur at any or all stages, and to be most effective should be equally ongoing, even if the type of impact differs from stage to stage. (Vogel *et al.*, 2007). At the same time, different moments in time require different types of knowledge and different modes of communication (Inter-American Institute for Global Change Research, 2005).

20.5 Changing worlds

It is not only science that is changing: The world of news is experiencing a serious tsunami that creates new challenges for journalists and the media. The traditional way in which people access information – press from newsagents, radio, television, and, more recently, free press – is being pushed aside by new channels and media (websites, blogs, podcasts, Google/news, etc.), as well as by a gradual change in the attitude of the public about how to consume information and, in general, culture (De Semir, 2010). As a consequence, in 2008, CNN cut its entire Science and Tech Team (Brainard, 2008). In March 2009, 'in response to the decline in journalistic coverage of sciences', a group of prominent universities and research centers in the United States decided to create Futurity, a scientific news portal that provides information directly from those who produce it (the scientific, medical and environmental community) to the general public. Futurity, which has now extended to more organisations and also to institutions in Great Britain, is a clear alternative to what used to be the most common way of communicating science: Intermediation of journalists. In other words, a 'bypass' has been created today that allows the world of science to skip the unavoidable collaboration or (for some) the obstacle represented by the media in their objective of circulating information to the general public, who also have the option of searching for information directly from the specialised sources (De Semir, 2010).

Futurity is a good example of the change being experienced by social communication of scientific, medical, and environmental information. Furthermore, it is an approach that is becoming more widespread and that is characterized by the fact that producers of knowledge directly contact the public via the many channels now offered by the internet, without requiring the media to act as intermediaries. The media and the traditional advertising model that allowed them to exist are thus immersed in a difficult adaptation to the society of information on the internet. The communicative system as a whole has been destabilized (De Semir, 2010, p. 18). Society, as a whole, is in a period of mutation.

20.6 A learning experience

For those scientists who find it difficult to communicate to non-specialized audiences, or who still rely on 'classical' ways of reaching the public (i.e., with the intervention of journalists and science communicators) there is good news: Contrary to the common perception that the relationship between scientists and the media is irremediably mined with obstacles, an international mail survey (conducted in 2008) of 1354 biomedical researchers from the USA, Japan, Germany, Great Britain, and France, showed that media contacts of scientists in top research and development countries were more 'frequent and smooth' than was previously thought (Peters *et al.*, 2008).

About 70% of the respondents had interacted with the media in the past 3 years (30% of them more than five times), and 75% rated their encounters with the media as mainly good, whereas only 3% evaluated them as mainly bad. The five countries varied little in that respect. The authors also suggest that science communication can be particularly effective if the scientist and the journalist manage to overcome their mutual distrust and cooperate. Their competencies are, in fact, perfectly complementary, as long as there is mutual respect for each other's role. The scientist should be responsible for the content and the journalist for the communication's format, while the selection of the content and the choice of the message can be made together. The task is not always self-evident: journalists must accept the role of the amateur and give up a bit of control over the texts. On the contrary, the researcher must accept the journalist's way of arranging the work and agree to introduce changes only for the sake of greater accuracy. The easiest formula for cooperating is a final review of texts, to ensure that factual errors have not been overlooked.

Be it by talking to a journalist or by writing their own blogs, scientists seem to be more and more convinced of the importance of communicating their research to wider audiences. What may be driving this change in scientists' behavior is that they now see the benefit of a greater public understanding of the scientific enterprise through news coverage of research. Another driver is the prospect of rewards. Science that is more visible appears more credible to potential donors, and news coverage may enhance an individual scientist's career prospects.

In spite of this complementation, concluding that scientists are satisfied with their interactions with the media is quite different from claiming that one need not worry about the science-media relationship (Peters *et al.*, 2008). The case of the Intergovernmental Panel on Climate Change (IPCC) Fourth Assessment Report (IPCC, 2007) and the Himalayan glaciers is a good reminder.

The IPCC reports are massive undertakings, involving thousands of scientists, thousands of pages, and thousands of references. Their public impact and the eagerness with which the media expect and reproduce them, confirm that they have succeeded in raising public awareness. However, the credibility of the reports fell into question when, in a chapter on climate change impacts in Asia, the IPCC's Fourth Assessment Report (IPCC, 2007) relied on an error-riddled online article to discuss the likely state of Himalayan glaciers in 2035. It did so despite questions raised by some reviewers.

Details about the incident have come to light since 2009, when the Indian government published a report that contradicted the IPCC. The error highlighted the need to strengthen the IPCC's review process, and its capacity to respond quickly and appropriately to such problems. Failure to do so may undermine public confidence in the IPCC and invite opportunistic attacks by those opposing meaningful action on climate change. The 'climate gate' mentioned at the beginning of this chapter, also contributed to eroding the prestige of the reports.

In spite of these difficulties, the IPCC reports can still be seen as an interesting learning experience about effective communications for global change science-policy linkage. Some of their characteristics may even inspire new science products aimed at a bigger, non-specialized audience. The first of them is that the reports are demand driven, with involvement in the assessment process of the full range of decision makers who would implement the responses.

They are also designed as an open, representative and legitimate process, with well-defined principles and procedures (Shaw, 2005) and further involve experts from all relevant stakeholder groups in the scoping, preparation, peer-review, and outreach/communication.

One of the key features of the IPCC reports is that they incorporate institutional as well as local and indigenous knowledge whenever appropriate. Another is that their conclusions are not only policy-relevant, but also non-policy-prescriptive. The media appreciate that their conclusions try to be evidence-based and not value-laden, recognizing that the assessment conclusions will be used within a range of different value systems.

Above all, the IPCC reports have appeared to successfully deal with the difficult issue of uncertainty in science: They cover risk assessment and management and they quantify, or at least qualify, the uncertainties involved while presenting different points of view.

These findings have enormous implications: scientific problems need to be meaningful and relevant; people need help to understand both causes and solutions; communicators must – despite uncertainty – create a sense of appropriate urgency (but not irrational fear); and they must enable and empower people to act in sustainable ways and support relevant public policy (Vogel *et al.*, 2007, p. 355). Importantly, scholars and other communicators must avoid (or abandon) the assumption that better information and understanding alone will lead to action or policy support.

Acknowledgements

With special thanks to Russell Morgan (IHDP) for his comments and copy-editing.

References

Baron, N. (2010) *Escape from the Ivory Tower: A Researcher's Guide to Making Your Science Matter*. Washington, DC: Island Press.

Bradshaw, G.A., Borchers, J.G. (1999) Using scientific uncertainty to shape environmental policy. In: IUFRO Task Force on Forest Science-Policy Interface, Occasional Paper Number 13, May. International Union of Forest Research Organizations.

Brainard, C. (2008) CNN cuts entire Science, Tech Team. Columbia Journalism Review. Available at: http://www.cjr.org/the_observatory/cnn_cuts_entire_science_tech_t.php?page=all

Briscoe, M. (2004) *Communicating Uncertainty in the Science of Climate Change: An Overview of Efforts to Reduce Miscommunication Between the Research Community and Policymakers & the Public*. International Center for Technological Assessment.

Burns, T.W., O'Connor, D.J., Stocklmayer, S.M. (2003) Science communication: a contemporary definition. *Public Understanding of Science* **12**: 183.

Chapman, G. (2009) Popular perception and climate change: mapping the varying experience of precipitation. In: Ostreng, W. (ed.), *Transference. Interdisciplinary Communications 2008/2009*. Oslo: CAS.

Christensen, L.L. (2007) *The Hands-On Guide for Science Communicators: A Step-by-Step Approach to Public Outreach*. New York: Springer.

Clark, R., Meidinger, E.E. (1998) Integrating science and policy in natural resource management: lessons and opportunities from North America. Gen. Tech. Rep. PNW-GTR-441. Portland, OR: U.S. Department of Agriculture, Forest Service, Pacific Northwest Research Station, 22 pp.

Collier, J.H., Toomey, D.M. (eds) (1997) *Scientific and Technical Communication: Theory, Practice, and Policy*. Thousand Oaks, CA: Sage Publications.

Court, J., Young, J. (2003) Bridging Research and Policy: Insights from 50 Case Studies. Working Paper 123, August. London: Overseas Development Institute.

De Semir, V. (2010) *Science Communication & Science Journalism. A Meta-Review*. Barcelona: Media For Science Forum. Science Communication Observatory, Pompeu Fabra University.

Dronkers, J. (2009) Science-Policy Interaction. Coastal Wiki. Available at: http://www.encora.eu/coastalwiki/Science-Policy_Interaction#_note-0 [accessed 19 May 2014].

European Commission (2006) *Communicating Science – A Scientist's Survival Kit*. Luxembourg: Office for Official Publications of the European Communities.

Fahnestock, J. (2004) Preserving the figure: Consistency in the presentation of scientific arguments. *Written Communication* **21**: 6–31.

Grorud-Colvert, K., Lester, S.E., Airame, S., Neeley, E., Gaines, S.D. (2010) Communicating marine reserve science to diverse audiences. Special Feature: Perspective. *Proceedings of the National Academy of Sciences of the United States of America* **107**: 18306–11.

House of Lords (2009) Communicating Uncertainty and Risk. Select Committee on Science and Technology. Third Report. Available at: http://www.publications.parliament.uk/pa/ld199900/ldselect/ldsctech/38/3806.htm [accessed 19 May 2014].

IHDP (International Human Dimensions Programme on Global Environmental Change) (2009) Concept Note of the 7th International Science Conference on the Human Dimensions of Global Environmental Change: 'The Social Challenges of Global Change'. Bonn, Germany, 26–30 April 2009.

Inter-American Institute for Global Change Research (IAI) (2005) Linking the Science of Environmental Change to Society and Policy: Lessons from Ten Years of Research in the Americas. (Group D on Communicating Science). Workshop report from a meeting in Ubatuba, Brazil, 27 November–2 December.

IPCC (Intergovernmental Panel on Climate Change) (2007) Fourth Assessment Report (AR4). Geneva: World Meteorological Organization.

Klotz, J.W. (1970) Assumptions in science and paleontology. In: Zimmerman, P.A. (ed.), *Rock Strata and the Bible Record*. St Louis, MO: Concordia Publishing House, pp. 24–39.

Lewin, R. (1987) *Bones of Contention*. New York: Simon & Schuster, pp. 18–19.

Mahlman, J.D. (1998) Science and nonscience concerning human-caused climate warning. *Annual Review of Energy and Environment* **23**: 100.

Manning, M.R. (2003) The difficulty of communicating uncertainty. *Climatic Change* **61**: 9.

Moser, S.C. (2008) What is asked of us? A clarion call to scientists at an urgent time. IHDP Update Magazine Extra on the Science-Policy Interaction Dialogue on Energy and Sustainability. University of California Santa Barbara. www.ihdp.org

Moss, R., Schneider, S. (2000) Uncertainties. In: Pachauri, R., Taniguchi, T., Tanaka, K. (eds), *Guidance Papers on the Cross Cutting Issues of the Third Assessment Report of the IPCC*. Geneva: Intergovernmental Panel on Climate Change.

Nisbet, M.C., Scheufele D.A. (2009) What's next for science communication? Promising directions and lingering distractions. *American Journal of Botany* **96**: 1767–78.

Peters, H.P., Brossard, D., Cheveigné, S.D., *et al.* (2008) Science communication: interactions with the mass media. *Science* **321**: 204–5.

Pittock, B.A., Jones, R., Mitchell, C. (2001) Probabilities will help us plan for climate change. *Nature* **413**: 249.

Rogers, C.L. (2000) Making the audience a key participant in the science communication process. *Science and Engineering Ethics* **6**: 553–7.

Shaw, A. (2005) Policy relevant scientific information: the co-production of objectivity and relevance in the IPCC. Breslauer Symposium. University of British Columbia. See http://escholarship.org/uc/ucias_breslauer

Stone, D., Maxwell, S., Keating, M. (2001) *Bridging Research and Policy*. Warwick University/UK Department for International Development.

Vogel, C., Moser, S.C., Kasperson, R.E, Dabelko, G.D. (2007) Linking vulnerability, adaptation, and resilience science to practice: Pathways, players, and partnerships. *Global Environmental Change* **17**: 349–64.

Warren, D.R., Weiss, M.S., Wolfe, D.W., Friedlander, B., Lewenstein, B. (2007) Lessons from science: communication training. *Science* **316**: 1122.

Weber, J.R., Word C.S. (2001) The communication process as evaluative context: What do nonscientists hear when scientists speak? *Bioscience* **51**: 487–95.

Webster, M. (2002) The curious role of 'learning' in climate policy. *Energy Journal* **23**: 97–119.

Webster, M. (2003) Communicating climate change uncertainty to policy-makers and the public. *Climatic Change* **61**: 1–8.

21 Planning for Resilience: the Quest for Learning and Adaptation

Fernando Teigao dos Santos
Instituto de Geografia e Ordenamento do Território (IGOT), Lisbon University, Lisbon, Portugal

21.1 Introductory insights

In recent years a significant amount of research has been conducted worldwide leading to the consolidation of what can be called resilience theories in socio-ecological systems, comprehending multiple concepts, prepositions and case studies, which are helping to bridge the gap between theoretical conception and practical application. Nevertheless, further knowledge and new approaches are needed in order to bring resilience closer to policy-making. Resilience as a framework can bring added value to the policy and planning processes, helping to understand how systems can be better prepared to adapt, to anticipate and to innovate when facing changing circumstances.

Disturbances and crises will become more frequent and intense in future, due to the aggravation of global problems like climate change, environmental degradation, energy demands or socio-economic inequities in a highly interdependent and interconnected world. Resilience is more than ever a critical property that reflects the system's capacity (nations, regions, communities, enterprises, families, etc.) to adapt to and resist these and other disturbances without collapsing.

The main objective of this chapter is to explore the relations and the implications of resilience theories for planning processes, bearing in mind that the global context is becoming more uncertain, unstable and undeniably demanding. This conceptual chapter will argue that resilience can be enhanced through planning processes more focused on 'learning and adaptation' instead of 'command and control', proactively enabling the

Vulnerability of Land Systems in Asia, First Edition. Edited by Ademola K. Braimoh and He Qing Huang.

systems to transform and auto-organise, which is crucial when the external context is constantly changing.

21.2 The global 'carousel' context

Recent years have brought several crises and multiple disturbances related to the falling of financial markets, the soaring prices of food or the skyrocketing costs of fuel in some periods, with worldwide impacts that affected the life of nations, regions, communities, enterprises and families, pushing the most vulnerable over the limit and leading to social ruptures. In this 'carousel' context, resilience – being the capacity of those systems to absorb disturbance and reorganize without disruption or collapse – becomes a critical property. Planning for resilience is gaining relevance from a capacity-building perspective, aiming to prepare organizations and societies to collectively learn how to adapt to change.

As global fluxes of people, resources, capital and information become stronger, more intense and faster, so do the systems become more complex and unpredictable, especially when faced with disturbances and changes. Peter Senge in 1990 wrote that mankind has the capacity to produce far more information than anyone can absorb, to foster far greater interdependency than anyone can manage, and to accelerate change far faster than anyone's ability to keep pace. Globalization is increasing the speed of interactions, with an intensification and multiplication of the linkages among elements of the system, stretching the impacts of human activities to the global scale, leading to a general decline in ecological and social diversity (Younga *et al.*, 2006).

Changes in drivers such as population growth, economic activities and consumption patterns are placing increased pressure on the environment (UNEP, 2007), which impacts coupled with the greater interconnectedness between human and natural systems can even lead to the occurrence of what Homer-Dixon (2007) called 'synchronous failures', meaning the occurrence of cascading disruptions and collapses, in the economic, environmental, social or political order.

The World Economic Forum (2009) highlights major categories of transnational risk with emphasis on systemic financial risk, food security, supply chains and energy. Interdependency has increased the probability that a disruption in any one region may have significant global repercussions. Globalization, technological complexity, interdependence, terrorism, climate and energy volatility, and pandemic potential are increasing the level of risk that societies and organizations now have to deal with.

The world faces a compounding series of crises spawned in part by human activity, which are outpacing the capacity of governments and institutions to deal with them. Energy, food and water crises, climate disruption, declining fisheries, ocean acidification, emerging diseases and increasing antibiotic resistance are examples of serious, intertwined, global-scale challenges spawned by the accelerating scale of human activity. And there are few institutional structures to achieve cooperation globally on the sort of scales now essential to avoid very serious consequences (Rockström *et al.*, 2009; Walker *et al.*, 2009).

If resilience continues to decrease in social-ecological systems as we strive to increase production efficiencies, the frequency of regional catastrophes will escalate. The ongoing climatic and ecological changes, together with population growth, rapid urbanisation, land-use change and globalisation, are key drivers of human vulnerability to natural disasters. In addition, human population growth has forced people and economic activities to settle in vulnerable areas, such as lowlands and coastal areas (Adger *et al.*, 2005).

According to the Human Impact Report on Climate Change (Global Humanitarian Forum, 2009), the impacts of climate change are happening right now. Events like weather-related disasters, desertification and rising sea levels affect individuals and communities around the world. They bring hunger, disease, poverty and lost livelihoods – reducing economic growth and posing a threat to social and even political stability. Many people are not resilient to extreme weather patterns and climate variability and they are unable to protect their families, livelihoods and food supply from negative impacts of seasonal rainfall leading to floods or water scarcity during extended droughts.

Complex adaptive systems (Holland, 1995) like the climate, countries, ecosystems or economies, are hard to predict, to understand and to manage, due to their dynamic character with many components acting in parallel and constantly reacting. Uncertainty, instability and unpredictability are inherent to our systems, despite the will and the ability of human society to organize and to deal with some levels of complexity. Crises and disturbances are nevertheless becoming more evident and severe, demanding greater capacity from our systems to absorb disturbances without collapsing. And that is what resilience is all about.

21.3 Looking at the resilience framework

The resilience framework is becoming increasingly relevant and revolutionary, bringing a more proactive and less resigned way of dealing with problems that affect socio-ecological systems. Resilience theories are focusing on understanding, managing, and governing complex linked systems of people and nature (Folke, 2006), bringing new perspectives and helping to strength the sustainability science frameworks. Resilience as a framework is a broad, multifaceted, and loosely organized cluster of concepts, each one related to some aspect of the interplay of transformation and persistence (Carpenter & Brock, 2008).

Strengthening the capacity of societies to manage resilience is critical to effective pursuit of sustainable development (Lebel *et al.*, 2006). Sustainability involves maintaining the functionality of a system when it is disturbed, or maintaining the elements needed to renew or reorganize if a large disturbance radically alters structure and function. Managing complex, coevolving social-ecological systems for sustainability therefore requires resilience as the ability to cope with, adapt to, and shape change without losing options for future development (Folke *et al.*, 2002). Building adaptive capacity is a prerequisite for sustainability in a world of rapid transformations (Gunderson & Holling, 2002) and resilience can be seen as an issue of environmental, social, and economic security (Germany Advisory Council on Global Change, 2000).

Resilience has been described as the capacity of a system to absorb disturbance and reorganize while undergoing change, so as still to retain essentially the same function, structure, identity, and feedbacks (Walker *et al.*, 2004). According to Carpenter *et al.* (2001) resilience is the amount of disturbance a socio-ecological system can absorb and still remain within the same state. It can also be seen as the degree to which the system is capable of self-organization (vs lack of organization, or organization forced by external factors), or even the degree to which the system can build and increase the capacity for learning and adaptation. The main objective of managing resilience is thus to prevent the system from moving to undesired configurations in the face of external stresses and disturbance (Walker *et al.*, 2002).

Understanding the loss, creation, and maintenance of resilience through the process of co-discovery (by scientists, policy makers, practitioners, stakeholders, and citizens) is at the heart of sustainability (Gunderson & Holling, 2002), especially when the systems are facing so much turbulence and when people tend towards resignation in face of overwhelming problems. The bottom line is that resilience can be taken as an opportunity to fight resignation, enhancing the capacities and the wills that are needed to face the threats of the future.

The loss of resilience means an increase in vulnerability, which can be defined as a measure of the extent to which a community, structure, service, or geographical area is likely to be damaged or disrupted by the impact of a particular disaster hazard (OECD, 1997) or the state of susceptibility to harm and stress associated with environmental and social change, resulting from the absence of capacity to adapt (Adger, 2000). Vulnerability is the flipside of resilience: when a social or ecological system loses resilience it becomes vulnerable to a change that could previously be absorbed (Kasperson & Kasperson, 2001). Vulnerability is not only a function of the physical characteristics of climate events, but more importantly an inherent property of a society determined by factors such as poverty, inequality, gender patterns, and access to health care and housing (Brooks, 2003).

Adaptive capacity reflects learning (individual, organizational, social), flexibility to experiment and adopt new solutions, and the development of generalized responses to broad classes of challenges (Walker *et al.*, 2002). Adaptive capacity also resides in aspects of memory, creativity, innovation, and diversity of ecological components and human capabilities. Adaptability in a resilience framework implies adaptive capacity not only to respond within the social domain, but also to respond to, and shape, ecosystem dynamics and change them in an informed manner (Berkes *et al.*, 2003). Adaptive capacity can also be defined as the ability to plan, prepare for, facilitate, and implement adaptation options (Klein *et al.*, 2003).

The resilience perspective shifts policies from those that aspire to control change in systems assumed to be stable, to ones that manage the capacity of social-ecological systems to cope with, adapt to, and shape change (Berkes *et al.*, 2003; Smit & Wandel, 2006). Resilience is an approach to managing systems that takes into account social, ecological, and economic influences at multiple scales, accepts continuous change, and acknowledges the level of uncertainty (The Resilience Alliance, 2007b).

According to Holling (2001) one approach to understanding current global change is by analogy with past events. A more systemic and deeper understanding of the system's past and present may help to shape a different vision for the future and to plan for an alternative trajectory. Focusing on disturbances and vulnerabilities, crisis and thresholds, and the solutions adopted over time, may provide lessons for planning and preparing the future.

Adaptive capacity is a prerequisite for sustainability in a context of rapid transformations (Gunderson & Holling, 2002), helping to keep resilience as the capacity to absorb disturbance and reorganize, without collapsing or changing considerably to a worse condition. For example, resilient communities with high adaptive capacity are more able to reconfigure themselves without significant declines in crucial functions in relation to social relations or economic prosperity (Folke *et al.*, 2002). In highly adaptable systems, the actors can reorganize and shape a desirable state in reaction to changing conditions and disturbances. To Luers *et al.* (2003) adaptive capacity is also the extent to which a system can modify its circumstances to move to a less vulnerable condition. Adaptive capacity is a key property based on the capacity of people, individually and collectively, to manage resilience by building social capital and trust and thereby reducing vulnerabilities in face of disturbances and changes (Folke *et al.*, 2003).

According to Tobin (1999), comprehensive planning for sustainability requires changes in the structure and thinking of society to accommodate hazards within the framework of day-to-day affairs. Handmer and Dovers (1996) also link resilience to planning for and adapting to hazards, observing that responses to environmental change are shaped by what is perceived to be politically and economically palatable in the near term rather than by the nature and scale of the threat itself. Klein *et al.* (2003) also note that, from an economic perspective, sustainability is a function of the degree to which key hazard impacts are anticipated.

Resilience is widely discussed and applied in many different contexts, under different disciplines and themes, and therefore with different focuses and specifications, but always maintaining common elements. For example, resilience in the context of disaster reduction is defined as the capacity of a system, community, or society to resist or adapt to change, and to achieve acceptable function and structure (UN/ISDR, 2002).

Resilience is being actively promoted as a management strategy by organizations and agencies, suggesting that planning for resilience can proactively reduce vulnerability by recognizing and accounting for the complex ecological, social, engineering, and economic links that exist in communities (Collini, 2008). Resilience of coastal communities is inherently complex, consisting of numerous linked ecological, social, engineered, and economic systems occurring at various temporal and spatial scales, which are vulnerable to a range of natural and man-made disturbances.

Another relevant example can be found in tourism in coastal and island destinations, which is highly vulnerable to direct and indirect impacts of climate change (such as storms and extreme climatic events, coastal erosion, physical damage to infrastructure, sea level rise, flooding, water shortages, and water contamination), given that most infrastructure is located within a short distance of the shoreline. This high vulnerability often couples with a low adaptive capacity, especially in coastal destinations of developing countries, leading to disasters and stressful events (WTO/UNEP, 2008). Coastlines are economically of outstanding importance not only for tourism, but also in large measure for coastal fisheries, providing a wide range of ecological goods and services on which coastal communities depend (Moberg & Folke, 1999).

The harm associated with climate change falls disproportionately on poorer nations and communities: whereas the wealthy may lose some wealth, the poor risk losing their livelihoods and lives. Poor people often tend to live in fragile or degraded environments and have livelihoods that are more dependent on ecosystem services. With fewer resources from which to draw during periods of stress or crisis, they are more vulnerable to increasing frequency and intensity of weather extremes, seasonal shifts in precipitation, sea level rise, and other observed or predicted effects of climate change. Supporting adaptation measures in poor communities is an urgent priority (Mcgray *et al.*, 2007).

Moving to a broader perspective we can consider the idea of strategic resilience proposed by Gary Hamel and Lisa Valikangas (2003), which is not about responding to a onetime crisis, nor about rebounding from a setback; it's about continually anticipating and adjusting to deep, secular trends that can permanently impair the future of a system. According to the authors it's about having the capacity to change before the case for change becomes desperately obvious. In that regard planning is an essential process where people make continual and systematic decisions about intended future outcomes, how outcomes are to be accomplished, and how success is measured and evaluated (Reinharth *et al.*, 1981).

In the perspective of Day and Schoemaker (2005) the biggest dangers are the ones you don't see coming. Understanding those threats – and anticipating opportunities – requires strong peripheral vision. Planning is needed to prepare for the future, to perceive emergent disturbances, and to identify early signals and solutions, instead of just reacting to crises or just to program interventions that might (or not) be applied if the context changes. In the face of the accelerating pace of change, the uncertainties of the future, and the increasing complexity of phenomena and interactions, an anti-fatalistic, pre-active (anticipating changes), and pro-active (provoking changes) attitude is essential (Godet & Roubelat, 1996). And effective systems are those in which the elements have a capacity to learn to predict changes in their environments, identify the influence of such changes, search for suitable strategies to cope with changes, and develop appropriate structures to implement those changes (Shrivastava, 1983).

21.4 Planning for resilience

Planning for resilience cannot be seen as a technical or closed procedure aiming to find the best solutions, but instead fostering social capacities and true strategic thinking. Static planning processes based on regulations, bureaucratic objectives, or reactive actions are not suitable for fostering strategic adaptation to such a dynamic context. Instead, planning must act as a catalyst for dealing with change, fostering anticipation and learning in order to generate adaptive capacity.

Planning frameworks are more likely to be successful when they have a long-term vision of sustainable development with transparent objectives, and when they include clear priorities upon which stakeholders agree (OECD, 2001a). Mintzberg (1994) also says that the key is not getting the right strategy but fostering strategic thinking. The associated challenges to planning are about institutional change, generating awareness, reaching consensus on values, building commitment, creating an environment with the right incentives, working on shared tasks – and doing so at a pace with which stakeholders can cope (Dalal-Clayton *et al.*, 2002). Stata (1989) argues that the benefits that accrue from planning are not just the strategies and objectives that emerge, but more importantly the social learning that occurs during the planning process.

Social learning can be seen as the synergistic relation between the concept of organizational learning and the resource of social capital. According to Robert Putnam (1995), social capital is the features of social life – networks, norms, and trust – that enable participants to act together and to more effectively pursue shared objectives. Social capital refers to social connections, institutions, relationships, and rules that shape the quality and quantity of a society's social interactions. Social capital is not just the sum of the institutions that underpin a society – it is the glue that holds them together. Peter Maskell (2000) emphasizes the idea that social capital is accumulated within the community through processes of interaction and learning. The word 'capital' implies that we are dealing with an asset. The word 'social' tells that it is an asset attained through membership of a community. To the OECD (2001b) social capital is of central importance for policy learning at the regional and local level. Therefore, planners and policy makers are encouraged to develop strategies to foster appropriate forms of social capital, as a key mechanism in promoting more effective individual, organisational, and social learning.

The impacts of climate change, although heavily shaped by global systems, are inherently local. As with impacts, options for adapting are also location specific but are shaped by economic, climatic, water resource, and other systems that operate from regional to global levels (Moench & Dixit, 2007). Building resilience is particularly important in areas such as coastlines, cities, agricultural land, and industrial zones, which are often the most impacted by humans. It is these same areas that people value highly, both economically and aesthetically, and upon which society often depends (The Resilience Alliance, 2007a). Shaped as they are by factors as diverse as topography, social stratification and gender, most risk vectors lie below the radar screen of national governments. This invisibility poses major challenges for the development of national or regional programmes because such approaches must be tailored to the specific local conditions (Moench & Dixit, 2007).

Society depends on a range of infrastructures and services, and preventing their interruption and restoring their operations becomes an important concern for public policies. From that perspective resilience can be defined (with variants) as the capacity of infrastructure, services, and social systems potentially exposed to hazards from events to adapt either by resisting system degradation or by readily restoring and maintaining acceptable levels of functioning, structure, and service following an event (Bruneau *et al.*, 2003). Today the economy is highly dependent on a complex system of production and distribution of resources and services, including the electricity grid, water supply, telecommunications, emergency services, and transportation network, which are critical infrastructures. Resilience is not necessarily about physical measures and it can be gained through changes in management, procedure, and awareness. It also does not necessarily involve spending large amounts of money – changes in how existing funds are distributed can be very effective in building resilience (Sivell *et al.*, 2008).

Planning as learning implies different assumptions and approaches, meaning a change in processes from being essentially rationalist, reactionary, or bureaucratic, to becoming more adaptable, anticipative, flexible, collaborative, and co-accountable. Planning should pay greater attention to the creation of the conditions for more productive engagement between planners and stakeholders (Counsell & Haughton, 2006). Regulatory frameworks and cultural values provide long-term stability, whereas flexibility and change are provided by learning and negotiation processes in dynamic actor networks in which the interpretation of rules may be substantially renegotiated or rules may even be changed (Pahl-Wostl *et al.*, 2007).

Planning as a learning process is fundamental to generate trust, awareness, and knowledge, to align ideas and goals, to shape leaderships, to produce consensus and co-accountability, to mobilize collective action, and to prepare people for managing change. The world is changing quickly and in order to grow and survive, organizations and communities must learn to adapt faster (Schein, 1993). Adaptive strategies involve mechanisms to encourage groups to learn from each other through social learning and to help policy reflect a range of different values and viewpoints (Stringer *et al.*, 2006). For example, adaptation in the context of climate change refers to the process of adjustment that takes place in natural or human systems in response to the actual or expected impacts of climate change, aimed at moderating harm or exploiting beneficial opportunities (Arnell *et al.*, 2002). Planning must therefore induce new relationships and enhance multidirectional information flows, leading people to develop flexible ways of managing their environments (Carpenter & Gunderson, 2001).

21.5 'Command-and-control' vs 'learning-and-adaptation'

Resilience implies different assumptions and approaches, meaning a change in planning processes from being essentially rationalist, reactionary, or bureaucratic, to becoming more adaptable, anticipative, and flexible. A shift in planning from 'command and control' to 'learn and adapt' is fundamental to address the challenges raised. Nevertheless, it is important to bear in mind that in reality, there aren't two pure extremes to be confronted or to be chosen. Laws and regulations, paper plans and static goals will always exist with their roles and redundancies. What is important is to understand what are the characteristics of each perspective and to explore their suitability for dealing with resilience (Table 21.1).

Planning focused on command and control assumes a more traditional and reductionist view of the systems and their problems, trying to understand them by individualizing their elements with a more 'thinking-in-boxes' approach, and paying less attention to interactions and interrelations. The approaches followed are based in regulations and laws, static goals and guidelines, bureaucratic frameworks, and technocratic procedures. Planning is

Table 21.1 Planning distinctive characteristics

Planning	Command-and-control	Learn-and-adapt
Vision	Reductionist, thinking-in-boxes	Integrated, systemic thinking
Approaches	Regulation, bureaucracy, rationalist, technocratic	Strategic, proactive, flexible, creative
Character	Mainly formal	Formal and informal
Main output	Paper plan	People in process
Participation	Restricted, punctual	Inclusive, ongoing
Actions planned	Many, dispersed, comprehensive	Few, focused, differentiated
World view	Change is gradual, main-equilibrium	Change can be sudden (thresholds), multi-equilibrium
Future focus	Short-term, predictable, one future	Long-term, uncertain, various futures
Leadership	Determinist, distant	Facilitator, catalyst
Institutional environment	Government, hierarchical, centralized	Governance, polycentric, networks
Dealing with disturbances	Reaction, coping capacity	Anticipation, adaptive capacity
Resilience	Low adaptability	High adaptability

assumed to be a highly rationalist and formal process aiming to produce a plan, which is more important than the process itself. Public participation and stakeholder engagement is restricted and punctual, if possible included in the final stages of the process, and seen as an obligation to address. Planning aims to identify as many actions and measures as possible, sometimes with lack of scale, substance, or relevance. The world is assumed as stable, evolution is gradual, and the future is considered predictable and amenable to forecasting, mainly in the short term, which is the most relevant timeframe for the process. Leadership of the process tends to be deterministic, decisional, and distant, acting in an institutional environment mainly hierarchical and centralized. When disturbances occur, the system tends to react based on its coping capacity in order to mitigate the impacts. Resilience tends to be weaker due to the lower flexibility and adaptability to change.

Planning focused on learn and adapt assumes a more integrated vision and a systemic thinking when looking to problems and structures, considering the interdependencies and interrelations between the different elements. The approaches followed tend to be more strategic, proactive, creative, and flexible to adopt new ideas. Planning can be seen as a formal or an informal exercise; what is important is the strategic thinking, the alignment of viewpoints, and the collective learning. The 'people in process' perspective is essential and participation tends to be ongoing, inclusive, and structured all along. The identification of actions and measures to be delivered or implemented tends to be focused, differentiated, and selective. The world is assumed as unstable, gradual evolution can interplay with sudden surprises, and thresholds might be crossed bringing substantial change and a different state of equilibrium. Assumptions about the future are flexible: various possible and plausible futures might occur according to different dynamics and alternative scenarios. Leadership in the process is mainly a facilitator of interactions and a catalyst of change. The institutional environment is more polycentric based on flexible networks working according to a logic of good governance. Anticipation of disturbances is essential and more attention is paid to macro and micro trends, wildcards, and early warning. Planning is above all a collective capacity-building process, aiming to strengthen the adaptive capacities. Therefore, resilience tends to be stronger due to the greater adaptability.

These are two metaphorical extremes of what can be seen as 'command and control' and 'learning and adaptation' in the perspective of planning for resilience. It is important to highlight that in reality what we find in planning processes is a conjugation of characteristics from the two trends. Nevertheless, it is obvious that planning for resilience is above all learning and adaptation, therefore it is important to develop and integrate those orientations into the procedures.

Despite those theoretical considerations there is a need for further knowledge but also for practical approaches in order to apply the concepts related to the resilience framework. An example is the framework SPARK – Strategic Planning Approach for Resilience Keeping – proposed by Santos and Partidario (2010), which intends to be an integrated conceptual and methodological framework for approaching resilience in the planning processes, aiming to bridge the theory with application.

21.6 The strategic SPARK example

The SPARK methodological framework proposed is structured in four sequential stages (Figure 21.1). In the first stage the aim is to understand the system (e.g. region, natural reserve, etc.) and to gain insight about its structure and dynamics, starting by focusing on the problem to address regarding the resilience of the system. The focus on the *process* must be clearly defined and be kept in mind throughout the whole process. Once the structure of the system is clear, the second stage moves into the *story*, to analyse the trajectory followed over time, to understand how the past shaped the present, to get a broader perspective of its

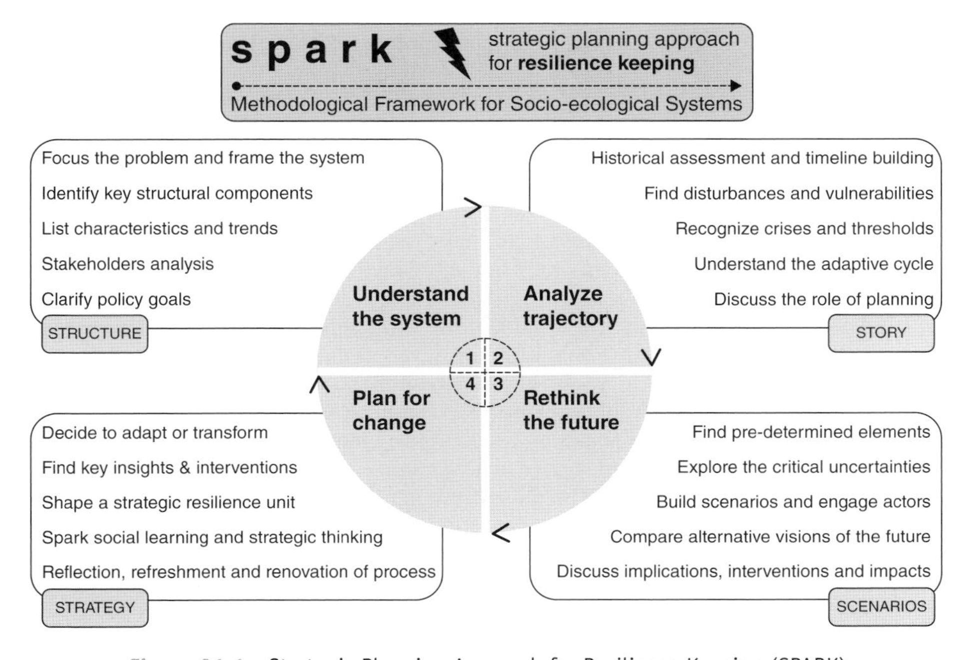

Figure 21.1 Strategic Planning Approach for Resilience Keeping (SPARK)

evolution. In this stage a systematic historical analysis is conducted, aiming at constructing the system's timeline, helping to find disturbances and vulnerabilities (but also answers and reactions), and to detect crises and thresholds that might have occurred.

If the first two stages of the process are more directed to understanding and getting to know the present and the past of the system, the third stage turns out to be more oriented to explore the *future* and to rethink different options according to distinct plausible scenarios. The intention is to involve multiple stakeholders and different perspectives over a scenario planning process, starting from the identification of critical uncertainties and predetermined elements. Different techniques and procedures may be followed, according to the needs of the process (see methodological procedures in Schwartz, 1991, or in Scearce & Fulton, 2004). The aim of the process is to find, if possible and relevant, alternative visions of the future of the system, exploring their potential implications, interventions, and impacts. Having gained insight about the present, the past, and the future of system, the fourth and last stage is more directed to define, and to clarify, *strategies* for resilience planning. The aim is to understand when, where, and how to intervene in order to adapt or to transform the system into a desirable configuration. This strategy builds on insights found and interventions detected along the rest of the process to keep resilience. The most important thing is to find ways to spark social learning and strategic thinking about the future, and to be prepared for the changes that may happen.

This methodological proposal, briefly discussed here, is an example of an approach that tries to integrate planning and resilience elements, in order to think (e.g. development processes) with a more systematic perspective about change, adaptation, and transformation.

21.7 Final considerations

Resilience is a framework for systematic thinking about the adaptive capacity of a region, a landscape, a local economy, or even a household, especially when facing disturbance and change, which are always elements shaping the system. Planning for resilience therefore is neither about creating new formal plans nor about introducing new orientations, but it is instead about how to raise awareness in relation to a more openminded-to-change perspective, which is critical for dealing with uncertain realities, even in more traditional planning contexts. The transition from 'command-and-control' to 'learn-and-adapt' also gives more importance to a 'put-people-in-the-process-perspective' and less relevance to a 'paper-plan-production-process', which is normally the end of it.

Sustainable development depends more and more on the capacity of social and ecological systems to support disturbances and persist in the long term, meaning that resilience is a critical property for sustainability. Sustainable development challenges societies, organizations and individuals to learn to deal with high levels of uncertainty, long time horizons, and the interaction of ecological, social, and economic systems as well as multi-level thinking to link local, regional, and global perspectives (Siebenhüner, 2005). The resilience lens accepts a world where multiple scales and stakeholders interact in time and space, where many pathways can be found and shaped, where change can be incremental or irregular, where several points of equilibrium might occur, regardless of its desirability for humans.

Planning for resilience is not about waiting for disturbances to happen, in order to cope with their impacts. Planning for resilience is all about being proactive, focusing on learning processes (individual, organizational, and social), and being less dependent on laws, regulations, strategic goals, and bureaucratic procedures, which can bring order and orientation but also inflexibility, inefficiency, and conflict, instead of adaptive capacity, anticipation, and motivation. Resilience can be enhanced through planning processes in order to prepare the systems (nations, regions, communities, enterprises, families, etc.) to adapt and to cope with disturbances and crises. The resilience framework is emerging as an alternative path for dealing with sustainable development issues, in a more interconnected world under greater pressure and turbulence. And that is why resilience is a crucial property and planning has a crucial role in sparking learning and adaptation.

Acknowledgements

The author is grateful to the Portuguese Foundation for Science and Technology (FCT) for the financial support granted (SFRH/BD/18366/2004/H39K) under a PhD fellowship.

References

Adger, N. (2000) Social and ecological resilience: are they related? *Progress in Human Geography* **24**: 347–64.

Adger, N., Hughes, T., Folke, C., Carpenter, S., Rockström, J (2005) Social-ecological resilience to coastal disasters. *Science* **309**: 1036–9.

Arnell, W., Cannell, M., Hulme, M., *et al.* (2002) The consequences of CO2 stabilisation for the impacts of climate change. *Climatic Change* **53**: 413–46.

Berkes, F., Colding, J., Folke, C. (2003) *Navigating Social-Ecological Systems; Building Resilience for Complexity and Change*. Cambridge: Cambridge University Press.

Brooks, N. (2003) Vulnerability, risk and adaptation: a conceptual framework. Tyndall Centre Working Paper No. 38. Tyndall Centre for Climate Change Research.

Bruneau, M., Chang, S., Eguchi, R., *et al.* (2003) A framework to quantitatively assess and enhance the seismic resilience of communities. *Earthquake Spectra* **19**: 733–52.

Carpenter, S., Brock, W. (2008) Adaptive capacity and traps. *Ecology and Society* **13**: 40.

Carpenter, S., Gunderson, L. (2001) Coping with collapse: ecological and social dynamics in ecosystem management. *BioScience* **6**: 451–7.

Carpenter, S., Walker, B., Anderies, J., Abel, N. (2001) From metaphor to measurement: resilience of what to what? *Ecosystems* **4**: 765–81.

Collini, K. (2008) Coastal community resilience: an evaluation of resilience as a potential performance measure of the Coastal Zone Management Act. NOAA Office of Ocean and Coastal Resource Management in collaboration with the Coastal States Organization.

Counsell, D., Haughton, G. (2006) Sustainable development in regional planning: the search for new tools and renewed legitimacy, *Geoforum* **37**: 921–31.

Dalal-Clayton, B., Swiderska, K., Bass, S. (2002) Stakeholder dialogues on sustainable development strategies. Lessons, opportunities and developing country case studies. *Environmental Planning Issues*, No. 26. London: International Institute for Environment and Development.

Day, G., Schoemaker, P. (2005) Scanning the periphery. *Harvard Business Review* November, pp. 1–12.

Folke, C. (2006) Resilience: the emergence of a perspective for social-ecological systems analyses. *Global Environmental Change* **16**: 253–67.

Folke, C., Carpenter, S., Elmqvist, T., *et al.* (2002) Resilience and sustainable development: building adaptive capacity in a world of transformations. Scientific Background Paper on

Resilience for the process of The World Summit on Sustainable Development on behalf of The Environmental Advisory Council to the Swedish Government.

Folke, C., Colding, J., Berkes, F. (2003) Synthesis: building resilience and adaptive capacity in social-ecological systems. In: Berkes, F., Colding, J., Folke, C. (eds), *Navigating Social-Ecological Systems: Building Resilience for Complexity and Change*. Cambridge: Cambridge University Press, pp. 1–30.

German Advisory Council on Global Change (2000) *World in Transition: Strategies for Managing Global Environmental Risks*. Berlin: Springer-Verlag.

Global Humanitarian Forum (2009) *Human Impact Report on Climate Change. The Anatomy of a Silent Crisis*. Geneva: Global Humanitarian Forum.

Godet, M., Roubelat, F. (1996) Creating the future: the use and misuse of scenarios. *Long Range Planning* **29**: 164–71.

Gunderson, L., Holling, C.S. (2002) *Panarchy: Understanding Transformations in Human and Natural Systems*. Washington, DC: Island Press.

Hamel, G., Valikangas, L. (2003) The quest for resilience. *Harvard Business Review* September, pp. 1–13.

Handmer, J., Dovers, S. (1996) A typology of resilience: rethinking institutions for sustainable development. *Industrial and Environmental Crisis Quarterly* **9**: 482–511.

Holland, J. (1995) Can there be a unified theory of complex adaptive systems? In: Morowitz, H., Singer, J. (eds), *The Mind, The Brain, and Complex Adaptive Systems*. Addison-Wesley, pp. 45–50.

Holling, C. (2001) Understanding the complexity of economic, ecological, and social systems. *Ecosystems* **4**: 390–405.

Homer-Dixon, T. (2007) *The Upside of Down. Catastrophe, Creativity and the Renewal of Civilisation*. London: Souvenir Press.

Kasperson, J., Kasperson, R. (2001) *Global Environmental Risk*. London: United Nations University Press/Earthscan.

Klein, R., Nicholls, R., Thomalla, F. (2003) Resilience to natural hazards: How useful is this concept? *Environmental Hazards* **5**: 35–45.

Lebel, L., Anderies, J., Campbell, B., *et al.* (2006) Governance and the capacity to manage resilience in regional social-ecological systems. *Ecology and Society* **11**: 19.

Luers, A., Lobell, D., Sklar, L., Addams, C., Matson, P. (2003) A method for quantifying vulnerability, applied to the agricultural system of the Yaqui Valley, Mexico. *Global Environmental Change* **13**: 255–67.

Maskell, P. (2000) Social capital, innovation and competitiveness. In: Field, J., Schuller, T., Baron, S. (eds), *Social Capital: Critical Perspectives*. Oxford: Oxford University Press, pp. 111–23.

Mcgray, H., Hammill, A., Bradley, R. (2007) *Weathering the Storm: Options for Framing Adaptation and Development*. World Resources Institute.

Mintzberg, H. (1994) *The Rise and Fall of Strategic Planning*. New York: The Free Press.

Moberg, F., Folke, C. (1999). Ecological goods and services of coral reef ecosystems. *Ecological Economics* **29**: 215–33.

Moench, M., Dixit, A. (2007) *Working with the Winds of Change. Toward Strategies for Responding to the Risks Associated with Climate Change and other Hazards*, 2nd edn. ProVention Consortium, Institute for Social and Environmental Transition-International and Institute for Social and Environmental Transition-Nepal.

OECD (Organization for Economic Cooperation and Development) (1997) Glossary of environment statistics. *Studies in Methods*, Series F, No. 67. New York: United Nations.

OECD (Organization for Economic Cooperation and Development) (2001a) *Strategies for Sustainable Development. Practical Guidance for Development Cooperation*. Paris: OECD.

OECD (Organization for Economic Cooperation and Development) (2001b) *Cities and Regions in the New Learning Economy*. Paris: OECD.

Pahl-Wostl, C., Craps, M., Dewulf, A., Mostert, E., Tabara, D., Taillieu, T. (2007) Social learning and water resources management, *Ecology and Society* **12**: 5.

Putnam, R. (1995) Bowling alone: America's declining social capital. *Journal of Democracy* **6**: 65–78.

Reinharth, L., Shapiro, J., Kallman, E. (1981) *The Practice of Planning: Strategic, Administrative and Operational*. New York: Van Nostrand Reinhold.

Rockström, J., Steffen, W., Noone, K., *et al.* (2009) A safe operating space for humanity. *Nature* **461**: 472–5.

Santos, F.T., Partidario, M.R. (2010) Strategic planning approach for resilience keeping. *European Planning Studies* **19**: 1517–36.

Scearce, D., Fulton, K. (2004) *What If? The Art of Scenario Thinking for Nonprofits*. Global Business Network.

Schein, E. (1993) How can organizations learn faster? The challenge of entering the green room. *Sloan Management Review* **34**: 85–92.

Schwartz, P. (1991) *The Art of the Long View: Planning for the Future in an Uncertain World*. New York: Doubleday.

Senge, P. (1990) *The Fifth Discipline: The Art and Practice of the Learning Organizations*. New York: Doubleday.

Shrivastava, P. (1983) A typology of organizational learning systems. *Journal of Management Studies* **20**: 7–28.

Siebenhüner, B. (2005) The role of social learning on the road to sustainability. In: Rosenau, N., Weizsäcker E., Petschow, U. (eds), *Governance and Sustainability*. Sheffield: Greenleaf, pp. 86–99.

Sivell, P., Reeves, S., Baldachin, L., Brightman, T. (2008) *Climate Change Resilience Indicators*. Transport Research Laboratory. Prepared for the South East England Regional Assembly.

Smit, B., Wandel, J. (2006) Adaptation, adaptive capacity and vulnerability. *Global Environmental Change* **16**: 282–92.

Stata, R. (1989) Organizational learning: the key to management innovation. *Sloan Management Review* Spring: 63–74.

Stringer, L., Dougill, A., Fraser, E., Hubacek, K., Prell, C., Reed, M. (2006) Unpacking "participation" in the adaptive management of social-ecological systems: a critical review. *Ecology and Society* **11**: 39.

The Resilience Alliance (2007a) *A Research Prospectus for Urban Resilience. A Resilience Alliance Initiative for Transitioning Urban Systems towards Sustainable Futures*. CSIRO (Australia), Arizona State University (USA) and Stockholm University (Sweden).

The Resilience Alliance (2007b) *Assessing and Managing Resilience in Social-Ecological Systems: A Practitioners Workbook*, Version 1.0. Available at: http://www.resalliance.org/index.php/resilience_assessment

Tobin, G. (1999) Sustainability and community resilience: the holy grail of hazards planning? *Environmental Hazards* **1**: 13–25.

UNEP (United Nations Environment Programme) (2007) *Global Environment Outlook: Environment for Development (GEO-4)*. Mriehel, Malta: Progress Press Ltd.

UN ISDR (United Nations International Strategy for Disaster Reduction) (2002) *Living with Risk: A Global Review of Disaster Reduction Initiatives*. Geneva: United Nations.

Walker, B., Carpenter, S., Anderies, J., *et al.* (2002) Resilience management in social-ecological systems: a working hypothesis for a participatory approach. *Conservation Ecology* **6**: 14.

Walker, B., Holling, C., Carpenter, S., Kinzig, A. (2004) Resilience, adaptability and transformability in social-ecological systems. *Ecology and Society* **9**: 5.

Walker, B., Barrett, S., Polasky, S., *et al.* (2009) Looming global-scale failures and missing institutions. *Science* **11**: 1345–6.

World Economic Forum (2009) Global Risks 2009: A Global Risk Network Report. Geneva: World Economic Forum.

WTO/UNEP (World Tourism Organization, United Nations Environment Programme) (2008) *Climate Change and Tourism – Responding to Global Challenges*. Madrid: WTO/UNEP.

Young, O.R., Berkhout, F., Gallopin, G.C., Janssen, M., Ostrom, E., van der Leeuw, S. (2006) The globalization of socio-ecological systems: An agenda for scientific research. *Global Environmental Change* **16**: 304–16.

22 Conclusion

He Qing Huang[1] and Ademola K. Braimoh[2]

[1] *Institute of Geographic Science and Natural Resources Research, Chinese Academy of Sciences, Beijing, China*

[2] *The World Bank, Washington DC, USA*

This book brings together diverse expertise on the vulnerability of land systems in Asia. It begins with a detailed account of the complexity of the coupled human-nature systems and the importance of the vulnerability assessment in land change science (Chapter 1 by Braimoh and Huang). To improve knowledge on the locality-based stressors, the causal processes and coping measures that affect the vulnerability of land systems in Asia, Chapters 2 to 21 present detailed case studies by scientists across the world. These case studies focus on different aspects of vulnerability and so are presented in three sections, namely 'hazards and vulnerability', 'land-use change modeling and impact assessment', and institutions. Table 22.1 provides a summary of the key issues addressed in each group of these case studies. We believe this is an important step towards a deeper understanding of the major issues and underlying processes determining the vulnerability of land systems in Asia in several aspects.

22.1 Improving understanding in areas lacking data

Asia has vast, physiographically diverse land systems, and due to physical, religious, political, financial and technical reasons, much of these land systems has not received detailed studies, particularly those located in the boreal climatic zone and the west and central parts, which are predominantly arid. Opportunities for scientists working in the region to communicate their study results are also few, although an increasing trend has been evident in recent years due to the influence of global change. As a result, a large part of the land systems has been essentially unknown to the rest of the world. Nevertheless, it can be seen in Table 22.1 that the case studies presented in this book are numerous and cover most of the land systems in Asia. Hence, this book provides a good opportunity for improving

Vulnerability of Land Systems in Asia, First Edition. Edited by Ademola K. Braimoh and He Qing Huang.

Table 22.1 A summary of the key issues examined by case studies in the book

Book section	Exposure	Sensitivity	Coping measures	Study area	Chapter (number)
Hazards and Vulnerability	Extreme cold winter, drought, climate warming, economic development, land use, urban growth, mining	Pastoral and agricultural systems, communities, ecosystems	Pastoral control, cultural landscape conservation, ecosystem restoration, water harvesting and tillage techniques, economic restructuring	Mongolia, China, India, Kazakhstan	Sternberg (2) Ojima *et al.* (3) Chuluun *et al.* (4) Liu *et al.* (5) Singh and Kumar (6) Winchester *et al.* (7)
Land-use Change: Modelling and Impact Assessment	Socio-economic development, urban growth	Land use/land-cover, agricultural productivity, ecosystem services	Land use planning and management	Southeast Asia, China, Iran, Indonesia	Fox *et al.* (8) Yan *et al.* (9) Haruyama *et al.* (10) Cui *et al.* (11) Shahbazi *et al.* (12) Rahajoe *et al.* (13) Yue *et al.* (14)
Institutions	Ecosystem conservation and restoration, socio-economic activities	Land use/land cover, ecosystems, communities, societies, knowledge, planning	Tradeoffs of ecosystem services, eco-compensation, community-based management, climate financing, science-policy linkage, planning	Thailand, China, Indonesia, India, general	Lebel and Daniel (15) Li *et al.* (16) Thapa *et al.* (17) Saharjo (18) Damodaran (19) Litre (20) dos Santos (21)

understanding of those land systems for which almost no information is available. This is also beneficial for the integration and synthesis analysis of land systems at a global scale, one of the major tasks of the Global Land Project co-funded by the IGBP and IHDP (GLP, 2005).

22.2 Highlighting the effects of scale

The vulnerability of land systems in Asia results from many causes and has specific forms varying from one place to another and from small to large scale. In most cases, both biophysical and human stressors are coupled in the land systems. Due to the complex behaviors of the two subsystems, some vulnerability problems may appear only at small scales, such as the one caused by an extreme cold winter climate as detailed in Chapter 2. The effects of such localized small-scale problems are minor at a large scale because they do not involve large human populations and are exhibited only in remote deserted areas. On the other hand, these localized small-scale vulnerability problems may be very sensitive to large-scale changes in climate but not to large-scale changes in socio-economy. Due to the small population, and the remote location, traditional cultures and customs can be preserved and there is no need to make significant changes in land use to satisfy the low level of demands on food and residential accommodation. To distinguish these effects of scale, one needs to examine vulnerability problems from both top-down and bottom-up directions.

22.3 Validating the conceptual framework for vulnerability assessment

The vulnerability assessment of a specific land system generally needs to be performed from three aspects: exposure, sensitivity and coping capacity (GLP, 2005; Turner *et al.*, 2007). These three aspects reveal the exposure of the land system to various stressors, the degree to which the land system is affected by the stressors, either adversely or beneficially, and the capacities of people and organizations who are able to adopt reasonable measures that prevent the land system from reaching thresholds beyond which the system is vulnerable. The detailed case studies presented in this book and outlined in Table 22.1 essentially follow this conceptual framework, although each group of them addresses a different key issue.

22.4 Land system vulnerability in other parts of the world

Over the last 150 years, US agriculture has exhibited a remarkable adaptive capacity to climate and natural resources variability, as well as dynamic changes in knowledge, technologies and markets (National Research Council, 2010). These adaptations were made during a period of relative climatic stability and abundant technical, financial and natural resources. However, multiple stressors, including climate change, increasingly compromise the ability of agricultural ecosystems to support productivity and other services. Future adaptation will be undertaken in a decision environment characterized by uncertainty driven by sensitivity of the USA's agricultural systems to climatic variability, the complexity of interactions between the land system, non-climatic stressors and the global climate system, and the increasing pace and intensity of climate change (United States Department of Agriculture, 2013).

In Europe, observed climate change and socioeconomic drivers have already led to a wide range of impacts on land systems and society. There is an earlier occurrence of spring seasonal events and a later occurrence of autumn seasonal events in plants and animals; lengthening of breeding seasons; and northward and uphill movement of many plant and animal species. Land suitability for crops is projected to expand northwards with earlier flowering and harvest dates in cereals; reduced yield of some crops due to heatwaves and droughts, mostly in central and southern Europe, but increased yields of other crops, mostly in northern Europe; and increased water demand for irrigation in southern and southwestern Europe (European Environment Agency, 2012). Existing socioeconomic vulnerabilities may be exacerbated by the impacts of climate change. Countries across Europe have different economic, technical and institutional capacity to cope with and adapt to climate change. An integrated assessment of European regions' vulnerability suggests that climate change, in addition to socioeconomic and demographic pressures, may negatively affect territorial cohesion in Europe by deepening existing socioeconomic imbalances (European Environment Agency, 2012).

Responsible for only less than 13% of the world's global greenhouse gas emissions, Latin America could be one of the regions most affected if temperatures were to rise 4°C by 2100, with the Caribbean and tropical regions shouldering the greatest burden and the region's poorest populations likely to suffer the most (World Bank, 2012). Already one of the region's most dangerous meteorological hazards, the frequency of high-intensity tropical cyclones is predicted to increase, with Central America and the Caribbean likely to bear the brunt of the resulting damage. Climate change will also lead to loss of arable land in the region with severe consequences for food security.

Sub-Saharan Africa relies heavily on agriculture as a source of food and income, but climate change may increase poverty and vulnerability in the region. Most African countries are the most vulnerable to the effects of climate change largely due to a low adaptive capacity. High levels of vulnerability and low adaptive capacity in the developing world are linked to high reliance on natural resources, limited ability to adapt financially and institutionally, low per capita GDP and high poverty, and a lack of safety nets (Thomas & Twyman, 2005). The impacts of a changing climate on agricultural production in a world that warms by 4°C or more are likely to be severe in Africa, compounding vulnerability, limiting crop and livestock production options, and making development goals more difficult to achieve (Thornton *et al.*, 2011; World Bank, 2013).

Irrespective of world region, government policies and programs are crucial to effective adaptation of the land system to climate change and other non-climatic stressors (Antle & Capalbo, 2010). The local nature of adaptation, however, complicates the implementation of adaptation programs because of the potential for complex cross-scale interactions between top-down policy decisions made at national or international scales, and bottom-up adaptive responses (Urwin & Jordan, 2008). Owing to the uncertainties associated with the impacts, vulnerability drivers, and the complexity of adaptation processes, adaptive governance strategies are recommended to implement, evaluate, and revise adaptation approaches (Biesbroek *et al.*, 2010).

22.5 Roads ahead

Among all the detailed case studies, it can be seen in Table 22.1 that a large number of them focus on institutions (7 out of 22 chapters). These studies were done in many countries, including Thailand, China, Indonesia, and India. As detailed in the corresponding chapters of this book and Section 22.4 above, institutions, typically governmental policies, are indeed a very important factor affecting the vulnerability as well as the adaptive capacity of the land systems. However, adaptation strategies to reduce land system variability may not always be transferable from region to region due to the local operating context of adaptation.

Next to institutions are case studies on land change modeling and impact assessment. The seven chapters in this section apply different approaches to model land-use change processes and to assess the potential vulnerabilities in land systems. As demonstrated clearly in the book, these chapters provide viable tools for perceiving the sensitivity of land systems to climate change and human activities. However, each case study applies a different model, making it difficult to evaluate if all of the models produce essentially consistent outcomes in the same region. In fact, the different models tend to complement each other in elucidating vulnerability.

The section on hazards and vulnerability presents six interesting chapters with exposure variables ranging from extreme temperatures to mining activities. As demonstrated by Turner *et al.* (2003b), the subsystems concepts and associated theory in the coupled human-environment land systems uncovered in detailed case studies are increasingly rewarding and significant for future research.

22.6 Final remarks

This book documents vulnerability in Asia through a range of case studies and integrative methods placed in the context of frameworks for addressing land systems' interrelationships. An abiding theme throughout this book is the need for place-based analysis to evaluate land system vulnerability and design adaptation responses. This leads to an improved understanding of the potential impacts of climate change and non-climate stressors, opportunities for adaptation, as well as more effective methods of management for land system sustainability.

References

Antle, J.M., Capalbo, S.M. (2010) Adaptation of agricultural and food systems to climate change: An economic and policy perspective. *Applied Economic Perspectives and Policy* **32**: 386–416.

Biesbroek, G.R., Swart, R.J., Carter, T.R., *et al.* (2010) Europe adapts to climate change: Comparing national adaptation strategies. *Global Environmental Change* **20**: 440–50.

European Environment Agency (2012) Climate change, impacts and vulnerability in Europe 2012. An Indicator Based Report. EEA Report No. 12/2012.

GLP (Global Land Project) (2005) GLP – Science Plan and Implementation Strategy. IGBP Report No. 53/IHDP Report No. 19. Stockholm: IGBP Secretariat.

National Research Council (2010) *Toward Sustainable Agricultural Systems in the 21st Century*. Washington, DC: National Research Council Committee on Twenty-First Century Systems, Agriculture; National Academies Press.

Thomas, D.S.G., Twyman, C. (2005) Equity and justice in climate change adaptation amongst natural resource dependent societies. *Global Environmental Change* **15**: 115–24.

Thornton, P.K., Jones, P.G., Ericksen, P.J., Challinor, A.J. (2011) Agriculture and food systems in sub-Saharan Africa in a 4°C+ world. *Philosophical Transactions of the Royal Society A* **369**: 117–36.

Turner, B.L. II, Matson, P.A., McCarthy, J.J., *et al.* (2003a) A framework for vulnerability analysis in sustainability sciences. *Proceedings of the National Academy of Sciences of the USA* **100**: 8074–9.

Turner, B.L. II, Matson, P.A., McCarthy, J.J., *et al.* (2003b) Illustrating the coupled human-environment system for vulnerability analysis: The case studies. *Proceedings of the National Academy of Sciences of the USA* **100**: 8080–5.

Turner, B.L. II, Lambin, E.F., Reenberg, A. (2007) The emergence of land change science for global environmental change and sustainability. *Proceedings of the National Academy of Sciences of the USA* **104**: 20666–71.

United States Department of Agriculture (2013) Climate change and agriculture in the united states: effects and adaptation. USDA Technical Bulletin 1935. Available at:http://www.usda.gov/oce/climate_change/effects.htm

Urwin, K., Jordan, A. (2008) Does public policy support or undermine climate change adaptation? Exploring policy interplay across different scales of governance. *Global Environmental Change* **18**: 180–91.

World Bank (2012) Turn down the heat. Why a 4°C warmer world must be avoided. A report for the World Bank by the Potsdam Institute for Climate Impact Research and Climate Analytics, Washington, DC.

World Bank (2013) Turn down the heat II. Climate extremes, regional impacts and the case for resilience. A report for the World Bank by the Potsdam Institute for Climate Impact Research and Climate Analytics, Washington, DC.

Index

Vulnerability of Land Systems in Asia, First Edition. Edited by Ademola K. Braimoh and He Qing Huang.